Computational Int1
Optimization Algorithms

Computational intelligence-based optimization methods, also known as meta-heuristic optimization algorithms, are a popular topic in mathematical programming.

These methods have bridged the gap between various approaches and created a new school of thought to solve real-world optimization problems. In this book, we have selected some of the most effective and renowned algorithms in the literature. These algorithms are not only practical but also provide thought-provoking theoretical ideas to help readers understand how they solve optimization problems. Each chapter includes a brief review of the algorithm's background and the fields it has been used in.

Additionally, Python code is provided for all algorithms at the end of each chapter, making this book a valuable resource for beginner and intermediate programmers looking to understand these algorithms.

Babak Zolghadr-Asli is currently a joint researcher under the QUEX program, working at the Sustainable Minerals Institute at The University of Queensland in Australia and The Centre for Water Systems at The University of Exeter in the UK. His primary research interest is to incorporate computational and artificial intelligence to understand the sustainable management of water resources.

Computational Intelligence-based Optimization Algorithms

From Theory to Practice

Babak Zolghadr-Asli

CRC Press
Taylor & Francis Group
Boca Raton London New York

CRC Press is an imprint of the
Taylor & Francis Group, an **informa** business

Designed cover image: Shutterstock

First edition published 2024
by CRC Press
2385 NW Executive Center Drive, Suite 320, Boca Raton FL 33431

and by CRC Press
4 Park Square, Milton Park, Abingdon, Oxon, OX14 4RN

CRC Press is an imprint of Taylor & Francis Group, LLC

Library of Congress Cataloging-in-Publication Data
Names: Zolghadr-Asli, Babak, author.
Title: Computational intelligence-based optimization algorithms :
from theory to practice / Babak Zolghadr-Asli.
Description: First edition. | Boca Raton, FL : CRC Press, 2024. |
Includes bibliographical references and index.
Identifiers: LCCN 2023019666 (print) | LCCN 2023019667 (ebook) |
ISBN 9781032544168 (hardback) | ISBN 9781032544151 (paperback) |
ISBN 9781003424765 (ebook)
Subjects: LCSH: Computer algorithms. | Computational intelligence.
Classification: LCC QA76.9.A43 Z65 2024 (print) |
LCC QA76.9.A43 (ebook) | DDC 005.13–dc23/eng/20230623
LC record available at https://lccn.loc.gov/2023019666
LC ebook record available at https://lccn.loc.gov/2023019667

ISBN: 978-1-032-54416-8 (hbk)
ISBN: 978-1-032-54415-1 (pbk)
ISBN: 978-1-003-42476-5 (ebk)

DOI: 10.1201/9781003424765

Typeset in Times
by Newgen Publishing UK

Contents

Figures

Foreword

This is a unique reference book providing in one place: information on the main meta-heuristic optimization algorithms and an example of their algorithmic implementation in Python. These algorithms belong to the class of computational intelligence-based optimization methods that have addressed one of the key challenges plaguing mathematical optimization for years – that of dealing with difficult and realistic problems facing any industry with resource restrictions. What do I mean by difficult and realistic? Instead of simplifying the problem that needs to be solved due to the limitations of the method, as was the case with many mathematical optimization algorithms, these meta-heuristics can now tackle large, complex, and previously often intractable problems.

The book includes 20 meta-heuristic algorithms, from the now-classical genetic algorithm to more "exotic" flower pollination or bat algorithms. Each of the algorithms is presented as far as possible using the same structure so the reader can easily see the similarities or differences among them. The Python code provides an easy-to-access library of these algorithms that can be of use to both novices and more proficient users and developers interested in implementing and testing some of the algorithms they may not be fully familiar with. From my own experience, it is much easier to get into a subject when somebody has already prepared the grounds. That is the case with this book, if I had it on my desk 30 years ago, I would've been able to try many more different ways of solving problems in engineering. With this book, I may still do it now!

Dragan Savic
Professor of Hydroinformatics
University of Exeter, United Kingdom
and
Distinguished Professor of Hydroinformatics
The National University of Malaysia, Malaysia

Preface

Computational intelligence-based optimization methods, often referred to as meta-heuristic optimization algorithms, are among the most topical subjects in the field of mathematical programming. This branch of optimization methods is basically an alternative approach to accommodate the shortcomings of conventional analytical-based approaches and unguided sampling-based methods. In a sense, these methods were able to bridge the gap between these two vastly different approaches and create a new school of thought to handle real-world optimization problems.

By the early 1990s, many researchers had started conceptualizing CI-based frameworks to tackle optimization problems. Hundreds of meta-heuristic optimization algorithms are out there, which could be overwhelming for beginners who have just started in this field. As such, in this book, we would not only provide a rock-solid theoretical foundation about these algorithms, but we tend to tackle this subject from a practical side of things as well. Any algorithm introduced in this book also comes with readily available Python code so the reader can implement the algorithm to solve different optimization problems. We strongly believe that this could help the readers to have a better grasp over the computational structure of these algorithms.

We have handpicked some of the literature's most exciting and well-known algorithms. Not only are these algorithms very efficient from a practical point of view, but they all also consist of through-provoking theoretical ideas that can help the reader better understand how these algorithms actually tend to solve an optimization problem. In this book, we will learn about *pattern search, genetic algorithm, simulated annealing, tabu search, ant colony optimization, particle swarm optimization, differential evolution algorithm, harmony search algorithm, shuffle frog-leaping algorithm, invasive weed optimization, biogeography-based optimization, cuckoo search algorithm, firefly algorithm, gravity search algorithm, plant propagation algorithm, teaching-learning-based algorithm, bat algorithm, flower pollination algorithm, water cycle algorithm,* and *symbiotic organisms algorithm.* Though the chapters are arranged chronically, there are some pedagogical reasoning behind this arrangement so that the readers can easily engage with the presented materials in each chapter. Note that the basic idea here is to ensure that each algorithm is presented in a stand-alone chapter. This means that after reading the first chapter, which we highly encourage you to do, you can go to a given chapter and learn all there is to understand and implement an algorithm fully. Each chapter also contains a brief literature review of the algorithm's background and showcases where it has been implemented successfully. As stated earlier, there is a Python code for all algorithms at the end of each chapter. It is important to note that, while these are not the most efficient way to code these algorithms, they may very well be the best way to understand them for beginner to intermediate programmers. As such, if, as a reader, you have a semi-solid understanding of the *Python* syntax and its numeric library *NumPy*, you could easily understand and implement these methods on your own.

1 An Introduction to Meta-Heuristic Optimization

Summary

Before we can embark upon this journey of ours to learn about computational intelligence-based optimization methods, we must first establish a common language to see what an optimization problem actually is. In this chapter, we tend to take a deep dive into the world of optimization to understand the fundamental components that are used in the structure of a typical optimization problem. We would be introduced to the technical terminology used in this field, and more importantly, we aim to grasp the basic principles of optimization methods. As a final note, we would learn about the general idea behind meta-heuristic optimization algorithms and what this term essentially means. By the end of this chapter, we will also come to understand why it is essential to have more than one of these optimization algorithms in our repertoire if we tend to use this branch of optimization method as the primary option to handle real-world complex optimization problems.

1.1 Introduction

What is optimization? That is perhaps the first and arguably the most critical question we need to get out of the way first. In the context of mathematics, *optimization*, or what is referred to from time to time as *mathematical programming*, is the process of identifying the best option from a set of available alternatives. The subtle yet crucial fact that should be noted here is that one's interpretation of what is "best" may differ from the others (Bozorg-Haddad et al., 2021; Zolghadr-Asli et al., 2021). That is why explicitly determining an optimization problem's objective is essential. So, in a nutshell, in optimization, we are ultimately trying to search for the optimum solution to find an answer that minimizes or maximizes a given criterion under specified conditions.

Optimization problems became an integrated part of most, if not all, quantitative disciplines, ranging from engineering to operations research and economics. In fact, developing novel mathematical programming frameworks has managed to remain a topical subject in mathematics for centuries. Come to think of it, there is a valid reason that optimization has incorporated itself into our professional and personal modern-day life to the extent it has. This is more understandable in the

DOI: 10.1201/9781003424765-1

context of engineering and management problems, where there are often limited available resources, and the job at hand is to make the best out of what is at our disposal. Failing to do so would simply mean that in the said procedure, whatever that may be, there is going to be some waste of resources. This could, in turn, imply that we are cutting our margin of profits, wasting limited natural resources, time, or workforce over something that could have been prevented if the process were optimized. So, in a way, it could be said that optimization is simply just *good common sense*.

There are several formidable approaches to go about mathematical programming. The traditional approach to solving optimization problems that are categorized under the umbrella term of *analytical approaches* is basically a series of calculus-based optimization methods. Often these frameworks are referred to as *derivate-based* optimization methods, given that they rely heavily on the idea of differential algebra and gradient-oriented information to solve the problem. As such, the core idea of these methods is to utilize the information extracted from the gradient of a *differentiable* function, often from the first- or second-order derivative, as a guide to find and locate the optimal solution. The main issue here is that this could not be a practical method to approach real-world optimization problems, as these problems are often associated with *high dimensionality*, *multimodality*, *epistasis*, *non-differentiability*, and *discontinuous search space* imposed by constraints (Yang, 2010; Du & Swamy, 2016; Bozorg-Haddad et al., 2017). As such, often, these methods are dismissed when it comes to handling intricate real-world problems as they are not by any means the ultimate practical approach to tackle such problems.

The alternative approach here would be to use a series of methods that are categorized under the umbrella term of *sampling*-based approaches. These, to some extent, use the simple principle of *trial-and-error* search to locate what could be the optimum solution. These methods are either based on *unguided* or *untargeted* search or the searching process that is *guided* or *targeted* by some criterion.

Some of the most notable subcategories of unguided search optimization methods are *sampling grid*, *random sampling*, and *enumeration*-based methods. The sampling grid is the most primitive approach here, where all possible solutions would be tested and recorded to identify the best solution (Bozorg-Haddad et al., 2017). In computer science, such methods are said to be based on *brute force* computation, given that to find the solution, basically, any possible solution is being tested here. As you can imagine, this could be quite computationally taxing. While this seems more manageable when the number of potential solutions is finite, in most, if not all, practical cases, this can be borderline impossible to implement such an approach to find the optimum solution. If, for instance, the search space consists of continuous variables, the only way to implement this method is to deconstruct the space into a discrete decision space. This procedure, known as *discretization*, transforms a continuous space into a discrete one by transposing an arbitrarily defined *mesh grid network* over the said space. Obviously, the finer this grid system, the better chance of getting closer to the actual optimum solution. Not only it becomes more computationally taxing to carry this task, but from a theoretical point of view, it is also considered impossible to locate the exact optimal solution for a continuous space with such an approach. However, it is possible to get a close approximation of the said value through this method.

Another unguided approach is random sampling. The idea here is to simply take a series of random samples from the search space and evaluate their performance against the optimization criterion (Bozorg-Haddad et al., 2017). The most suitable solution found in this process would then be returned as the optimal solution. Though this process is, for the most part, easy to execute, and the amount of computational power needed to carry this task can be managed by limiting the number of samples taken from the search space, as one can imagine, the odds of locating the actual optimum solution is exceptionally slim. This is, of course, more pronounced in complex real-world problems where there are often numerous continuous variables.

The other notable approach in the unguided search category is enumeration-based methods (Du & Swamy, 2016). These methods are basically a bundle of computation tasks that would be executed iteratively until a specific termination criterion is met, at which point the final results would be returned by the method as the solution to the optimization problem at hand. Like any other unguided method, here, there is no perception of the search space and the optimization function itself. As such, the enumeration through the search space would be solely guided by the sequence of computational tasks embedded within the method. In other words, such a method could not learn from their encounter with the search space to alter their searching strategies, which is in and of itself the most notable drawback of all the unguided searching methods.

Alternatively, there are also targeted searching methods. One of the most notable features of this branch of optimization is that they can, in a sense, implement what they have learned about the search space as a guiding mechanism to help navigate their searching process. As such, they attempt to draw each sample batch from what they learned in their last attempt. As a result, step by step, they are improving the possibility that the next set of samples is more likely to be better than the last until, eventually, they could gradually move toward what could be the optimum solution. It is important to note that one of the distinctive features of this approach, like any other sampling method, is that they aim to settle for a close-enough approximation of the global optima, better known as *near-optimal solutions*. The idea here is to possibly sacrifice the accuracy of the emerging solution to an acceptable degree to find a close-enough solution with considerably less calculation effort. One of the most well-known sub-class of the guided sampling methods is meta-heuristic optimization algorithms. However, before diving into what these methods actually are and what they are capable of doing, it is crucial that we improve our understanding of the structure of an optimization problem and its components from a mathematical point of view.

1.2 Components of an Optimization Problem

As we have discovered earlier, the optimization problem's main idea is to identify the best or optimum solution out of all possible options. Thus, the core idea of an optimization problem is to create a search engine that enumerates all possible solutions to locate what could be the optimum solution. As we have seen in the previous section, there are different approaches to solving an optimization problem, which, for the most part, comes down to how the said approaches tend to search through the possible solutions to, ultimately, locate what could be the optimal solutions. Analytical approaches, for instance, resort to calculus-based methods that use the obtained gradient-oriented information to solve the problem at hand. Alternatively, there are also sampling-based approaches, which, at their core, use the simple principle of trial-and-error search to locate the potential optimal solutions.

Regardless of what approach is selected to tackle the problem at hand, there is a fundamental requirement here to represent the optimization problem in a mathematical format. There are certain standard components that commonly help shape this mathematical representation of an optimization problem. To understand how these optimization methods proceed with their task, it is essential to learn how an optimization problem can be expressed through these components. With that in mind, the first step is to see what a standard optimization model is like.

From a mathematical standpoint, an optimization problem can be formulated as the following generic standard form:

$$\underset{X \in \Re^N}{Optimize\ f(X)} \tag{1.1}$$

Subject to

$$g_k(X) \le b_k \qquad \forall k \tag{1.2}$$

$$L_j \le x_j \le U_j \qquad \forall j \tag{1.3}$$

in which $f()$ represents the objective function, X is a point in the search space of an optimization problem with N decision variables, N denotes the number of decision variables, $g_k()$ is the kth constraint of the optimization problem, b_k denotes the constant value of the kth constraint, x_j represents the value associated to the jth decision variable, and U_j and L_j represent the upper and lower feasible boundaries of the jth decision variable, respectively. Note that in an optimization problem with N decision variables, an N-dimension coordination system could be used to represent the *search space*. In this case, any point within the search space, say X, can be represented mathematically as a $1{\times}N$ array as follows:

$$X = \left(x_1, x_2, x_3, \ldots, x_j, \ldots, x_N \right) \tag{1.4}$$

With this basic representation of an optimization model in mind, we can continue dissecting the structure of the said problem to isolate and study each component that is in play here.

1.2.1 Objective Function

As one's interpretation of what is best may differ from others, it is crucial in an optimization problem to explicitly define a mathematical procedure to evaluate how desirable an option actually is. The idea here is to establish a robust mathematical framework that quantifies the desirability of a potential solution. By doing so, it becomes possible to evaluate and compare the tentative options against one another so that, ultimately, the best solutions can be identified among all the possible options. In mathematical programming terminology, the *objective function* constitutes the goal of an optimization problem (Bozorg-Haddad et al., 2017). Within the context of an optimization problem, one could attempt to *minimize* or perhaps *maximize* the objective function value. In the former case, the idea is to identify a solution that yields the lowest objective function value possible, while in the latter case, the ultimate goal is to find a solution that is associated with the objective function's highest value. For instance, the objective function may be defined in a way to minimize the amount of risk imposed on a system (e.g., Issa, 2013; Zolghadr-Asli et al., 2018; Capo & Blandino, 2021) or perhaps maximize the profit of an operation (e.g., Husted & de Jesus Salazar, 2006; George et al., 2013; Kamrad et al., 2021). It should be noted that, from a theoretical standpoint, any given maximization can be expressed as a minimization problem and vice versa with a simple mathematical trick that is to multiply the said function by -1 (Bozorg-Haddad et al., 2017).

1.2.2 Decision Variables

In an optimization problem, there are so-called *variables* that, by changing their values, you are effectively creating new solutions. Naturally, each solution is associated with an objective function value. As such, the optimization can be seen as identifying the most suitable values for these variables. Whether this is a designing, operation, layout, or management problem, the critical feature associated with these variables is that they can be controlled through this process. In fact, that is why in mathematical programming terminology, these variables are referred to as *decision variables*, given that the whole point behind optimization is to decide which variable would be deemed the most suitable choice.

From a mathematical standpoint, as we have seen earlier, these variables could be bundled together to form an array. This bundle represents a solution to the said optimization problem, and the idea here is to find the right array that yields the best objective function value.

A decision variable could take the form of an *integer number*. For instance, the objective of an optimization problem could be to find the most economical way to place a number of filters to refine an industrial site's wastewater. In this case, the number of filters, which is an integer number, is the decision variable. Note that the problem may not even have a numeric variable. For instance, in the previous case, we might also want to determine the type of filters as well. In such case, for each filter, we have another decision variable that is by nature a *nominal* or *categorical* variable; that is to say, we want to figure out which type of filter should be installed to get the best result. The variable may also be *binary* in nature. This means that only two possible values can be passed for that variable. For instance, if we want to figure out whether an industrial site should be constructed in a place where we tend to maximize the margin of profits. Here the variable could be either going ahead with the project or shutting the project down. Mathematical programming terminology refers to all three cases as *discrete variables*. Alternatively, a decision variable may also be a *float number*, which is a number drawn from the *real number set*. An example of this would be when you want to determine the maximum amount of partially refined industrial site wastewater that can be released back into the stream without violating the environmental regulatory thresholds set to protect the natural ecosystem. In mathematical programming terminology, such a case is an example of a *continuous variable*. Of course, in real-world optimization problems, we may have a combination of discrete and continuous variables. These are said to be *mixed-type* optimization problems.

1.2.3 State Variables

In most practical, real-world optimization problems, such as those revolving around engineering or management-oriented problems, in addition to decision variables, we also deal with another type of variable called *state variables*. These are a set of dependent variables whose values would be changed as the decision variables' values are being changed. In a sense, these variables reflect how the decision variables' values affect the system and, in turn, its performance. As such, these would help get additional information about the system's state. It is important to note that these variables are not being controlled directly through the optimization process, but rather any change in these values is a product of how decision variables are selected. In the example we have seen earlier, where we wanted to determine the maximum amount of released wastewater to the stream, the stream's heavy metal concentration measures, for instance, are the state variables. Notice that the decision variable in this problem is the amount of outflow from the site, which can be controlled through the optimization problem. But as a consequence of such releases, the heavy metal concentration measures of the streamflow would be changed, making them state variables of this problem.

1.2.4 Constraints

Usually, optimization problems are set up in a way that decision variables cannot assume any given value. In other words, an optimization problem can be limited by a set of *restrictions* or *constraints* that bounds them between often two acceptable thresholds. Often, this is because resources are limited, and as such, it is impossible to pour unlimited supplies into a process or an operation. For instance, if you intend to optimize the company's workflow, there are budget and human resources limitations that need to be accounted for. In addition to this, there are some legal or physical restrictions that pose some limitations to the problem. For instance, in optimizing the schedule of a power plant, some safety measures restrict how you can operate the plant, even if there are resources available to pour into the system. Another example of this would be optimizing the design of an infrastructure, where tons of regulatory and physical restrictions must be considered in the optimization process.

From a mathematical standpoint, two types of constraints can be found in an optimization problem. The first type of restriction is what in mathematical programming dialect is referred to as *boundary conditions*. Here the restrictions are directly imposed on the decision or state variables themselves. As such, a decision or state variable should always be within a specified boundary. This could either mean that the said variable should always assume a value between a lower and upper boundary or that the boundary is only applied in one direction. A general representation of such boundary conditions can be seen in Equation (1.3).

The other form of restriction is *non-boundary conditions*, which pretty much sums up any other form of restriction that can be imposed on an optimization problem. From a mathematical standpoint, this means that a function of two or more decision or state variables is bound to hold a specific condition. This could mean that the said function should be equal to, greater, or less than a specified constant. A general representation of such boundary conditions can be seen in Equation (1.2). Note that while an optimization problem may have no condition, it may also be restricted by multiple conditions of different sorts.

In mathematical programming lingo, a solution that can hold all the possible constraints of the optimization problem is referred to as a *feasible solution*. On the other hand, if a solution violates even one condition, it is labeled an *infeasible solution*. It is important to note that the ultimate goal of optimization is to find a feasible solution that yields the best objective function value. This simply means that an infeasible solution can never be the answer to an optimization problem, even if it yields the best result.

1.2.5 Search Space

The idea behind optimization is to identify the so-called optimum solution out of all available solutions. As we have seen earlier, in an optimization problem with N decision variables, any solution can be mathematically expressed as an N-dimensional array. By the same token, the combination of all the possible solutions to an optimization problem can be interpreted mathematically as an N-dimensional *Cartesian coordinate system*, where each decision variable denotes an axis to the said system. This would create a hypothetical N-dimensional space, which in mathematical programming dialect is referred to as the *search space*. The idea is that any solution is a point within this coordination system, and the main point of optimization is to search through this space to locate the optimum solution.

As the previous section shows, not all solutions can be valid answers to an optimization problem. In fact, any answer that violates at least one constraint is considered an infeasible solution and, as such, cannot be passed as a viable result to the optimization process. With that in mind, the search space could be divided into

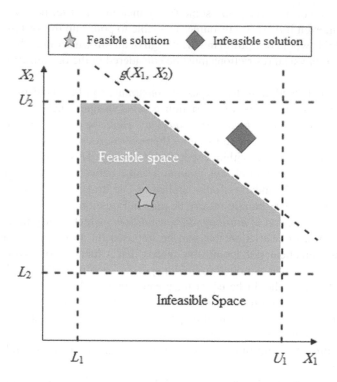

Figure 1.1 Search space of a standard constrained two-dimensional optimization problem.

two mutually exclusive sections that are *feasible space* and *infeasible space*. The former constitutes the portion of the search space where all constraints can be held, while one or more constraints are violated in the latter area. Naturally, the solution to the optimization problem must be selected from the former space. Note that the constraints can be assumed as a *hyperplane* that divides the search space into two mutually exclusive sections. Based on the condition of the said constraint, either one of these parts constitutes the feasible space, or the hyperplane itself denotes this space. If more than one condition is involved, the intersection of all the feasible portions will denote the feasible search space. In other words, the feasible space should be able to satisfy all the conditions of the optimization problem at hand. Note that, in mathematical programming lingo, the portion of the search space in which all the boundary conditions are met is called the *decision space*. Figure 1.1 illustrates the search space of a hypothetical constrained two-dimensional optimization problem.

1.2.6 Simulator

As we have seen earlier, state variables play a crucial role in the optimization process. Their role is often more pronounced in intricate real-world optimization problems where there is a form of dependency between the variables; or in cases where there is detectable *autocorrelation* in decision variables, that is to say, the value of the said variable in a given time step is statistically dependent to the values of the same variable in previous time steps; or lastly in cases where there are restrictions on some of the state variables. In such situations, which is often the case for complex real-world problems, capturing the underlying process that governs the system through mathematical equations is an absolute necessity. The said mathematical model, which in mathematical programming lingo is referred to as the *simulation model* or the *simulator*, for short, is an integrated part of the optimization model and the optimization process, as it enables the process to unravel the intricate relationship between the variables and compute their corresponding values at any given time. This mathematical representation would help compute the state and decision variables under any condition. If need be, the problem's condition can also be checked through these computed values. Therefore, these models are to emulate the nature of the system for the optimization problem, and as such, they are an inseparable part of the optimization models. In other words, you cannot have an optimization model without some form of a simulation model.

A notable thing about the simulation portion of the optimization models is that while this is the low-key part of the computational process for most benchmark problems, as you dive deeper into the real-world optimization problems, these simulators are naturally getting more intricate. This could, and often in complex real-world problems will, get to the point that the simulation portion of the optimization problem becomes the most computationally taxing part of the whole process. As such, one of the main goals of the optimizer would be to limit the number of times the simulator is called upon during the optimization process.

1.2.7 Local and Global Optima

The optimization problem's main task is locating the optimal solution in the search space. This could be either identifying the point greater than the other solutions, which is the case in the maximization problems or locating the point that yields the lowest value of the objective function, which is the task of the minimization problems. All the maxima and minima of an objective function are collectively known as *optima* or *extrema*. In mathematical analysis, we have two types of extrema or optima.

The first case is where the extremum is either the absolute largest or smallest value in the entire domain of the search space. In mathematical analysis lingo, these are referred to as the *absolute* or *global extrema*. Note that a function may have more than one global optimum. From a mathematical standpoint, a point X^* can be considered an absolute optimum of a minimization problem if the following condition can be held:

$$f\left(X^*\right) \le f\left(X\right) \qquad \forall X \qquad\qquad (1.5)$$

In a maximization problem, a global optimum can be defined as follows:

$$f(X^*) \geq f(X) \qquad \forall X \tag{1.6}$$

In addition to global or absolute extrema, mathematical analysis recognizes another form of optimum that is called *local* or *relative extrema*. A point is considered a local or relative extrema if it has either the largest or smallest value in a specified neighboring range. In other words, these values are either the maximum or minimum points in a specific section of the search domain. Again it is important to note that a function may have more than one local optimum. From a mathematical standpoint, a point X' can be considered a local optimum of a minimization problem if the following condition can be held:

$$f(X') \leq f(X) \qquad X - \varepsilon \leq X \leq X + \varepsilon \tag{1.7}$$

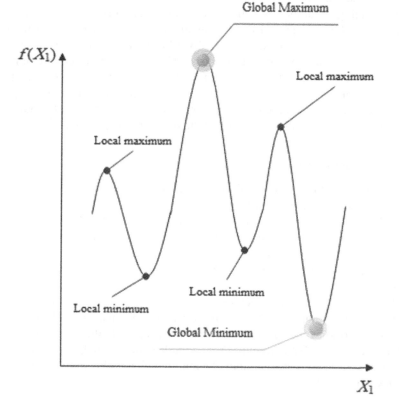

Figure 1.2 The relative and absolute extrema of a generic one-dimensional optimization problem.

in which ε denotes a very small positive value.

In a maximization problem, a local optimum can be defined as follows:

$$f(X') \geq f(X) \qquad X - \varepsilon \leq X \leq X + \varepsilon \tag{1.8}$$

Figure 1.2 depicts the relative and absolute extrema of an optimization problem with a single decision variable.

All the local and global maxima and minima of an optimization problem are collectively known as the extrema of the said problem. As we have seen, it is possible for an optimization problem to have more than one extremum. In mathematical programming lingo, an optimization problem with such characteristics is referred to as a *single-modal* or *unimodal* optimization problem. On the other hand, if a problem has more than one extremum, it is called a *multimodal* optimization problem. Figure 1.3 depicts the schematic form of the single-modal and multimodal maximization problem with one decision variable.

While in some practical problems, you may also need to identify all the extrema of the multimodal optimizing problem at hand, often, as one can imagine, the main objective of the optimization process is to identify all the possible global or absolute optima of a given problem. This, however, as we would explore through this book, is a bit challenging for sampling-based optimization methods. In this branch of optimization, it is often hard to distinguish local and global optima from one another. More importantly, there is no guarantee that you have encountered them all through the searching process or whether the point is, in fact, an extrema or we just did not happen to run into a point with better properties in our stochastic sampling process. So in optimization methods that are established on the idea of sampling, it is crucial to be aware of these so-called pitfalls as we strive to locate the absolute optimum solutions.

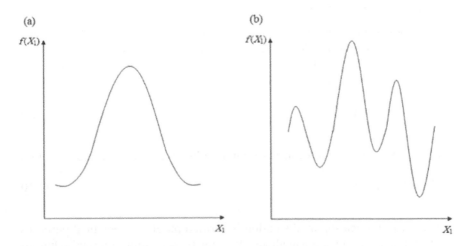

Figure 1.3 The generic scheme of (a) single-modal and (b) multimodal one-dimensional maximization problem.

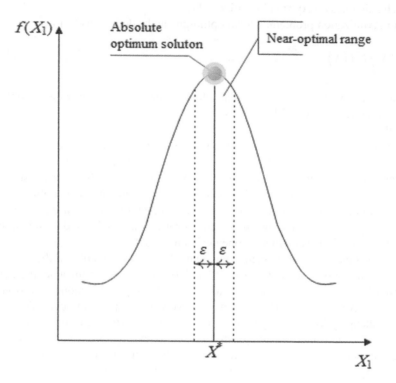

Figure 1.4 The generic scheme of near-optimal solutions in a one-dimensional maximization problem.

1.2.8 Near-Optimal Solutions

A *near-optimal solution* refers to an approximate feasible solution that is within the vicinity of the global solution, but it is not the absolute solution itself. From a mathematical point of view, a near-optimal solution, here denoted by P, of a minimization problem can be defined as follows:

$$f(P) \leq f(X^*) + \varepsilon \qquad (1.9)$$

In the maximization problem, an ε-optimal solution can be expressed as follows:

$$f(P) \geq f(X^*) + \varepsilon \qquad (1.10)$$

The idea of a near-optimal solution is formed purely for practical purposes. Most analytical optimization methods can identify the absolute optimal solutions, but as seen earlier, there are some inherent problems with these approaches when it comes to handling complex real-world problems. As such, sampling methods were introduced as an alternative approach to address these shortcomings. But the innate structure of these models poses some serious challenges for them as they tend to locate the absolute optimal solutions. In other words, while often they can get a close-enough approximation of the optimal solution to the problem at hand, as the problems get more intricate, they fail to identify the absolute optimal solutions. However, in most practical, real-world optimization cases, this can be passed as an acceptable compromise. In other words, often, the time and computational labor needed to find the absolute optimal solutions to such intricate problems are so overwhelming that it justifies settling down for a passable approximation of the optimal solution. Figure 1.4 depicts the central concept of the near-optimal solution in a maximization problem with one decision variable.

1.3 The General Theme of Meta-Heuristic Algorithms

As we have seen, using the gradient to detect the optimum point could pose serious computational challenges when dealing with real-world problems. Sampling-based optimization methods were an answer to these noted shortcomings. However, most of these methods were too random in nature to be considered a formidable option to handle complex optimization problems. For the most part, this is due to the fact that for most of these methods, there was no guiding system to navigate the search process. And this is where the *computational intelligence* (CI)-based optimization methods enter the picture.

CI-based optimization methods, which are often referred to as *meta-heuristic optimization algorithms*, are a subset of guided sampling methods. This branch of optimization methods is basically an alternative approach to accommodate the shortcomings of conventional analytical-based approaches and unguided sampling-based methods. In a sense, these methods were able to bridge the gap between these two vastly different approaches and create a new school of thought to handle real-world optimization problems.

Here, the optimization method is an algorithm composed of a series of iterative computational tasks that enumerate through the search space over a sequence of trial-and-error and often converge to a close-enough approximation of the optimum solution. In computer science lingo, the term algorithm refers to a finite sequence of explicitly defined, computer-implementable instructions to either solve a class of specific problems or perhaps to perform a predefined computational task (Yanofsky, 2011).

The term meta-heuristic is rooted in the Greek language, where "meta" translates to "upper level" or "higher level," and the phrase "heuristic" means "the art of discovering new strategies." In computer science terminology, the term *heuristic* refers to experience-derived approaches that are used to either solve a problem or gather additional information about a subject. Often, however, these methods can be seen as *ad hoc* problem-solving approaches given that they are primarily problem-dependent and hatched solely with a particular problem in mind. And as such, despite the fact that such approaches could often lead to reasonably close-enough solutions with adequate computational time, they cannot be seen as a universal, multi-purpose framework.

It is hard to pinpoint the exact origin of meta-heuristic algorithms accurately. A simple review of the literature on this subject would reveal that some algorithms and optimization methods were proposed over the years that defiantly resembled some of the characteristics of CI-based methods to some extent. But perhaps the most notable attempt to deviate from the principles of traditional optimization methods was the introduction of the pattern search algorithm. Theorized by Hooke and Jeeves (1961), the pattern search algorithm, also known as *random search*, represents, arguably, one of the pivotal points in the history of optimization techniques, as the principles used in this algorithm closely resembled those from the next generation of optimization methods that become known as meta-heuristic optimization algorithms.

By the early 1990s, many researchers had started conceptualizing CI-based frameworks to tackle optimization problems. These methods were based on the theoretical foundation of random sampling. The distinctive feature of these methods was that they were conducting a guided search that could be adjusted by the user to be tailored for a specific optimization problem. This is why Glover (1986) coined the phrase "meta-heuristic" to distinguish this branch of optimization methods. The idea is that these methods were to be conceptually ranked above heuristic methods in the sense that they can alter and control the structure of a heuristic framework, elevating them from the *ad hoc* method to a more universally applicable framework. It is important to note that these are still sampling methods in nature, and as such, they would inevitably inherit some of the generic drawbacks often associated with such methods. But ultimately, the innate characteristics of these methods enable them to go over a large number of feasible tentative solutions with fairly reasonable computation effort to locate what could be the near-optimal solution.

Often, meta-heuristic optimization algorithms are designed to be fine-tuned to better handle the search space and the problem at hand, hence the name meta-heuristic algorithm. The idea is that through a series of trial-and-error, the algorithm could guide us toward the vicinity of the optimum solution. If need be, it is possible to alter the algorithm to reflect and adapt to the unique characteristics of the search space. Again the backbone of these algorithms is to resort to approximation, hence being categorized under the umbrella of CI.

The other notable feature of these algorithms is rooted in their guiding mechanism. The search engine of these algorithms enables them to navigate the search space. But what distinguishes them from other known sampling-based optimization methods is that the searching process is guided by the information obtained by the algorithm during its searching phase. As such, these algorithms are, in a sense, learning from their encounter with the search space and utilizing the gained information to adjust their performance. But what is important here is that, unlike calculus-based optimization methods, here the guiding information is not derived from gradient-based information. Instead, these methods use a direct search to gain information about the search space and ultimately navigate their search process. In this context, a direct search indicates that the algorithm would utilize the function values directly in its searching process rather than resorting to information from the derivation of a function.

In order to bring together these ideas in a practical way, these algorithms need to be designed so that they can execute a series of commands repeatedly until they can confidently return what could be passed as the near-optimal solutions. In fact, the iterative process seen in all meta-heuristic algorithms allows these algorithms to thoroughly search the space by adjusting their performance in each attempt.

As a result of this unique computational formation, the structure of meta-heuristic algorithms can often be classified into three main stages, namely, *initiation*, *searching*, and *termination* stage. In the initiation stage, the algorithm would be set up to conduct the search. Here the algorithm would take all the input information, and the user would need to tune the algorithm via its *parameters*. The parameters of an algorithm are the mathematical instruments through which you can change the properties of an algorithm. After the initiation stage, the algorithm executes the search process. This is, for the most part, where the algorithm differs from one another. Meta-heuristic algorithms often look at the outside world as a source of inspiration to formulate a search strategy that helps them build an efficient search engine. This engine would help the algorithm to enumerate through the search space to locate what could be the near-optimal solution. As we have seen earlier, this is an iterative process, meaning that the engine would keep repeating the same strategy until it can finally stop the process. At this point, the algorithm enters the last phase, the termination stage. From a computational standpoint, this means that a termination criterion has been met, which signifies that the search process can come to an end. Note that without such a termination stage, the algorithm could be executed in an infinite loop. In effect, the termination stage determines whether the algorithm has reached what could be the optimum solution. While some algorithms are equipped with uniquely designed termination criteria, most meta-heuristic algorithms lack such features. As such, in order to implement these algorithms in practice, one would need to ensemble a termination criterion into their computational structure. Some of the most commonly cited options in this regard are limiting the number of iterations, run time, or perhaps monitoring the improvement made to the best solution in consecutive iterations. Among these options, limiting the number of iterations is arguably the most known mechanism to create a termination stage for meta-heuristic algorithms. The idea is that the predefined procedures of the algorithm would be executed only a specified number of times before coming to a halt and returning the best solution found in the process.

As a side note on the searching process, it is essential to know that often in this stage, the search engines of meta-heuristic optimization algorithms would ultimately need to go over two distinctive phases that are *exploration* and *exploitation* phase. In the exploration phase, also known as the *diversification phase*, the algorithm usually tends to explore the search space with more random-based moves that are more lengthy and sporadic by nature. After this phase, the algorithm would tend to focus more on specific regions of the search space where it deemed them more likely to house the optimum solution. This is the task of the exploitation phase, which is also referred to as the *intensification phase*. Having a search strategy that emphasizes adequacy on both these pillars is crucial for an efficient and thorough search. Often, the algorithms would tend to transition smoothly from the exploration phase to the exploitation. However, as we would see later in this book, there are those algorithms that are designed in a way that they can execute both phases simultaneously.

Today, meta-heuristic algorithms have established themselves as one of the most fundamental and pragmatic ways to tackle complex real-world optimization problems. These methods could theoretically handle high-dimensional, discontinuous, constrained, or non-differentiable objective functions. This does not mean that these methods are without their drawbacks. Firstly, the algorithm would pursue the near optima rather than the global optimum solutions. This, of course, does not pose a serious challenge to most practical cases. Secondly, given the stochastic nature of this process, and the trial-and-error component that is the inseparable part of meta-heuristic optimization methods, there are no guarantees that these algorithms could converge to the optimum solution. Also, there is always the possibility that the algorithm could be trapped in local optima, as there is no sure way here to distinguish the relative and absolute extrema. Of course, one can always try to play with the algorithm's parameter and, through a series of trial-and-error experiences, find a more suitable setting for the problem. The bottom line, however, is that while fine-tuning the parameters of a meta-heuristic optimization algorithm can undoubtedly have an impact on the final solution, and there are certain measures to see if an algorithm is performing the way it should, there can never be any guarantee that the emerged solution is indeed the global optimum. All in all, these shortcomings should not undermine the importance of these methods and their capacity to handle complex real-world problems.

1.4 Different Types of Meta-Heuristic Algorithms

Meta-heuristic algorithms have become one of the most topical subjects in the field of mathematical programming. Nowadays, there are hundreds of meta-heuristics available for those interested in employing these types of optimization methods. While these meta-heuristic algorithms are built upon the very same foundations we explained earlier, naturally, they differ from one another in numerous ways. From a pedagogical standpoint, it is never a bad thing to see how different methods are similar to one another, understand the common denominator in their computational structure, or dissect their general theoretical foundation. As such, in this section, we tend to explore how these methods are categorized in the literature.

1.4.1 Source of Inspiration

It became a common practice in the field of meta-heuristic optimization to ponder about the outside world as a source of inspiration to establish the computational foundation of a search engine. If you come to think of it, nature has evolved through millions of years to finally establish the perfect way to perform a specific task. In a sense, through countless iterations of trial-and-error, natural selection, and even the evolutionary process, nature has found the optimum solution for a given task, which may range from the foraging behavior of a group of animals, the reproductive procedure of a plant species, chemical, physical, or biological events, or perhaps even the evolutionary process itself. As such, meta-heuristic optimization algorithms find a way to imitate an abstract and idealized interpretation of these events with the hope that they could get the same effect as they search for the optimum solution within the search space. It used to be the case where algorithms with akin inspiration sources, such as those inspired by nature, were often based on similar analogy, terminology, and even computational structure. Modern meta-heuristic optimization algorithms tend to implement more intricate algorithmic architectures, making these classifications obsolete (Bozorg-Haddad et al., 2017). That said, understanding the inspiration source of an algorithm, or those that are somewhat related to it, could help you appreciate the computational intricacies of an algorithm and, in turn, become handy as one tends to implement the said algorithm.

1.4.2 Number of Search Agents

Meta-heuristic optimization algorithms are based on taking samples from the search space. In effect, they tend to enumerate through the search space via an iterative process. The algorithm would use the information obtained from previous encounters to adjust its performance and the searching process in each iteration. The act of searching would be conducted through what is known in CI lingo as a *search agent*. Although on the surface, the search agent has a mathematical structure of a solution that is an N-dimensional array, it plays a more crucial role in the computational structure of the algorithm. A search agent, from a computer science standpoint, is, in fact, a *mutable object*. The idea is that the algorithm recognizes a search agent as a unique object throughout the search process, even though this object may assume different values. In other words, the values of the array that is the mathematical representation of the search agent would be altered throughout the searching process, but the algorithm would always recognize this as the same computational object. This would, in turn, enable the algorithm to keep track of the search agent's movement during the run time of the searching stage. As such, the information obtained from this repositioning of the agent within the search space could provide the algorithm with additional information about the optimization problem. The algorithm would then use this feedback to suggest new positions for the search agent. And this process would be repeated until the algorithm reaches its termination point.

Some meta-heuristic algorithms would employ a single search agent to conduct the search. These are often referred to as *single-solution-based algorithms*. The other known terminologies for them are *trajectory methods* and *encompass local search-based meta-heuristic algorithms*. The reason is that the algorithm effectively traces an object's moving trajectory as it enumerates through the search space. Given that the searching process solely relies on tracing a single object throughout the iterations, these algorithms are often less computationally taxing. Alternatively, an algorithm may use multiple search agents. These algorithms are often referred to as *population-based algorithms*. Most modern meta-heuristic algorithms are, in fact, population-based. Having multiple search agents that simultaneously enumerate the search space would, in effect, increase the odds of locating the optimum solution while reducing the chance of being trapped in local optima, which happens to be one of the main pitfalls of meta-heuristic optimization. What is important here is that the unique computational structure of these algorithms often enables them to be less computationally taxing than one might initially think. The algorithm could carry this task with less computational effort through *parallel computation*, a common feature of most modern programming languages, including Python.

1.4.3 Implementation of Memory

An exciting feature of CI-based methods is the implementation of memory. While this is by no means a mutually exclusive or universal feature, when it comes to computation time, algorithms with this feature embedded in them would certainly have the edge over the competing methods. A memory-based feature could help the algorithm trace previously encountered information and, in turn, use this data to its advantage. Of course, having such capabilities would come at a price, primarily if not used efficiently, given that this requires more memory storage and computation power. That said, with the advancement of computational science and modern-day technologies, this is something that, at least to some extent, could be overshadowed by the many benefits that having such a feature would bring to the table.

By the same token, meta-heuristic optimization algorithms are divided into major groups when it comes to utilizing a memory-based feature within the computational structure of these algorithms, namely, *memory-based* and *memory-less* algorithms. In a memory-based algorithm, it is possible to keep track of the search agents' movements in the previous iteration. Notice that while it is always possible to trace all the moves that occurred during the searching process, it is certainly not the most efficient way to do things from a computational standpoint. In most real-world optimization problems, resorting to these practices may not even be possible, given the overwhelming memory use and computation power needed to keep up with the search agents, as this could quickly get out of hand. As such, the most particle way to implement a memory-based feature is to limit the memory's capacity to a certain number of events. After reaching its capacity, the memory would start dumping old data to make room for new information. Alternatively, in a memory-less algorithm, a *Markov chain*-like procedure would help update the position of the search agents. The idea here is that the algorithm's current stage only depends on its previous stage. As such, to reposition an agent, the algorithm would only need to know where a given agent is at the moment. Again, given the many advantages of having a properly designed memory-like feature in the structure of an algorithm, most modern-day meta-heuristic optimization algorithms are designed with this feature embedded in their algorithmic architecture (Bozorg-Haddad et al., 2017).

1.4.4 Compatibility with the Search Space's Landscape

As we have seen in the previous sections, there are two general types of decision variables; those that are of a continuous type and, of course, discrete decision variables. By the same token, there are two general types of search spaces, namely, *continuous* and *discrete search spaces*. What is important to note here is that, from a computational standpoint, searching through these search spaces would require different strategies. While some algorithms are innately designed to handle discrete search spaces, others can only tackle problems with continuous search spaces. That said, there are certain tricks to switch these two so that an algorithm would be compatible with both search spaces. The most notable idea here is discretization, where a continuous space would be transformed into a discrete one by transposing an arbitrarily defined mesh grid system over the said space. By doing so, you can use a discrete meta-heuristic algorithm to enumerate through a continuous search space. That, as you can imagine, is not without its drawbacks. The main problem is that you essentially lose some information in these transitions. As such, you cannot be sure if the optimum solution is, in fact, in the newly formed grid system, no matter how fine of a mesh grid you use for this transition. Often, however, the researchers would tend to use the governing computational principles of an algorithm to rebuild a *revised* or *modified* version of the algorithm that is compatible with the other type of search space. Note that researchers may also use this strategy to add new features or tweak the structure of an algorithm to create a more efficient modified algorithm (Yaghoubzadeh-Bavandpour et al., 2022). While the new algorithm is still built on the same principles, it can usually outperform its original standard version in one way or the other.

1.4.5 Random Components

As we have seen thus far, the structure of the meta-heuristic algorithms is built on the idea of guided random sampling. As the random sampling goes, we have three main ideas to draw a sample from a set; one being to select the samples based on pure randomness; we can select the samples using a systematic deterministic structure; and finally, there is the stochastic selection, which is something in between the previous approaches, meaning that while there is a random component to the selection procedure, it also follows some deterministic instruction as well. By the same token, meta-heuristic optimization algorithms can be categorized into two classes that are *stochastic* and *deterministic* algorithms.

In stochastic algorithms, a more common practice in this field, the algorithm has some random components within its computational structure. Often, you would find the random component in the initiation stage of the algorithms, where they randomly place the search agents within the feasible boundaries of the search space. These algorithms also introduce some random motion to the movement of the search agents. The idea is that doing so would reduce the odds of being trapped in local optima, as this component expands the search agents' exploration capacities. This also means that, due to the random nature of some of the computational procedures used in these algorithms, you may get different results each time you execute the algorithm. Though if appropriately set, these differences should not be so pronounced. Alternatively, the algorithm may be categorized as deterministic. This is not a common practice in this field by any means; in fact, there are only a handful of these algorithms. That said, the idea is that the computational structures of these algorithms, from the initiation stage down to the termination stage, are all based on deterministic-based instructions. As such, unlike stochastic algorithms, running the algorithm at any time should lead to the same result. But the main problem with these algorithms is that falling into the trap of local optima is always a real possibility. And if that happens to be the case, there is no way for the algorithm to get out of the said trap, no matter how many times you run the algorithm. And the only way to address this is, perhaps, to reset the algorithm's parameters, yet there is no guarantee that this could also solve the issue. Notice that we do not have a pure random meta-heuristic algorithm because we need a mechanism in the algorithm to *converge* the solutions into what could be the near-optimal solution. This is done through transiting between the exploration and exploitation phase. An algorithm that is built solely of random components would not have the necessary means to make this transition.

1.4.6 Preserving Algorithms

As we have seen, a meta-heuristic algorithm tends to trace the reposition of one or multiple search agents to gain information about the search space. In a sense, the algorithm is constantly updating the position of these agents within the search space, and by doing so, it is, in effect, drawing new samples from it. By doing so, the search agent would need to assume the new values, which, from the mathematical point of view, simply means that the array that represented the said agent would change in this update. There are two different schools of thought when it comes to perceiving these transitions by meta-heuristic optimization algorithms. One way of seeing things is to let any repositioning take place, regardless of whether it is an improving move. Often, the algorithms that follow this idea have a better exploration property. However, the sampling nature of meta-heuristic algorithms dictates that there are no guarantees that we are preserving the absolute best solutions within the search. This is significantly more pounced in memory-less algorithms.

Alternatively, there are those algorithms that are built on the idea that the best solutions must be somehow preserved during the search. From a practical point of view, there are two known ways to create such an effect. One is that the algorithm always preserves the best or even some of the best solutions as it iterates to locate the optimum solution. These ideas are computer science lingo called *elitism*, and an algorithm based on this idea is called an *elitist algorithm*. The other known approach here is only to permit the improving moves to occur and reject the non-improving ones. This idea in computer science lingo is known as the *greedy strategy*. Both these ideas can preserve the best solutions encountered through the search. But this also means that they would lose some of their exploration capacities. More importantly, the algorithm is more likely to be trapped in local optima, as they tend to only move toward better positions in the search space. These problems are more pronounced for single-solution-based algorithms and greedy strategy, given that in the first case, the search only relies on a single search agent, and in the second case, non-improving moves are not permitted through the search, meaning that there are no escaping the local optima if search agents assume these values. Notice that elitism and greedy strategy for a single-solution-based algorithm are, in effect, telling the same story, as in the end, they would have the same overall effect. Also, it is essential to bear in mind that either strategy can be later added to an algorithm to, perhaps, enhance its performance.

1.5 Handling Constraints

As we have seen, a solution that can satisfy all the constraints of optimization problems is referred to as a feasible solution. On the other hand, violating even one constraint would deem the solution infeasible. The main objective of the optimization methods is to locate the best possible feasible solution within the search space. In other words, an infeasible solution can never be seen as the answer to an optimization problem, even if it yields the best objective function value. With that in mind, given the exploratory sampling-based nature of the meta-heuristic algorithm, they must avoid returning an infeasible solution at any cost. There are basically three main strategies cited in the literature to handle this computational challenge: *removal, refinement*, and *penalty function* (Bozorg-Haddad et al., 2017).

Implementing the removal strategy to address the issue of passing infeasible solutions as the optimal solutions is perhaps the most primitive way to handle the given situation. According to this strategy, an infeasible solution must be eradicated from the search process at any given time it is presented by the search engines. As can be seen, this is a simple enough strategy to be executed or embedded in the structure of an algorithm. But there are a few notable shortcomings attributed to this method, which makes it not an ideal option when handling constraints in an optimization process. First and foremost, the algorithm must go the extra mile to create another solution, which might, again, be infeasible. The problem is more pronounced in deterministic algorithms as the new solution would undoubtedly be infeasible. So there is a possibility here that the algorithm may be stuck in a long, or even infinite, computational loop. The other problem is that by straight-up tossing the infeasible solutions from the search, you might lose some valuable information about the search space's landscape. This is mainly due to the fact that we are not making any distinction between a solution that slightly violates the constraints or a case where there are severe violations of the set restrictions. In the former case, the infeasible solution could have been used as a guide to help the search agent locate the optimum solution. All in all, while this is undoubtedly an easy way to handle constraints, it certainly is not the most efficient way to do so.

Alternatively, based on the refinement strategy, you can save the infeasible solution by the process as you adjust the variables to be deemed feasible again. As such, unlike the removal strategy, here, you are not directly eradicating any infeasible solution from the search process. While this is a manageable strategy when the violations are restricted to the boundary conditions, any other type of violation is rather hard to be refined by this method. In the former case, you can simply replace the value of the violating variable with one of the threshold values, which is a reasonably simple task. However, if the constraints involve multiple decision variables, there is no universal or straightforward approach to alter the solution in a way that transforms it into a feasible solution. And more importantly, even if you can make this transition, there is no way to tell if this is the best way to change an infeasible solution into a feasible one when it comes to preserving additional information about the search space, which is the whole point behind these strategies. As such, while this is an acceptable remedy to keep the solutions within the decision space, it is not necessarily the best option for constraints that involve multiple decision or state variables.

The last option to address constraints is implementing the idea of penalty functions. The idea here is to penalize the objective function value proportional to how much the said solution has violated the constraints of the problem at hand through a mechanism called *the penalty function*. In effect, a penalty function determines the severity of these penalties based on how much a solution has violated a said constraint. Naturally, the more violation forms a solution, the more severe these penalties would get. So in a maximization problem, as a penalty, a positive value would be subtracted from the objective function value rendering it less suitable. In contrast, a positive value would be added to the objective function value in a minimization problem. It is important to note that the amount of this penalty function must be directly proportional to how much a solution violates an optimization problem's constraints. The other notable thing is that each constraint could have its own penalty function mechanism. The overall penalty imposed on a solution is the accumulated values of all the said penalties. In mathematical programming lingo, a penalized objective function is referred to as the *cost function* or *fitness function*. This can be mathematically expressed as follows:

$$F(X) = f(X) \pm p(X) \tag{1.11}$$

where $F()$ represents the fitness function, and $p()$ denotes the penalty function. Note that we would subtract the penalty values from the objective function in the maximization problem, while in minimization problems, we would add the penalty values to the objective function.

The generic form to create the penalty function for a standardized constraint (i.e., a constraint that is written as a *less than or equal to* expression), such as the one depicted in Equation (1.2), can be formulated as follows:

$$p_k(X) = \begin{cases} \vartheta_k & \text{if} \quad g_k(X) > b_k \\ 0 & \text{if} \quad g_k(X) \le b_k \end{cases} \qquad \forall k \tag{1.12}$$

$$\vartheta_k = \alpha_k \times \left[g_k(X) - b_k \right]^{\beta_k} + \gamma_k \qquad \forall k \tag{1.13}$$

in which $p_k()$ denote the penalty function for the kth constraint; ϑ_k represents the amount of penalty that is associated with the kth constraint; and α_k, β_k, and γ_k are all constant values for adjusting the magnitude of the penalty function for the kth constraint. Using these constants, you can control the amount of penalty that would be applied to the objective function. In other words, increasing these values would apply a more severe penalty to the objective function. However, using the proper penalizing mechanism is a critical subject if you want to get a good result out of these algorithms. Because the point is, applying acute penalties would be, in effect, equivalent to omitting the infeasible solutions. Applying mild penalties could also have no visible result on the search algorithms' emerging solution. So it is crucial to keep a balance between two extreme situations. However, finding the proper values for these parameters would need some prior experience and perhaps sensitivity analysis.

By the same token, you can create a penalty function for any other constraint type. Of course, the general idea is to apply the penalty whenever the said solution violates the said constraint. All these penalties would be collectively applied to the objective function to get the fitness function value of the solution. Through this mechanism, in effect, you would map the objective function to the fitness function values, which would be used from that point onward as the evaluation mechanism. As such, the algorithm would use these values to determine how good a solution is and, in turn, select the optimum solution based on these values. Again, it is essential to note that, like meta-heuristic algorithms, this is also, by nature, a trial-and-error procedure, meaning that there could never be an absolute guarantee that the final solution is never infeasible or there is no better feasible solution in the search space.

1.6 Performance of a Meta-Heuristic Algorithm

As we have seen thus far, the whole idea behind meta-heuristic algorithms is to use the principle of guided sampling to locate the optimal solutions. But due to the nature of these algorithms, they often settle for identifying a near-optimal solution. What happens behind the scene is that the algorithm would continually enumerate through the search space by tracing the movement of the search agents. And if the algorithm is set up correctly, little by little, it gets closer and closer to a specific position in the search space until there is no room for any improvement. At this point, it is said that the algorithm has *converged* to the optimum solution. If, however, the algorithm's searching process terminates before this point, it is said that there is *premature convergence*; That is to say, the emerged solution from this search is not actually the optimum solution, and the search should have continued for a bit longer. So, as can be seen here, the final solution of a meta-heuristic algorithm, in and of itself, cannot be sufficient to interpret the quality of the result. Instead, you need to trace the progression of the search process to see how reliable the emerging solution actually is. Analyzing the *convergence rate* of an algorithm is one of the primary tools for understanding the performance of an algorithm and the reliability of the outcome solutions.

One of the simplest ways to tackle this matter is to plot the algorithm's convergence rate throughout the search. Here, like any data-driven science, *visualization* could offer reliable guidance to unravel the intricacies of a problem at hand. In order to do that, either the objective or the fitness function values would be plotted against either the run time, number of iterations, or, more commonly, a technical measure called the *number of function evaluations*, or NFE for short. NFE is a technical term that refers to the number of times the objective function has been called during the search. To understand this concept, we must first understand how the algorithm actually converges throughout the search.

Figure 1.5 illustrates how the algorithm's solutions converge as the search progresses for a minimization problem. For simplicity, let us assume that we are tracing the progression of the search against the iteration or the run time metrics. This should be reasonably straightforward for the single-solution-based algorithms, as we deal with a single search agent in this type of algorithmic structure. However, in population-based algorithms, we could select the best solution in the population set as the sole representative of the said set. As such, the said values would be plotted against time or the iteration count to trace the progression of the search process. As can be seen here, often, the algorithm experiences a sudden and massive drop within the first few iterations. But as the algorithm progresses, these changes become more subtle and less pronounced until, eventually, there are no visible changes in the values of the objective function. Again, it is essential to note that if there are notable changes in the tail of the convergence graph; this means that there is premature convergence, and you need to recalibrate the algorithm to get the optimum result. This can often be addressed by extending the termination process so the algorithm can do more iterations. It is also essential to note that there are two generic types of convergence graphs. In meta-heuristic algorithms that are based on the idea of preservation strategies, say elitism or greedy strategy, the graphs are *monotonic*, meaning that if the algorithm's returned values are decreasing, there are no sudden upticks in the graph, and if the values are increasing, there cannot be a decreasing section in the plot. This is mainly because these sorts of algorithms always preserve the best solutions in each iteration, so the algorithm is either continuously improving or at least staying the same. The graph of algorithms that do

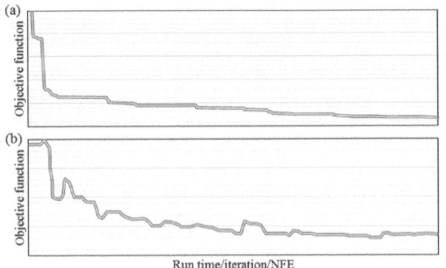

Figure 1.5 Convergence of an optimization problem toward the minimum solution of (a) an algorithm with a preservation strategy and (b) an algorithm with no preservation strategy.

not resort to preservation strategies is non-monotonic, meaning that there can be sudden, often mild, jumps in the graph. In either case, as the algorithm transits from the exploration phase to the exploitation phase, you would see a shift in the tone of the graph as it becomes more static.

Now let us expand on the idea of convergence rate in the context of different termination metrics. The first idea is to use the idea of run time, that is, to plot the objective function against the time that took the algorithm to reach the said value. In data science lingo, any graph that is plotted against the time variable is called a *time series*. While this graph has helpful information about the algorithm's performance in terms of how fast an algorithm actually is, there are a few things that need to be understood when you want to analyze these types of plots. First and foremost, you need to understand that the computation speed of an algorithm is not solely a reflection of the innate computational structure of an algorithm, but it is also a product of the quality of the executed code, the programming language itself, and of course the system that is doing the heavy lifting of the computational process. For instance, *compiled* programming languages such as C++ are inherently faster than *interpreted programming languages* such as Python. As such, it is crucial to mention which programming language was used to execute the meta-heuristic optimization algorithm. Naturally, the system that is used to carry the computation also plays a crucial role in how fast the code is executed. Again, that is why it is essential to report the *system configuration* with the obtained result, including the technical properties of the system's *random-access memory* (RAM), *central processing unit* (CPU), or *graphics processing unit* (GPU), depending on which processor was used to do the computation, and even the *operating system* (OS) can play a role here. It is important to note that comparing the convergence speed of two given meta-heuristic optimization algorithms or different parameter settings for a given algorithm would only make sense if the systems that executed them were identical. Lastly, there is the quality of the code in which the algorithm was written down in the given system. Naturally, while this is indeed a crucial factor in how fast or smoothly an algorithm handles a given optimization problem, it is quite challenging actually to quantify this factor in an objective way. The programmers' skills and knowledge of a programming language could help them find a more efficient way to implement an algorithm. For instance, it is entirely plausible for a more experienced programmer to find an elegant way to improve the performance of a coded algorithm by a beginner programmer in terms of the speed and efficiency of the program. And while this is indeed an essential factor in the convergence speed, it is impossible to report any objective information on this matter. All in all, while it is not a bad idea to have some idea about the run time of an algorithm, as we have seen here, it is not necessarily the best metric to use as an analytical tool to study the convergence of an algorithm.

The other option here is the number of iterations that took the algorithm to reach an objective function. Here, we could plot the objective function values against the iteration numbers. While, in this approach, we are certainly excluding the innate problems that we have named earlier, which are the system's computational power, the technical properties of the programming language, and, more importantly, the programmers' skill set from our analysis, there is still one major subtle problem with the way that we are breaking down the performance of the algorithms.

As we have seen in the previous section, some algorithms use one agent to enumerate through the search agent, while others may employ multiple agents to carry out this task. So in a single iteration, the former group is arguably doing less computation than the latter group. This is important to note, given that in most real-world optimization problems, the most computationally taxing part of the process is calling the simulator that gets activated as you call the objective function in your algorithm. Some algorithms may even call this function multiple times within the same iteration for each search agent, which requires additional computational power. Of course, this would also mean that such algorithms may have a better chance of converging to near-optimal solutions. So, the metric that should be monitored here is not the number of iterations per se but rather the number of times the objective function is called during the search. This is, in fact, the definition of the NFE. By using this metric, you are factoring in the number of search agents and, more importantly, the innate efficiency of the algorithm itself and not the programmer's skill set. For instance, an algorithm that often calls the objective function would consequently have a greater NFE than a more efficiently designed algorithm that rarely calls this function during the search. As such, plotting the objective function against the NFE values is a more reasonable way to analyze or even compare the convergence of meta-heuristic algorithms.

Aside from the visualization of the convergence rate, we need other quantifiable metrics to analyze the performance of an algorithm. These would not only be used in the fine-tuning procedure to get the best result out of an algorithm for a specific optimization problem, but they are also helpful when comparing the performance of two or even more algorithms. Again, to apply these measures, we must first run the algorithm with the same setting for a specified number of times. This is mainly due to the fact that often these algorithms are based on stochastic sampling. This means that the algorithm would return a different solution that may slightly differ for each run. So the idea is after fully running the algorithms numerous times, you would use these metrics to evaluate their performance. As such, in addition to reporting the solutions obtained in each run, you can report these metrics to give a more wholesome idea about the performance of the algorithms. There are four general areas to evaluate an algorithm's performance that are *efficacy*, *efficiency*, *reliability*, and, finally, *robustness*.

In this context, efficacy-based metrics measure how an algorithm is performing regardless of its convergence speed. Three known metrics here are *mean best fitness* (MBF), the *best solution in all runs* (BSAR), and the *worst solution in all runs* (WSAR) (Du & Swamy, 2016; Bozorg-Haddad et al., 2017). MBF denotes the average of the best objective or fitness function value of all runs. BSAR denotes the best objective or fitness function value that was found among all the runs. By the same token, WSAR denotes the worst objective or fitness function value that was found among all the runs.

The idea behind reliability-based metrics is to measure the ability of an algorithm to provide acceptable results. The most notable measure here is the *success rate* (SR) (Du & Swamy, 2016). To work on this measure, you must first identify the best solution among all the runs. Any time the solution comes within an acceptable range of this value, meaning that you also need to set a predefined threshold here as well, the said runs are deemed successful. SR is the ratio of successfully executed runs over the total number of runs. As seen here, the SR has a probabilistic nature; in a way, it can be interpreted as the probability of executing a successful run.

Efficiency-based measures tend to quantify the speed of an algorithm in identifying the optimum solution. The most notable measure here for this is the *average number of evaluations of a solution* (AES) (Du & Swamy, 2016). As the name suggests, this metric measures the average number of evaluations it takes for the algorithm to be deemed successful. If an algorithm did not have a so-called successful run, it is said that ASE is undefined.

Robustness-based measures determine how persistent an algorithm is in different runs. A so-called robust algorithm would always return reasonably similar solutions, while a non-robust algorithm's solutions vary significantly in each run. Figure 1.6 depicts the difference between robust and non-robust algorithms. As can be seen here, the non-robust algorithm final solution would rely heavily on the initiation stage, which is naturally not a good feature to have in a meta-heuristic algorithm. The most notable measures for this metric are *standard deviation* and the *coefficient of variation* of solutions in different runs (Bozorog-Haddad et al., 2017).

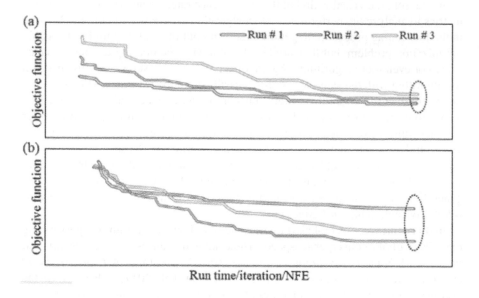

Figure 1.6 Convergence rate of an algorithm that is considered (a) robust and (b) non-robust.

1.7 No-Free-Lunch Theorem

So, the final yet very crucial question we need to answer here is why we have so many meta-heuristic optimization algorithms these days, which are seemingly all producing good results. And in fact, by the same token, do we actually need to have multiple algorithms in our repertoire to tackle optimization problems? And the answer to both questions is a resounding yes due to a known theory in data science called the *no-free-lunch theorem*.

No-free-lunch theorem is a known impossibility theorem with fundamental application across data-driven models such as meta-heuristic optimization. The gist of this idea is that if we have no prior assumption about the problem at hand, no data-driven strategy, model, or method would have a pre-advantage over the other, given that there could never be a guarantee that one approach here could outperform the others (Culberson, 1998). So this theory is actually telling us that when it comes to data-driven models such as meta-heuristic algorithms, a general-purpose, universal optimization strategy can never exist (Wolpert & Macready, 1997). So the no-free-lunch theorem is the epitome of the idea "if anything is possible, then nothing can be expected" (Ho & Pepyne, 2002).

But there is another implication that comes from this theorem. If there can be no universally superior algorithm, if there is no prior assumption about the given problem, there can also never be an ideal setting for a given algorithm that can handle all the optimization problems. In other words, if we want to get the best performance out of a meta-heuristic optimization algorithm, we have to always look for the best parameter setting for the said algorithm for that specific problem at hand. This means that a universally best parameter setting for an algorithm could never exist, and as such, fine-tuning an algorithm is an inseparable part of meta-heuristic optimization algorithms. Of course, it is always possible to use our intuition, experience, and default values suggested for an algorithm's parameters as a good starting point. That said, one should bear in mind that fine-tuning these parameters is, more than anything, a trial-and-error process. Thus, while it is possible to get a good enough result by having an educated guess for setting the parameters of these algorithms, to get the best possible performance, it is necessary to go through this fine-tuning process.

1.8 Concluding Remarks

In order to embark upon our journey to the fascinating realm of meta-heuristic optimization algorithms, we first had to learn about the general concepts of mathematical programming or what we commonly know as optimization. We learned the general approaches available to conduct optimization and the merits and drawbacks of each idea. For instance, we saw that analytical methods are guided approaches that use gradient-based information to locate the optimum solutions. The main problem with these methods is that they have difficulty handling real-world optimization problems. These problems are often associated with high dimensionality, multimodality, epistasis, non-differentiability, and discontinuous search space imposed by constraints. Alternatively, we also have sampling-based optimization methods. These methods are based on the idea of searching among the possible solutions until we come across a solution that can be passed as the optimum solution. We have two general groups of sampling-based methods that are unguided and guided methods. Unguided methods could not learn from their encounter with the search space to alter their searching strategies, which is, in and of itself, the most notable drawback of this branch of optimization. This is not the case for guided or targeted methods. The most notable example of this is meta-heuristic optimization methods. These methods would search the optimization problems and are not inherently bound by the drawbacks we commonly associate with analytical-based methods. However, one thing that should be noted about these methods is that there can never be a guarantee that these methods can always locate the optimal solution, and more often than not, they return the near-optimal solutions rather than the absolute optimal solution.

We also reviewed the main components of optimization problems and the generic representation of an optimization model. We have come to learn about the concepts of objective functions, decision variables, state variables, constraints, search space, simulator, local and global optima, and near-optimal solutions. We briefly reviewed the history of meta-heuristic algorithms and different known classifications for these algorithms. We have also learned a few tricks to handle constraints in the context of meta-heuristic algorithms. Then we reviewed the metrics used to analyze these algorithms' performance. And as a final note, we have discussed the idea of the no-free-lunch theorem. In a nutshell, the two implications of this theorem in the context of meta-heuristic optimization algorithms are that if there are no prior assumptions about the problem at hand, there can never be a universally superior algorithm, nor can be a universally ideal parameter setting for a given algorithm. This means we have to have multiple meta-heuristic optimization algorithms in our repertoire and do parameter fine-tuning to get the best performance out of the algorithm. That said, we can finally start learning about different algorithms in this field.

References

Bozorg-Haddad, O., Solgi, M., & Loáiciga, H.A. (2017). *Meta-heuristic and evolutionary algorithms for engineering optimization.* John Wiley & Sons. ISBN: 9781119386995

Bozorg-Haddad, O., Zolghadr-Asli, B., & Loaiciga, H.A. (2021). *A handbook on multi-attribute decision-making methods.* John Wiley & Sons. ISBN: 9781119563495

Capo, L. & Blandino, M. (2021). Minimizing yield losses and sanitary risks through an appropriate combination of fungicide seed and foliar treatments on wheat in different production situations. *Agronomy*, 11(4), 725.

Culberson, J.C. (1998). On the futility of blind search: An algorithmic view of "no free lunch". *Evolutionary Computation*, 6(2), 109–127.

Du, K.L. & Swamy, M.N.S. (2016). *Search and optimization by metaheuristics: Techniques and algorithms inspired by nature.* Springer International Publishing Switzerland. ISBN: 9783319411910

George, M., Kumar, V., & Grewal, D. (2013). Maximizing profits for a multi-category catalog retailer. *Journal of Retailing*, 89(4), 374–396.

Glover, F. (1986). Future paths for integer programming and links to artificial intelligence. *Computers & Operations Research*, 13(5), 533–549.

Ho, Y.C. & Pepyne, D.L. (2002). Simple explanation of the no-free-lunch theorem and its implications. *Journal of Optimization Theory and Applications*, 115(3), 549–570.

Hooke, R. & Jeeves, T.A. (1961). "Direct Search" solution of numerical and statistical problems. *Journal of the ACM*, 8(2), 212–229.

Husted, B.W. & de Jesus Salazar, J. (2006). Taking Friedman seriously: Maximizing profits and social performance. *Journal of Management Studies*, 43(1), 75–91.

Issa, U.H. (2013). Implementation of lean construction techniques for minimizing the risks effect on project construction time. *Alexandria Engineering Journal*, 52(4), 697–704.

Kamrad, B., Ord, K., & Schmidt, G.M. (2021). Maximizing the probability of realizing profit targets versus maximizing expected profits: A reconciliation to resolve an agency problem. *International Journal of Production Economics*, 238, 108154.

Wolpert, D.H. & Macready, W.G. (1997). No free lunch theorems for optimization. *IEEE Transactions on Evolutionary Computation*, 1(1), 67–82.

Yaghoubzadeh-Bavandpour, A., Bozorg-Haddad, O., Zolghadr-Asli, B., & Gandomi, A.H. (2022). Improving approaches for meta-heuristic algorithms: A brief overview. In Bozorg-Haddad, O., Zolghadr-Asli, B. eds. *Computational intelligence for water and environmental sciences*, Springer Singapore, 35–61.

Yang, X.S. (2010). *Nature-inspired metaheuristic algorithms.* Luniver Press. ISBN: 9781905986286

Yanofsky, N.S. (2011). Towards a definition of an algorithm. *Journal of Logic and Computation*, 21(2), 253–286.

Zolghadr-Asli, B., Bozorg-Haddad, O., & Loáiciga, H.A. (2018). Stiffness and sensitivity criteria and their application to water resources assessment. *Journal of Hydro-Environment Research*, 20, 93–100.

Zolghadr-Asli, B., Bozorg-Haddad, O., & van Cauwenbergh, N. (2021). Multi-attribute decision-making: A view of the world of decision-making. In Bozorg-Haddad, O. ed. *Essential tools for water resources analysis, planning, and management*. Springer, 305–322.

2   Pattern Search Algorithm

Summary

The pattern search algorithm is arguably one of the earliest, if not the first, meta-heuristic optimization algorithms. The ideas and novelties used in this algorithm would help shape the next generation of optimization algorithms. In this chapter, we will dig deep and explore the mechanisms used in this algorithm. We would then see how one can implement the pattern search algorithm in the Python programming language. Finally, we will explore the potential merits and drawbacks of this algorithm.

2.1 Introduction

The core principle of *derivate-based optimization* methods is to utilize the information extracted from the gradient of a *differentiable* function, often from the first- or second-order derivative, as a guide to find and locate the optimal solution within the *search space*. As it turned out, this could not be a practical method to approach real-world optimization problems, as these problems are often associated with *high dimensionality*, *multimodality*, *epistasis*, *non-differentiability*, and *discontinuous search space* imposed by constraints (Yang, 2010; Du & Swamy, 2016).

One of the main characteristics of conventional derivative methods, primarily calculus-based optimization techniques, is that they are designed to find the *exact* solution or *global optimum* to the problem at hand. This, as stated, could be a challenging task, if not impossible, in some practical cases. Alternatively, one could settle for a close-enough approximation of the global optima, better known as *near-optimal solutions*. Methods that are designed based on this paradigm are known as *approximate algorithms*. The idea here is to possibly sacrifice the accuracy of the emerging solution to an acceptable degree in order to find a close-enough solution with considerably less computation effort. But this is not a novel approach to solving mathematical problems per se, as this notion describes the fundamental foundation of most enumeration-based methods. These methods are, at their core, a bundle of computation tasks that would be executed iteratively until a certain termination criterion is met, at which point the algorithm would return the final results as the approximated solution. But these algorithms are what is known as

DOI: 10.1201/9781003424765-2

unguided or untargeted sampling-based methods, that is, the algorithm would have no perception of the search space and the function, and as such, the enumeration through the search space would be solely carried out by executing a sequence of computational tasks embedded within the algorithm. This contrasts calculus-based methods, where the function's gradient would be used as a guide for the optimization process. As we have seen, using gradient could be problematic in real-world problems, but one should also bear in mind that using some properties of the function could potentially enhance the searching capabilities of an optimization algorithm. And this is where computational intelligence (CI)-based optimization methods enter the picture.

CI-based optimization methods, often known as meta-heuristic algorithms, are basically alternatives to accommodate the shortcomings of conventional derivate-based optimization methods. Here the optimization method is, in fact, an algorithm composed of a series of iterative computational tasks that, over a sequence of trial-and-errors, enumerates through the search space and often converges to a close-enough approximation of the optimum solution. In most cases, these algorithms are designed so that they can be fine-tuned to better handle the search space and the problem at hand, hence the name meta-heuristic algorithm. The idea is that through a series of trial-and-error, the algorithm could guide us toward the vicinity of the optimum solution, and if need be, it is possible to alter the algorithm to better match the characteristics of the problem at hand. Again the backbone of these algorithms is to resort to approximation, hence being categorized under the umbrella of CI.

It is hard to pinpoint the exact origin of meta-heuristic algorithms accurately. A simple review of the literature on this subject would reveal that some algorithms and optimization methods were proposed over the years that defiantly resembled some of the characteristics of CI-based methods to some extent. But perhaps the most notable attempt to deviate from the principles of traditional optimization methods was the introduction of the pattern search algorithm. Theorized by Hooke and Jeeves (1961), the pattern search algorithm, also known as random search, represents, arguably, one of the pivotal points in the history of optimization techniques, as the principles used in this algorithm closely resembled those from the next generation of optimization methods that become known as meta-heuristic.

Pattern search is a single-solution optimization method, as the computational procedures are based on a single agent that moves through the search space. The position of the said agent would be adjusted through the optimization process, where the information of nearby points to the said agents would be used as a guide to creating a trajectory that would likely converge to the near-optimal solution. As such, the said method is categorized as a *local search*, for in each iteration, the agent's trajectory is guided solely by the information gathered from the points in the vicinity. The said algorithm is also considered stochastic, given that the agent's initial position is designated randomly within the search space confinements. But what distinguished the pattern search algorithm from its precursors was how it guides the agent through the search space. The pattern search algorithm is categorized as a direct search algorithm. In this context, a direct search indicates that the algorithm would utilize the function values directly in its computation process rather than resorting to information from the derivation of a function.

This was undoubtedly a drastic shift in how optimization methods were perceived to explore the search space. The algorithm could theoretically handle high-dimensional, discontinuous, constrained, or non-differentiable objective functions. This was not, unfortunately, without its drawbacks. Firstly, the algorithm would pursue the near optima rather than the global optimum. Secondly, given the stochastic nature of this process, which will be discussed in the later section, and the trial-and-error component that is the inseparable part of meta-heuristic optimization methods, there are no guarantees that the algorithm could converge to the optimum solution. This turned out to be one of the main characteristics of meta-heuristic optimization. Of course, one can always try to play with the algorithm's parameters and, through a series of trial-and-error-based experiences, find a more suitable setting for the problem. The bottom line, however, is that while fine-tuning the parameters of a meta-heuristic optimization algorithm can undoubtedly have an impact on the final solution, and there are certain measures to see if an algorithm is performing the way it should, there can never be any guarantee that the emerged solution is indeed the global optimum.

Despite being one of the earliest meta-heuristic algorithms out there, the pattern search algorithm demonstrated great potential as it was used to solve a variety of problems such as data engineering (e.g., Shehab et al., 2007), energy industry (e.g., Alsumait et al., 2010; AlHajri et al., 2012), environmental assessments (e.g., Khorsandi et al., 2014), food industry (e.g., Smitabhindu et al., 2008), geo-hydrological simulation (e.g., Bozorg-Haddad et al., 2013), human resources management (e.g., Swersey, 1994), hydrological modeling (e.g., Tung, 1985; Moradkhani & Sorooshian, 2009), oil industry (e.g., Bellout et al., 2012), quality control (e.g., Banerjee & Rahim, 1988; Carlyle et al., 2000), structural engineering (e.g., Evins, 2013; Machairas et al., 2014), and system monitoring and analysis (e.g., Tillman et al., 1977; Caponio et al., 2007), to name a few. Even more importantly, the fundamental ideas that shaped this algorithm would later be a source of inspiration for other meta-heuristic optimization algorithms, such as the tabu search algorithm (Glover, 1989, 1990), some of which will be discussed in the following chapters. With that said, we can explore the computational structure of the pattern search algorithm.

2.2 Algorithmic Structure of Pattern Search Algorithm

The pattern search algorithm, as stated, follows a fundamental principle called direct search. The idea is that the objective function values could be used as a guide for the navigation of the algorithm through the search space. To do that, the algorithm initiates by selecting a random point within the search space. The said point, which in pattern search algorithm terminology is referred to as the *base point*, would serve as a *search agent* for the algorithm. Due to the random nature of the algorithm initiation stage, running the pattern algorithm with a similar setting for a specific optimization problem may lead to different results. Although if the algorithm's parameters are fine-tuned carefully, the difference between emerging solutions should be statistically negligible.

After the initiation stage, the algorithm would conduct a series of local searches and select a set of *trial* or *tentative points* within the vicinity of the search agent. During this search process, any time the algorithm comes across a point with better properties (i.e., better objective function) than the current base point, the said point would replace the current base point. This is often referred to as a *greedy strategy*, where the search agent's position would be updated on the sole condition that the "new place" has better properties. This process would be repeated until a certain termination criterion is met, upon which the current base point would be returned as the potential optimum solution. Again, it is crucial to bear in mind that there is no guarantee that the emerged solution is, in fact, the global optima.

The pattern search algorithm's flowchart is depicted in Figure 2.1. As can be seen, the computational structure of the pattern search algorithm is fairly simple and straightforward. Upon closer investigation, one can deduce that the pattern search algorithm consists of three main stages that are the *initiation, searching*, and *termination* stages. The pattern search algorithm is equipped with two main search mechanisms, namely, *exploratory* and *pattern moves*. The following subsection will discuss each of these stages and their mathematical structures.

2.2.1 Initiation Stage

An *N*-dimension coordination system could represent the search space in an optimization problem with N decision variables. In this case, any point within the search space, say X, can be represented mathematically as a $1 \times N$ array as follows:

$$X = \left(x_1, x_2, x_3, \ldots, x_j, \ldots, x_N \right) \tag{2.1}$$

where X is a point in the search space of an optimization problem with N decision variables, and x_j represents the value associated with the *j*th decision variable.

As stated, the pattern search algorithm generates a random point within the *feasible* area of the search space (i.e., the decision space). Given that the algorithm has no previous encounter with the search space at the initiation stage, the randomly generated point is temporarily set as the base point. Naturally, this is a placeholder until the algorithm stumbles upon a point with more desirable properties, where the base point would be replaced. Note that tentative or trial points that would be subsequently generated and tested through the computational process of the pattern search algorithm have the same mathematical structure.

2.2.2 Searching Stage

After the initiation stage, the algorithm would enumerate through the search space, with the focal point being the current base point. The algorithm would

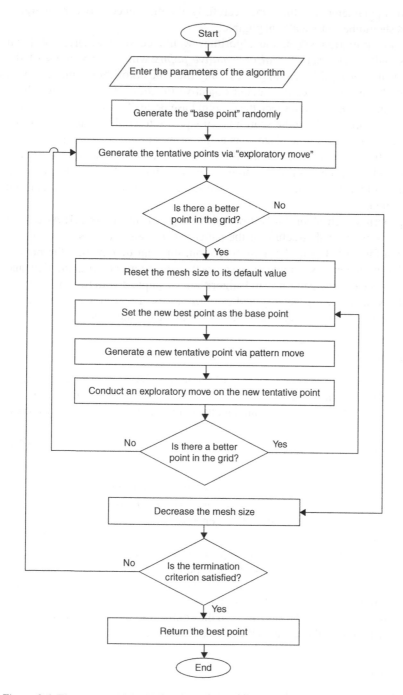

Figure 2.1 The computational flowchart of the pattern search algorithm.

proceed by selecting and evaluating a set of tentative or trial points placed within the vicinity of the search agent. At any given point during this process, a tentative point that can outperform the base point in the objective function would be replaced as the new base point. Pattern search is equipped with two searching mechanisms, namely, exploratory and pattern moves. An exploratory move tends to serve as a local search where the vicinity of the search agent (i.e., current base point) would be explored to locate a potentially more suitable base point. While different alternatives are suggested in the literature to conduct this task, the core idea is to form a mesh grid around the search agent. The points within this grid would then be evaluated to identify the next potential base point. Next, we have the pattern move, which amplifies the moving pattern detected by a successful exploratory move. Here the algorithm tends to test whether following the trajectory of a successful exploratory move would still yield a more desirable solution. The following subsections explore the computational procedure to conduct these search mechanisms.

2.2.2.1 Exploratory Move

As stated, the main idea behind the exploratory move is to navigate through the search space and create a path toward what is potentially the optimum solution. This movement is associated with two defining characteristics. Firstly, the exploratory move is what is known as a local search. The idea is that the search premise should be limited to the vicinity of the search agent. Secondly, the movement resulting from such a search should be guided via information obtained from the search space and the positioning of the objective function within this space. Although guided search is also commonly associated with calculus-based optimization methods, the main difference here is that, unlike the said methods where the gradient information is used as a guide, the objective function values are directly employed as a navigation tool for the searching process. Methods that resemble such characteristics are known as direct search methods.

While there are different approaches to implementing an exploratory move in the pattern search algorithm, they all follow the same computational principles. A gridded mesh would be formed surrounding the premise of the search agent, which is always located at the current base point. The size of this mesh grid, denoted by μ, is one of the parameters of the pattern search algorithm that needs to be fine-tuned by the users based on the characteristics of the search space. As one could intuitively deduce, a more extensive search space would probably require a larger mesh size, while smaller search spaces could benefit more from smaller mesh girds. The shape of the objective function can also influence the suitable value for this parameter. When dealing with a ragged objective function, one may need to consider a larger mesh size to avoid being trapped in local optima. One should note that the mesh grid size would be dynamically adjusted throughout the search process. For that, whenever the tentative or trial solutions within the mesh grid cannot outperform the current base point, the algorithm reduces the size of the mesh grid. By doing so, the algorithm would tend to focus more on the exploitation phase rather than exploration. The idea is if an algorithm cannot identify a more suitable point, the point in the closer vicinity of the searching agent that cannot be examined due to the current size of the grid may have more suitable properties. As such, the algorithm would concentrate on a more targeted portion of the search space by reducing the search area.

To that end, if the exploration move is deemed a failure, using a contraction coefficient, denoted by δ, the mesh grid size would be reduced. This can be mathematically expressed as follows:

$$\mu^{new} = \mu - \delta \qquad (2.2)$$

where μ^{new} is the new mesh grid size. Note that the contraction coefficient is another parameter of the pattern search algorithm that needs to be calibrated based on the properties of the search space and the objective function at hand. Opting for a smaller size would lead to a more refined and thorough search while increasing searching and computational time. On the other hand, choosing a more significant value for this parameter would crank up the search speed, leading to premature convergence.

Based on the pattern search algorithm's instruction, any time during the searching process when an exploratory move is found to be a success, the value for the mesh grid size should be restored to its initial default value. This is mathematically expressed as follows:

$$\mu^{new} = \mu \qquad (2.3)$$

The idea here is that if the algorithm executes a successful exploratory move, the base point would be updated and moved into a new uncharted territory of the search space. As such, it seems logical to search this new area with the same diligence that led the search agent to this point. The effect is that the mesh grid size needs to be reset to its initial value.

As for the gridding pattern, as stated earlier, a few options are suggested in the literature. The common denominator is that they create a finite number of tentative options within the vicinity of the search agent. These tentative points are all at an equal distance from the current base point, which is numerically equivalent to the mesh grid size parameter. While each method suggests a different number of tentative or trial points to be evaluated at this stage, the said number is proportional to the number of decision variables of the optimization problem. Naturally, the more decision variables we have, the more number of tentative points need to be evaluated in this stage. Two of the most acceptable methods to create these mesh grids are generalized pattern search (GPS) and mesh adaptive direct search (MADS) (Bozorg-Haddad et al., 2017). In GPS, only one element of the base point would be either increased or decreased proportionately to the mesh size grid parameter. As such, in the GPS method, the tentative points created in each exploratory move equal 2×N. This can be mathematically expressed as follows:

$$X_1^{new} = \mu.[1, 0, 0, ..., 0] + X \tag{2.4}$$

$$X_2^{new} = \mu.[0, 1, 0, ..., 0] + X \tag{2.5}$$

$$\vdots$$

$$X_N^{new} = \mu.[0, 0, 0, ..., 1] + X \tag{2.6}$$

$$X_{N+1}^{new} = \mu.[-1, 0, 0, ..., 0] + X \tag{2.7}$$

$$X_{N+2}^{new} = \mu.[0, -1, 0, ..., 0] + X \tag{2.8}$$

$$\vdots$$

$$X_{2N}^{new} = \mu.[, 0, 0, ..., -1] + X \tag{2.9}$$

where X_i^{new} is the *i*th tentative solution.

On the other hand, the MADS method suggests a more restricted number of tentative points. Based on this method, only $N+1$ tentative points would be generated. This can be done as follows:

$$X_1^{new} = \mu.[1, 0, 0, ..., 0] + X \tag{2.10}$$

$$X_2^{new} = \mu.[0, 1, 0, ..., 0] + X \tag{2.11}$$

$$\vdots$$

$$X_N^{new} = \mu.[0, 0, 0, ..., 1] + X \tag{2.12}$$

$$X_{N+1}^{new} = \mu.[-1, -1, -1, ..., -1] + X \tag{2.13}$$

As can be seen, the GPS method conducts a more thorough search as it generates a much more comprehensive mesh grid network in comparison to the MADS

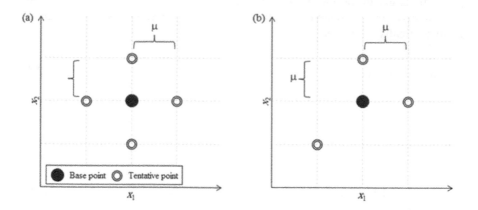

Figure 2.2 The mesh grid networks of tentative points are generated by (a) GPS and (b) MADS methods in a two-dimensional search space.

method's grids. This would naturally create a smoother transition through the search space. However, as one could imagine, this is more computationally taxing as the algorithm needs to evaluate each tentative point within this grid network. This could be problematic when the optimization problems contain numerous decision variables or the simulation stage of the optimization process (i.e., objective function evaluation) becomes more intricate than a simple mathematical equation. These conditions, which describe many real-world optimization problems, could pose serious challenges to the computation time of the algorithm. MADS provides a more practical alternative to the GPS method in such cases. Figure 2.2 depicts the schematic structure of the gird networks generated by both GPS and MADS methods in a two-dimensional search space.

2.2.2.2 Pattern Move

Pattern move is basically an extension to the exploratory move by amplifying the trajectory found by a successful exploratory move. The idea behind the pattern move is to imitate the pattern recognized in a successful exploratory move. By doing so, the algorithm can explore the search space with less computation effort, which, in turn, can potentially reduce the algorithm's convergence time and the probability of being trapped in local optima.

If an exploratory move is deemed successful, the search agent would be relocated to what is the new base point, creating a trajectory that links the previous base point to the new one. In mathematics, this can be described as a *vector* that bridges two points in the search space, here the previous and current base points. The pattern move recognizes this pattern and tries to follow along such a trajectory to see if, by doing so, it can find a more suitable solution. This can be mathematically expressed as follows:

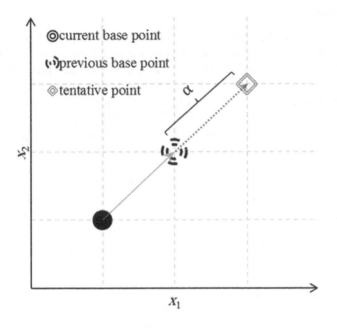

Figure 2.3 The schematic structure of the pattern moves in a two-dimensional search space.

$$X^{new} = X' + \alpha.(X - X')$$ (2.14)

where X' is the previous base point, X denotes the new base point, α represents a positive acceleration factor which happens to be another parameter of the pattern search algorithm, and X^{new} is a new tentative or trial point. Similar to the exploration move, the tentative point would be evaluated at this stage and would replace the current base point if it has more suitable properties; otherwise, it gets rejected. If a pattern move is deemed successful, rather than applying it once, alternatively, one can keep on executing the same pattern until it can no longer generate more suitable tentative points (Bozorg-Haddad et al., 2017). One can also control the length of these pattern moves by adjusting the acceleration factor parameter. Assigning a proper value to this parameter would, similarly, depend on the search space characteristic and the structure of the objective function. The schematic structure of the pattern move in a two-dimensional search space is depicted in Figure 2.3. As can be seen, unlike exploratory move, pattern search generates only one tentative point in each attempt. As such, the algorithm may need less computation time to evaluate the suitability of this trial point. This effect is more pronounced in real-world optimization problems where function evaluation takes a good chunk of computation time. Resultantly, a successful pattern move can potentially save a lot of computing time and effort.

2.2.3 Termination Stage

Meta-heuristic algorithms are basically a bundle of computational procedures that would be executed iteratively to enumerate through the search space and locate what is probably the optimum solution. These iterative computational procedures would be executed in an infinite loop without a termination stage. Thus, meta-heuristic algorithms would need a mechanism to determine whether the algorithm has successfully converged to the final optimum solution. Given that we do not have any prior information about the search space and the positioning of the objective function, the termination mechanism becomes quite essential for how these meta-heuristic optimization algorithms perform.

Often, there is a variety of options available for termination mechanisms, where the user could opt for the one that seems more fitting to a given problem, in which case the selection procedure can be perceived as a parameter for the optimization algorithm. Remember that these strategies may have other parameters that need to be fine-tuned for the problem. As for the pattern search algorithm, it is equipped with an internal mechanism by which it can terminate the search process.

As demonstrated in Figure 2.1, the algorithm conducts a series of patterns and exploratory moves to enumerate through the search space. But the mesh grid size used in the exploratory move is dynamically changed throughout the process. Each time the exploratory move fails to generate a tentative option that can outperform the current base point, the grid size decreases accordingly. On the other hand, locating a better point in the search space through the exploratory move would reset the mesh grid size to its initial default value at any given time. As such, upon consecutive failure execution of the exploratory move, the searching premise could drop below a certain termination threshold, denoted by γ, in which case the search would be terminated, and the current base point would be returned as the optimum solution. The neat thing about this termination mechanism is that we are not explicitly telling the algorithm the termination point, unlike most commonly used termination mechanisms. Alternatively, one can simply abandon this mechanism and replace it with other popular termination mechanisms, such as limiting the number of iterations, run time, or even monitoring the improvement made to the best solution in consecutive iterations.

2.3 Parameter Selection and Fine-Tuning the Pattern Search Algorithm

Based on the *no-free-lunch theorem*, fine-tuning an algorithm is a critical step in implementing a meta-heuristic algorithm to ensure that the said algorithm is equipped to handle a given optimization problem. While it is possible to use our intuition, experience, and default values suggested for an algorithm's parameters as a good starting point, this is more than anything a trial-and-error process. Indeed,

```
Begin
        Set the algorithm's parameter and input the data
        Generate a random point within the search space and set it as the base point
        While the mesh size is greater than the termination threshold
                Generate tentative points by exploratory move
                If tentative points contained a better point than the current base point
                        Reset mesh size
                        Implement a pattern move
                        If the pattern move is successful
                                Replace the base point
                                Conduct another exploratory move, and replace the base point if need be
                        End if
                Otherwise
                        Decrease the mesh size grid using the contraction coefficient
                End if
        End while
        Report the base point as the solution
End
```

Figure 2.4 Pseudocode for the pattern search algorithm.

it is possible to get a good enough result by having an educated guess for setting the parameters of these algorithms to get the best possible performance. However, more often than not, it is necessary to go through this fine-tuning process. As for the pattern search algorithm, these parameters are mesh grid size (μ), contraction coefficient (δ), mesh grid generation methods (GPS and MADS), acceleration factor (α), termination threshold (γ), or possibly opting for other termination strategies, in which case parameters that are associated with each of these strategies can be added to this list. The pseudocode for the pattern search algorithm is shown in Figure 2.4.

2.4 Python Codes

The code to implement the pattern search algorithm can be found below:

```python
import numpy as np

def init_generator(num_variable, min_val, max_val):
    return np.random.uniform(low=min_val,high=max_val,size=num_
    variable)

def exploratory_move(X, mu, func, method='GPS'):
    '''method = {'GPS, 'MADS'}'''

    if method == 'GPS':
        array_pos = mu*np.eye(len(X)) + X
        array_neg = -mu*np.eye(len(X)) + X
        mesh = np.concatenate((array_pos, array_neg), axis=0)
        eval_results = np.apply_along_axis(func1d=func, axis=1, arr=
        mesh)
        return mesh, eval_results
```

```python
    elif method == 'MADS':
        array_pos = mu*np.eye(len(X)) + X
        array_neg = np.reshape(-mu*np.ones(len(X)) + X, (1,len(X)))
        mesh = np.concatenate((array_pos, array_neg), axis=0)
        eval_results = np.apply_along_axis(func1d=func, axis=1, arr=
        mesh)
        return mesh, eval_results

def exploratory_move(X, mu, func, method='GPS'):
    '''method = {'GPS, 'MADS'}'''
    if method == 'GPS':
        array_pos = mu*np.eye(len(X)) + X
        array_neg = -mu*np.eye(len(X)) + X
        mesh = np.concatenate((array_pos, array_neg), axis=0)
        eval_results = np.apply_along_axis(func1d=func, axis=1, arr=
        mesh)
        return mesh, eval_results
    elif method == 'MADS':
        array_pos = mu*np.eye(len(X)) + X
        array_neg = np.reshape(-mu*np.ones(len(X)) + X, (1,len(X)))
        mesh = np.concatenate((array_pos, array_neg), axis=0)
        eval_results = np.apply_along_axis(func1d=func, axis=1, arr=
        mesh)
        return mesh, eval_results

def pattern_move(current_base, previous_base, alpha, func):
    current_base = np.array(current_base)
    previous_base = np.array(previous_base)
    X_new = previous_base + alpha*(current_base-previous_base)
    return X_new, func(X_new)
def PS_algorithem(mu_const, alpha, delta, obj_func,
                  final_step_size, num_variable, min_val,
                  max_val, meshing_method='GPS', minimizing = True,
    full_result=False):
    results_list = list()
    NFE_list = list()
    NFE_value = 0
    X = init_generator(num_variable, min_val, max_val)
    NFE_value += 1
    best_of = obj_func(X)
    results_list.append(best_of)
    NFE_list.append(NFE_value)
    mu = mu_const
    if minimizing:
        while mu > final_step_size:
            mesh, values = exploratory_move(X, mu, obj_func,
                                            method=meshing_method)
            if meshing_method == 'GPS':
                NFE_value += 2*num_variable
            else:
                NFE_value += (num_variable+1)
```

```
        if np.min(values) < best_of:
            mu = mu_const
            best_of = np.min(values)
            results_list.append(best_of)
            NFE_list.append(NFE_value)
            current_base = mesh[np.argmin(values)]
            previous_base = X
            while True:
                X_new, of_new = pattern_move(current_
                base, alpha,
                            previous_base, obj_func)
                NFE_value += 1
                if of_new < best_of:
                    previous_base = current_base
                    current_base, best_of = X_new, of_new
                    results_list.append(best_of)
                    NFE_list.append(NFE_value)
                    mesh, values = exploratory_move(current_
                    base, mu, obj_func,
                        method=meshing_method)
                    if meshing_method == 'GPS':
                        NFE_value += 2*num_variable
                    else:
                        NFE_value += (num_variable+1)
                    if np.min(values) < best_of:
                        best_of = np.min(values)
                        results_list.append(best_of)
                        NFE_list.append(NFE_value)
                        previous_base = current_base
                        current_base = mesh[np.argmin(values)]
                else:
                    mu -= delta
                    X = current_base
                    results_list.append(best_of)
                    NFE_list.append(NFE_value)
                    break
        else:
            mu -= delta
            results_list.append(best_of)
            NFE_list.append(NFE_value)

else:
    while mu > final_step_size:
        mesh, values = exploratory_move(X, mu, obj_func,
                                method=meshing_method)
        if meshing_method == 'GPS':
            NFE_value += 2*num_variable
        else:
            NFE_value += (num_variable+1)
        if np.max(values) > best_of:
            mu = mu_const
            best_of = np.max(values)
```

```
                    results_list.append(best_of)
                    NFE_list.append(NFE_value)
                    current_base = mesh[np.argmax(values)]
                    previous_base = X
                    while True:
                        X_new, of_new = pattern_move(current_base, alpha,
                                                     previous_base,
                                                     obj_func)
                        NFE_value += 1
                        if of_new > best_of:
                            previous_base = current_base
                            current_base, best_of = X_new, of_new
                            results_list.append(best_of)
                            NFE_list.append(NFE_value)
                            mesh, values = exploratory_move(current_
                                        base, mu,
                                                            obj_func,
                                                    method=meshing_
                                                    method)
                            if meshing_method == 'GPS':
                                NFE_value += 2*num_variable
                            else:
                                NFE_value += (num_variable+1)
                            if np.max(values) > best_of:
                                best_of = np.max(values)
                                results_list.append(best_of)
                                NFE_list.append(NFE_value)
                                previous_base = current_base
                                current_base = mesh[np.argmax(values)]
                        else:
                            mu -= delta
                            X = current_base
                            results_list.append(best_of)
                            NFE_list.append(NFE_value)
                            break
                else:
                    mu -= delta
                    results_list.append(best_of)
                    NFE_list.append(NFE_value)

    if not full_result:
        return X, best_of
    else:
        return X, best_of, results_list, NFE_list
```

2.5 Concluding Remarks

The pattern search algorithm is among the first, if not the first, meta-heuristic optimization algorithm introduced in the literature. In fact, one could argue that it pioneered the next generation of optimization algorithms. The novelty incorporated in the searching mechanisms of this algorithm laid the ground-work for many other meta-heuristic optimization algorithms. One of the most fundamental characteristics of this algorithm was that it introduced a way to conduct a guided search to find a close-enough optimum solution without bothering with the gradient information of the objective function. To do that, the algorithm would directly look into the values of the objective functions as a guide to navigating the search space. This would come to be known as a direct search method. The same idea enabled the pattern search algorithm to tackle optimization problems previously perceived to be challenging, if not impossible, due to high dimensionality, multimodality, epistasis, non-differentiability, and discontinuous search space imposed by constraints.

The pattern search algorithm benefits from a relatively simple computational structure that would be executed iteratively until a certain termination criterion is met. At this point, the algorithm would return what is probably the near-optimal solution. The search would be initiated by generating and locating a search agent at random points through the search space. The algorithm would navigate within the search space using two moving mechanisms, namely, exploratory and pattern moves. Having a single searching agent that follows simple computational instructions would help make the algorithm easy to understand and, more importantly, easy to execute. As such, this is not a particularly computationally taxing optimization algorithm. But the problem with this algorithm is that it is heavily centered on a local search. This is the problem with most single-agent optimization algorithms, where exploitation is more pronounced than exploration. As such, algorithms may converge to a local optimum solution rather than the global point. Thus, such algorithms might not be the best option to solve intricate search spaces or too ragged objective functions. The other problem with this algorithm is that it has not an internal mechanism to ensure the search agent does not bounce out of the boundaries imposed by the feasible range of decision variables. Algorithms, such as pattern search that use the idea of following a trajectory should be embedded with such a mechanism, otherwise, the algorithm may return an infeasible solution. This issue can be addressed by treating these boundaries as a regular constraint or creating a mechanism that replaces the infeasible values for the decision variables throughout the search process. That said, while the pattern search algorithm may strike you first as a rudimentary method due to its simplistic computational procedure, it is anything but that. In fact, it is still considered a viable optimization method with a lot of potentials to solve real-world problems if the parameters are fine-tuned correctly and with great care.

References

AlHajri, M.F., El-Naggar, K.M., AlRashidi, M.R., & Al-Othman, A.K. (2012). Optimal extraction of solar cell parameters using pattern search. *Renewable Energy*, 44, 238–245.

Alsumait, J.S., Sykulski, J.K., & Al-Othman, A.K. (2010). A hybrid GA–PS–SQP method to solve power system valve-point economic dispatch problems. *Applied Energy*, 87(5), 1773–1781.

Banerjee, P.K. & Rahim, M.A. (1988). Economic design of – control charts under Weibull shock models. *Technometrics*, 30(4), 407–414.

Bellout, M.C., Ciaurri, D.E., Durlofsky, L.J., Foss, B., & Kleppe, J. (2012). Joint optimization of oil well placement and controls. *Computational Geosciences*, 16(4), 1061–1079.

Bozorg-Haddad, O., Solgi, M., & Loáiciga, H.A. (2017). *Meta-heuristic and evolutionary algorithms for engineering optimization*. John Wiley & Sons. ISBN: 9781119386995

Bozorg-Haddad, O., Tabari, M.M.R., Fallah-Mehdipour, E., & Mariño, M.A. (2013). Groundwater model calibration by meta-heuristic algorithms. *Water Resources Management*, 27(7), 2515–2529.

Caponio, A., Cascella, G.L., Neri, F., Salvatore, N., & Sumner, M. (2007). A fast adaptive memetic algorithm for online and offline control design of PMSM drives. *IEEE Transactions on Systems, Man, and Cybernetics, Part B (Cybernetics)*, 37(1), 28–41.

Carlyle, W.M., Montgomery, D.C., & Runger, G.C. (2000). Optimization problems and methods in quality control and improvement. *Journal of Quality Technology*, 32(1), 1–17.

Du, K.L. & Swamy, M.N.S. (2016). *Search and optimization by metaheuristics: Techniques and algorithms inspired by nature*. Springer International Publishing Switzerland. ISBN: 9783319411910

Evins, R. (2013). A review of computational optimisation methods applied to sustainable building design. *Renewable and Sustainable Energy Reviews*, 22, 230–245.

Glover, F. (1989). Tabu search—part I. *ORSA Journal on Computing*, 1(3), 190–206.

Glover, F. (1990). Tabu search—part II. *ORSA Journal on Computing*, 2(1), 4–32.

Hooke, R. & Jeeves, T.A. (1961). "Direct Search" solution of numerical and statistical problems. *Journal of the ACM*, 8(2), 212–229.

Khorsandi, M., Bozorg-Haddad, O., and Mariño, M.A. (2014). Application of data-driven and optimization methods in identification of location and quantity of pollutants. *Journal of Hazardous, Toxic, and Radioactive Waste*, 19(2), 04014031.

Machairas, V., Tsangrassoulis, A., & Axarli, K. (2014). Algorithms for optimization of building design: A review. *Renewable and Sustainable Energy Reviews*, 31, 101–112.

Moradkhani, H. & Sorooshian, S. (2009). General review of rainfall-runoff modeling: model calibration, data assimilation, and uncertainty analysis. In Sorooshian, S., Hsu, K.L., Coppola, E., Tomassetti, B., Verdecchia, M., and Visconti, G. ed. *Hydrological modelling and the water cycle*. Springer, 1–24.

Shehab, M., Bertino, E., & Ghafoor, A. (2007). Watermarking relational databases using optimization-based techniques. *IEEE Transactions on Knowledge and Data Engineering*, 20(1), 116–129.

Smitabhindu, R., Janjai, S., & Chankong, V. (2008). Optimization of a solar-assisted drying system for drying bananas. *Renewable Energy*, 33(7), 1523–1531.

Swersey, A.J. (1994). The deployment of police, fire, and emergency medical units. *Handbooks in Operations Research and Management Science*, 6, 151–200.

Tillman, F.A., Hwang, C.L., & Kuo, W. (1977). Determining component reliability and redundancy for optimum system reliability. *IEEE Transactions on Reliability*, 26(3), 162–165.

Tung, Y.K. (1985). River flood routing by nonlinear Muskingum method. *Journal of Hydraulic Engineering*, 111(12), 1447–1460.

Yang, X.S. (2010). *Nature-inspired metaheuristic algorithms*. Luniver Press. ISBN: 9781905986286

3   Genetic Algorithm

Summary

The genetic algorithm is arguably one of the most well-known and prevalent meta-heuristic optimization algorithms. The novelties in this algorithm created a tectonic shift in how these algorithms were generally perceived, to the point that most of these ideas were implemented in shaping the next generation of meta-heuristic algorithms. In this chapter, we will dig deep and explore the mechanisms used in this algorithm. We get familiar with the genetic algorithm's terminology and see how one can implement this algorithm in the Python programming language. Finally, we will explore the potential merits and drawbacks of this algorithm.

3.1 Introduction

The creation of intelligent machines has long been a fascinating subject for scholars and researchers, inspiring early efforts to build modern computers (Turing, 1980). This objective has yet to be fully achieved despite numerous attempts, making it a relevant and ongoing study area to this day. However, over the years, there were some notable attempts in which an artificial system or a computer was able to behave in a manner that, in a way, it resembled an intelligence-like behavior. In such cases, the system would imitate the structure or behavior of an entity, a creature, or even a system that we know to be of significant intelligence. Arguably, one of the earliest attempts that capitalized on this idea was perhaps the introduction of the genetic algorithm. The genetic algorithm, to this day, remains one of the most well-received optimization algorithms. It revolutionized the field of meta-heuristic optimization and, in turn, inspired generations of neo-meta-heuristic optimization methods. But even more importantly, it laid the groundwork for a new field that utilizes what can be passed as computational intelligence to tackle complex problems that were perceived to be challenging to solve, if not impossible, in a practical way.

At its core, the genetic algorithm tries to imitate evolution, a gradual process through which a given species can evolve and adapt to a new environment. Technically speaking, the fundamental idea of the genetic algorithm was not novel per se, as the likes of Nils Aall Barricelli and Lawrence J. Fogel have already

DOI: 10.1201/9781003424765-3

implemented the idea of *evolutionary computing* to solve complex problems (Baricelli, 1962; Fogel et al., 1966). But when it comes to the modern algorithmic structure to imitate the evolution process, it is believed that the works of John Holland laid such theoretical groundwork that shaped the idea of the genetic algorithm in the 1960s and 1970s (Yang, 2010). Holland's (1975) groundbreaking work officially introduced the foundation of what became known as the genetic algorithm for the first time. This version of the evolutionary computation algorithm is embedded with operators such as *crossover*, *mutation*, and *selection*, by which one could capture the core mechanism of the evolutionary process. For the following years, however, most genetic algorithm-related research remained largely theoretical until the mid-1980s, when scholars and researchers tried to find new implementations for the said algorithm (e.g., Goldberg, 1989).

As a meta-heuristic optimization algorithm, like its predecessors, the genetic algorithm was basically a series of algorithmic instructions that would benefit from a guided direct search to navigate through the search space to locate what is potentially an acceptable approximation of the optimum solution. Being based on the idea of a direct search would relinquish this algorithm from the limitations of using gradient information as a guiding mechanism. But what was interesting about the computational architecture of this algorithm was that through reinforcing the search process, the search agents would gradually evolve into solutions with better properties. This would later become one of the defining characteristics of meta-heuristic optimization algorithms, where an idealized representation of the surrounding world inspires the search strategy. This would help create a general-purpose optimization algorithm that can converge to a near-optimal solution without being bound by the limitation that is commonly associated with conventional calculus-based models while being guided through this search process, a characteristic that gives these methods an edge over enumeration-based methods.

The genetic algorithm's inspiration for the search strategy comes from the evolution process. As such, taking a closer look at the generic idea behind evolution could be a good starting point for this chapter. In general, the adaption of a living being to a new environment may be rooted in two sets of phenomena that are *learning* and *evolution*. Learning is the adaptive process through which individuals acquire a new skill set that makes them more equipped to survive or thrive in the said environment. These skill sets can be learned and passed to the community members and, more importantly, through generations to come. Others, whether contemporary or individuals who acquired these skills from their predecessors, could constantly build on these skills and try to improve upon their knowledge pull. Thus, it is essential to note that in learning, knowledge gained by an individual could potentially have an impact not only on the upcoming generations but can also alter the behavior of the individuals of the current generation. Such interactions can be captured in the connectionist model of the human brain (Du & Swamy, 2016).

On the contrary, evolution tends to operate on biological entities rather than the behavior of individuals. Here, an individual with better genetic properties would have a better chance of survival, as such properties would help them better adapt to the new environment. Often, such individuals would also have a better chance of finding a more suitable mate. Here, the offspring could potentially inherit the genetic characteristics of the parents. Note that, unlike learning, in this process, the emphasis is mainly on the inheritable characteristics that can be passed to the next generation. Natural selection gives individuals with superior genetic characteristics a better ability to adapt to the environment. Mutation is another phenomenon that is associated with the evolution process. Here the mutation represents a sudden, random, and rare change in the biological structure of an individual. This could either improve an individual's odds of survival, or it can decrease them drastically. Thus, the evolution process can be described as a gradual stochastic process through which natural selection creates an environment that is more tolerant of suitable biological properties and harsher to harmful ones. As such, the species in a given community would gradually inherit these suitable characteristics and improve their odds of surviving in the given environment. Note that the evolution process can be captured through the *Darwinian model* and what is known as the *principle of natural selection* or *survival of the fittest* (Du & Swamy, 2016).

The genetic algorithm is a general-purpose meta-heuristic algorithm that is based on *the neo-Darwinian paradigm* to capture the essence of the natural evolution process that is associated with biological systems (Du & Swamy, 2016). At its core, the genetic algorithm is a stochastic process that utilizes the idea of direct search to enumerate through the search space. One of the main novelties in the computational structure of this algorithm is that, unlike its processor meta-heuristic optimization algorithms, it uses multiple search agents simultaneously in each iteration. The advantage of this feature of *parallel computation* can mainly be felt in today's programming languages that can use this at its full potential to improve search speed drastically. Meta-heuristic algorithms based on this searching paradigm are known as population-based optimization algorithms. In a population-based algorithm, rather than having a single search agent, there is a set of agents commonly known as a population in meta-heuristic terminology. A population is also referred to as a *generation* in the genetic algorithm lingo. The main idea is that through an iterative process, a randomly generated generation would eventually evolve to what could potentially be a superior population, which, in all likelihood, could contain an individual search agent with the properties of the optimum solution.

From a mathematical point of view, a population is typically represented as a *matrix*, a bundle of arrays stacked vertically where each row represents the properties of a search agent. *Population size* is the term used to describe the number of search agents in the said matrix. A search agent is often called a *chromosome* or a *genome* in the genetic algorithm terminology. Mathematically speaking, given that a chromosome or a genome is basically a search agent, it can also be represented via the structure of an array, where each element of it contains the numeric properties of the decision variable. Like search agents, a chromosome represents a point within the search space. A string of these elements, or more often a singular element of these arrays, is called a *gene*. The entire premise of the genetic algorithm is built upon two main drivers that are *natural selection* and *genetic drift* (Du & Swamy, 2016).

In natural selection, the biological properties of a species are altered gradually and over a generation by creating an environment that rewards more desirable genetic properties. In this environment, individuals with more suiting characteristics have a better chance to survive and thrive. In this case, such individuals have a better chance of finding a partner or mate and passing their genetic properties to the next generation. Thus, the offspring that form the next generation would probably inherit these distinctive good genetic properties. As such, the end result would be a mechanism that pushes the said species toward inheriting more suitable characteristics that help them survive in a given environment. The problem, however, is that this process does not encourage diversification of the said population. This could, in turn, means that a population could prematurely converge into local optimum solutions. Thus, this process needs a mechanism that encourages the exploration of other biological properties that could potentially increase the survival odds of a given species. This can be done through genetic drift. This is basically a stochastic process that could introduce new biological properties to a given population by mutating the genes of certain randomly selected chromosomes of a search agent. To capture the essence of this phenomenon, the genetic algorithm proposed three main operators that are *selection*, *crossover*, and *mutation*, which we will explore in the upcoming sections.

Bear in mind that the genetic algorithm, at its core, is still a meta-heuristic optimization algorithm. As such, it can overcome the issues related to high dimensionality, multimodality, epistasis, non-differentiability, and discontinuous search space (Yang, 2010). This is because, like other meta-heuristic algorithms, the genetic algorithm consists of simple-enough computational instructions that are not limited to gradient information. But, as stated earlier, two main novelties in the genetic algorithm distinguish it from its processors. Firstly, this is the earliest notable attempt to imitate a nature-based phenomenon as the core of the computational architect. As we would see in the later chapters, this eventually became the new standard for the next generation of meta-heuristic optimization algorithms. Secondly, this algorithm introduced the concept of *parallelism* that employs multiple search agents in each iteration. The idea of parallelized computation would increase the algorithm's searching capabilities while drastically improving the odds of converging to the near-global optima rather than a local one. Both these contributions significantly impacted the optimization field and heavily influenced many algorithms that followed the genetic algorithm. That said, implementing this algorithm is not without its challenges. The main issue with the genetic algorithm is that it is riddled with numerous parameters that are not often intuitively understandable, at least without a good deal of practice and experience. This makes parameter selection and fine-tuning the algorithm a really challenging task for untrained and inexperienced users. Bear in mind that inapt parameter values could make the algorithm's performance sub-optimal, to the point that it may not converge to the actual optimum solution of the given problem (Yang, 2010).

That said, to this very day, the genetic algorithm remains one of the most famous and revered options for optimization. Aerospace engineering (e.g., Adeli & Cheng,

1993; Bayley et al., 2008), bioinformatics (e.g., Jarvis & Goodacre, 2005), building design (e.g., Wang & Wei, 2021), data science (e.g., Abualigah & Dulaimi, 2021; Hamdia et al., 2021), energy industry (e.g., Iba, 1994), environmental engineering (e.g., Adams et al., 2004), geohydrology (e.g., Fallah-Mehdipour et al., 2014), healthcare (e.g., Wager & Nichols, 2003), hydrology (Jahandideh-Tehrani et al., 2021), irrigation and drainage engineering (e.g., Ines et al., 2006), mechanical engineering (e.g., Chen et al., 2021), nuclear engineering (e.g., Kleedtke et al., 2021), structural engineering (e.g., Gazonas et al., 2006), and water resources planning and management (e.g., Hınçal et al., 2011; Yaghoubzadeh-Bavandpour et al., 2022) are merely a few examples where the genetic algorithm has proven to be a viable option to tackle real-world optimization problems. Remember that researchers and scholars have attempted to modify the genetic algorithm to enhance its performance over the years. Some of the most notable variations of the genetic algorithm are the elitist version (De Jong, 1975), the messy genetic algorithm (Goldberg et al., 1990), the breeder genetic algorithm (Mühlenbein & Schlierkamp-Voosen, 1993), and the genetic algorithm with varying population size (Michalewicz, 1996). That said, the standard genetic algorithm is still considered a viable option to tackle complex real-world problems. As such, in this chapter, we mainly focus on the original computational architecture of the genetic algorithm.

3.2 Algorithmic Structure of the Genetic Algorithm

As stated, the genetic algorithm searching strategy is based on mimicking the evolutionary process. To capture the neo-Darwinian paradigm, the algorithm tracks the positions of multiple search agents placed within the feasible boundaries of the search space. The said population would evolve through each iteration to the point that they eventually are guided toward what could potentially be the optimum solution.

The genetic algorithm's flowchart is depicted in Figure 3.1. A closer look would reveal that the genetic algorithm actually consists of three main stages that are the *initiation*, *reproduction*, and *termination* stages. To conduct the reproduction stage, the genetic algorithm would utilize three main evolutionary operators that are *selection*, *crossover*, and *mutation*. The following subsection will discuss each of these stages and their mathematical structures.

3.2.1 Initiation Stage

The genetic algorithm is a population-based meta-heuristic optimization algorithm. As such, it works with multiple search agents, here called chromosomes or genomes, that would enumerate through the search space. As we have seen, in an optimization problem with N decision variables, an N-dimension coordination system could be used to represent the search space. In this case, any point within the search space, say X, can be represented mathematically as a $1 \times N$ array as follows:

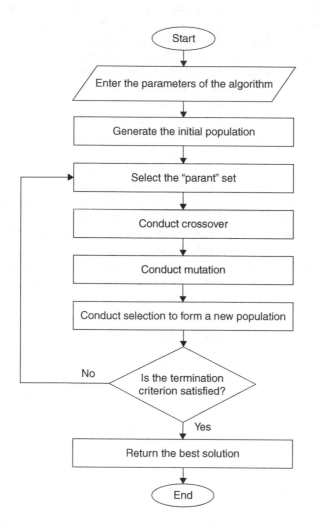

Figure 3.1 The computational flowchart of the genetic algorithm.

$$X = \left(x_1, x_2, x_3, \ldots, x_j, \ldots, x_N \right)$$ (3.1)

where X represents a chromosome in the search space of an optimization problem with N decision variables, and x_j represents the value associated with the jth decision variable, or what is technically known as a gene.

The genetic algorithm starts with randomly placing a series of chromosomes within the feasible boundaries of the search space. This bundle of arrays, which in the genetic algorithm's terminology is referred to as the population, can be mathematically expressed as $M \times N$ matrix, where M denotes the number of chromosomes or what is technically referred to as the population size. In such a structure, each row represents a single chromosome. Note that population size is one of the many parameters of the genetic algorithm. A population, denoted by *pop*, can be represented as follows:

$$pop = \begin{bmatrix} X_1 \\ X_2 \\ \vdots \\ X_i \\ \vdots \\ X_M \end{bmatrix} = \begin{bmatrix} x_{1,1} & x_{1,2} & \cdots & x_{1,j} & \cdots & x_{1,N} \\ x_{2,1} & x_{1,2} & \cdots & x_{2,j} & \cdots & x_{2,N} \\ & & \vdots & & & \\ x_{i,1} & x_{i,2} & \cdots & x_{i,j} & \cdots & x_{i,N} \\ & & \vdots & & & \\ x_{M,1} & x_{M,2} & \cdots & x_{M,j} & \cdots & x_{M,N} \end{bmatrix} \tag{3.2}$$

where X_i represents the ith chromosome in the population, and $x_{i,j}$ denotes the jth gene of the ith chromosome.

The initially generated population represents the first generation of chromosomes. As we progress, the values stored in the *pop* matrix will be altered according to the computational structure of the genetic algorithm's reproduction stage. By the end of this iterative computation process, when the termination criterion is met, one or possibly multiple chromosomes will converge to the optimum solution.

3.2.2 Reproduction Stage

The reproduction stage is a general term used here to refer to a procedure in the genetic algorithm by which the older generation would evolve into a new one. The reproduction stage consists of three main pillars that are selection, crossover, and mutation operators. These operators are used as alteration tools to ultimately increase the odds of surviving for the next generation. In the genetic algorithm's terminology, the potential next generation is called the *child population*, while the previous generation is commonly known as the *parent population*. The main idea is to create a stochastic mechanism by which the child population would, in all likelihood, inherit suitable biological properties from the parent population. These operators are explored in the following subsections.

3.2.2.1 Selection Operators

A selection operator is basically a procedure by which a number of the chromosomes, say R, would be picked from a larger pool of chromosomes. Note that the selection operator would be called upon in several places within the computational structure of the genetic algorithm. More importantly, there are several different approaches to creating a selection operator. One may opt for different selection operators based on the optimization problem's numeric structure, the type of objective function, and the computational task at hand. Among the most cited methods to assemble a selection operator, three methods would stand out, which are *ranking selection*, *tournament selection*, and *proportionate selection* or the *roulette wheel method*. Note that selecting of these methods as the selection operator for the genetic algorithm can be seen as one of the model's parameters.

Ranking selection is based on opting for solutions with better properties. There are two main approaches to implementing ranking selection, one being *deterministic*, while the other is *stochastic*. In both approaches, the algorithm needs to first rank the solutions in descending order based on their objective or cost function in case we deal with constraints in the optimization problem. As such, the best solution would assume the first rank, while the worst solution would be ranked last. For instance, in a minimizing problem, the chromosome with the minimum objective function would be placed on the top, while the chromosome with the maximum value would be placed at the bottom of the list. In the deterministic ranking selection method, the top R chromosomes are selected. This is a great approach when you want to merge two population sets or in cases where you must ensure that the *elite* chromos with the most desirable biological properties actively participate in the reproduction process. The computational procedure of this method is straightforward and can be easily implemented whenever the selection operator is called. However, the problem with this method is that it may lead to premature convergence.

Alternatively, you could first compute a probability distribution system that assigns the highest probability to the best chromosome and the lowest likelihood to the worst chromosome. These distributions can be defined as follows (Du & Swamy, 2016):

$$Pr_i = \frac{1}{Z}\left[\beta - 2(\beta-1)\frac{i-1}{Z}\right] \quad \forall i \tag{3.3}$$

in which Pr_i represents the probability of selecting the ith chromosome, Z is the total number of chromosomes in the selection poll, and β is a user-defined parameter that can range from [0, 2].

The tournament selection follows the same structure as the probabilistic ranking selection, with the notable exception that the probability distribution is uniform here. This means that all the chromosomes in the selection poll have the same chance of being selected in this process. The operator would continually try to pull out chromosomes randomly from the pool until R chromosomes were selected, at which point the selection process would come to an end. In each attempt, a random number would be generated if the said value is less than the selection probability for the chromosome, it would be selected. As can be seen, this is a simple and straightforward idea to select chromosomes, though the main problem is that it gives no advantage to elite chromosomes (Du & Swamy, 2016). As such, employing this method as a selection operator could potentially decrease the chance of converging the optimum solution.

Finally, proportionate selection, sometimes referred to as the roulette wheel, is another probabilistic-based selection operator. To implement this selection method, a hypothetical roulette wheel would be divided into different segments, each associated with a chromosome in the selection pool. The area of each segment would be proportional to each chromosome's relative objective function value. This can be mathematically expressed as follows (Holland, 2000):

$$Pr_i = \frac{f(X_i)}{\sum_{i=1}^{Z} f(X_i)} \qquad \forall i \tag{3.4}$$

in which $f()$ denotes the objective or cost function if there are constraints in the optimization problem. Similar to what we have seen, the algorithm would generate a random number ranging from 0 to 1. If the probability associated with a chromosome exceeds this random value, the said chromosome will be selected. This process would be repeated until all the R chromosomes were selected from the chromosome pool. Employing this method would ensure that better-performance chromosomes have a better chance of being selected. The problem with this method is that, like the ranking method, it may lead to premature convergence. Bear in mind that this method cannot be technically used for minimization because of the *scaling issue*. Furthermore, it also cannot be used when there are negative or null finesses values, as the probability computed by Equation (3.4) would be negative or null, respectively.

3.2.2.2 *Crossover Operators*

Crossover is one of the operators embedded in the genetic algorithm to create a new chromosome that inherits its properties from the parent population. To do that, the algorithm would first need to identify the parent population by dividing the current generation into *effective* and *ineffective* population sets through a stochastic procedure. Here the user must first assign a value to the crossover probability parameter denoted by P_c. This is another user-defined parameter of the genetic algorithm that ranges from 0 to 1. Through this parameter, we could see which of the chromosomes would be used in the process of creating the new offspring or what is known as the child population. Using one of the probabilistic selective methods, such sets could be drawn from the pool of the current population set. For instance, one could generate a random number in the range [0, 1] for each chromosome in the current generation. If the randomly generated value associated with a chromosome is below the crossover probability parameter, the said chromosome would be deemed effective; otherwise, the algorithm would identify it as an ineffective parent. The effective parent set would then be paired and used as an input to the crossover operator to create the offspring population.

Similar to what we have seen in the selection operator, there are several viable options when it comes to creating a crossover operator for the genetic algorithm. The most notable are *one-point crossover, two-point crossover*, and the *uniform crossover* method (Goldberg, 1989; Michalewicz, 1996). Again, opting for one of these methods can be seen as one of the parameters of the genetic algorithm.

In the one-point crossover, each time the crossover operator is called, a random point, say point *C*, would be selected to cut down each of the coupled selected effective parent chromosomes into two sections with respect to the said point. The genes stored in these sections would be swapped to create two child chromosomes (Figure 3.2a). The two-point crossover method follows a similar principle, but here, we have two randomly selected points, say points *C* and *C'*. The genes between these two crossover points are swapped to create two child chromosomes (Figure 3.2b). Lastly, in the uniform crossover method, rather than swapping a chunked segment, the crossover method would swap randomly selected individual genes between the parents to create two child chromosomes (Figure 3.2c).

The selected crossover method would generate a set of child chromosomes that inherited their properties from the former population set by enumerating through the coupled effective parent chromosome. The child set would be later concatenated

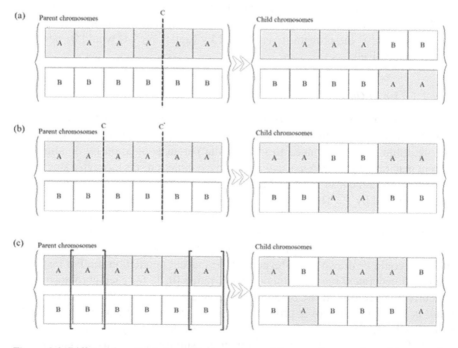

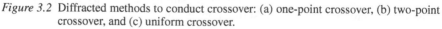

Figure 3.2 Diffracted methods to conduct crossover: (a) one-point crossover, (b) two-point crossover, and (c) uniform crossover.

to the pool of available chromosomes. As discussed in the upcoming section, the new generation will be selected from this newly formed pool.

3.2.2.3 Mutation Operators

The mutation is the random-based operator that simply handles the diversification of the population by introducing a random component to the search process. The mutation operator ensures a more thorough search so that algorithm does not stick in local optima by reintroducing some lost values for a gene. The mutation is considered a *unary* operator, which in computer science lingo refers to an operation that has only one *operand*. This simply means that the operation would be conducted on a singular chromosome. Like previous operators, several options are available to compose this operator, the most notable ones are *uniform* and *non-uniform* methods. Again, remember that opting for one of these options can be considered one of the parameters of the genetic algorithm.

Both these methods follow the same computational structure principle. The main idea is to randomly locate a set of genes that needs to be altered by replacing their assigned values. To do that, we first need to assign a proper value to the mutation probability parameter, denoted by P_m. This is another user-defined parameter of the genetic algorithm that should be selected from the range [0, 1]. Similar to what we have seen earlier, a randomly generated number within the range of 0–1 would be assigned to any gene in the population set. If the said value is below the threshold that is defined by the mutation probability parameter, the said gene will be replaced by the mutation operator.

In the uniform method, the replacing value for the selected genes would be randomly generated within their feasible range's lower and upper limits. This gives the missing values an equal opportunity to be reintroduced into the mix. This can be mathematically expressed as follows:

$$x_{i,j}^{new} = L_j + Rand \times \left(U_j - L_j\right) \tag{3.5}$$

in which $x_{i,j}^{new}$ represents the jth gene of the ith chromosome that has been selected randomly for mutation, *Rand* is a random number ranging from 0 to 1; U_j and L_j represent the upper and lower feasible boundaries of the jth gene, respectively.

While the aforementioned method could introduce a random component to the search and thus decrease the possibility of being trapped in local optima, it may prolong the search if a critical gene is selected for mutation. Furthermore, the logical idea is to downgrade the impact of randomness as we get closer to the end of the search process. This is because the last generations are converging toward what could be potentially the optimum point, and in such a case, a more local and narrow search could be more beneficial than a more exploratory-based search. This is the basic idea behind the non-uniform mutation method (Michalewicz, 1996). Note that this method is only applicable when the number of iterations is explicitly expressed, for instance, when the termination criterion is set by limiting the number of iterations. In such a case, the mathematical procedure to randomly generate a new gene can be expressed as follows:

$$x_{i,j}^{new} = x_{i,j} - d_t + Rand \times (2d_t)$$

(3.6)

$$d_t = d_0 \times \frac{T-t}{T}$$

(3.7)

in which $x_{i,j}$ is the current value for the jth gene of the ith chromosome; d_t is the maximum length of mutation at the tth iteration; T is the maximum number of iterations, which is a user-defined parameter; and finally, d_0 is the initial value for maximum mutation length, which as another user-defined parameter.

The mutation operator would alter the parent population through the procedure described above to create a new set of child chromosomes, which would be later concatenated to the pool of available chromosomes to form the new generation.

It should be noted that a high mutation rate would, in effect, transform the genetic algorithm into a random search. By increasing the chance of altering the desirable properties of elite chromosomes, it may also prolong the convergence process. On the other hand, assigning a quite low value for the mutation probability parameter could lead to premature convergence of the algorithm, which may cause the algorithm to be trapped in local optima. As such, it is essential to fine-tune this parameter cautiously. If the population size is large enough, experts believe that the crossover operator plays a more crucial role in the algorithm's performance. In contrast, in smaller populations, it is the mutation operator that could have a more significant impact on the algorithm's performance (Du & Swamy, 2016).

3.2.3 Termination Stage

Based on the instruction in Figure 3.1, the crossover and mutation operators would form new child chromosomes. These sets would then be merged with the current population to form a pool of potentially viable chromosomes for the next generation. A selection operator, say a deterministic ranking selection method, would be called upon at this stage to pick the next generation from this pool. The next generation would then be relabeled as the current generation to go through a similar computation process.

As can be seen, like other meta-heuristic algorithms, the sequence of operational structures of this algorithm needs to be executed iteratively until a certain termination criterion is met. At this point, the execution of the algorithm would be terminated. The best chromosome in the last generation, or depending on how the algorithm is coded, perhaps the best one observed thus far, would be reported as the solution to the optimization problem. Note that without such a termination stage, the algorithm could be executed in an infinite loop. In effect, the termination stage would determine whether the algorithm has reached what could be the optimum solution.

As the genetic algorithm is not equipped with an explicitly defined, unique termination mechanism, one could implement the commonly available options, most notably limiting the number of iterations, run time, or perhaps monitoring the improvement made to the best solution in consecutive iterations. Among these options, limiting the number of iterations is arguably the most cited mechanism to create a termination stage for the genetic algorithm. The idea is that the process would be executed only for a specified number of times, a parameter known as the maximum iteration. In any case, it should be noted that selecting the termination mechanism is also considered one of the algorithm's parameters. Bear in mind that in most cases, these termination mechanisms may require setting up additional parameters.

3.3 Parameter Selection and Fine-Tuning of the Genetic Algorithm

One of the main conclusions that one can derive from the *no-free-lunch theorem* is that fine-tuning an algorithm is quite essential to get the best performance out of a meta-heuristic algorithm. This would basically ensure that an algorithm is equipped to handle the unique characteristics of a given optimization problem. Of course, it is possible to use our intuition, experience, and default values suggested for an algorithm's parameters as a good starting point, one should bear in mind that fine-tuning these parameters is, more than anything, a trial-and-error process. Thus, while it is possible to get a good enough result by having an educated guess for setting the parameters of these algorithms, to get the best possible performance, it is necessary to go through this fine-tuning process. In the case of the genetic algorithm, these parameters are population size (M), crossover probability parameter (P_c), mutation probability parameter (P_m), and opting for a selection, mutation, crossover operators, termination criterion, and of course, all the parameters that are embedded in these operators. For instance, if limiting the number of iterations has been selected as a termination criterion, the maximum iteration (T) is another parameter that needs to be defined by the user.

As can be seen, the genetic algorithm has a considerable number of parameters that can be tweaked to control its performance. On the one hand, this gives us a lot of flexibility as we have many options to opt for when we intend to tune the algorithm's performance. But from a practical standpoint, this means that fine-tuning could be a challenging and time-consuming task. The other main challenge is that while some parameters have a noticeable effect on the emerged solution that one could intuitively understand, say the population size or the maximum number of iterations, the effects of others are not so intuitively clear, say crossover or mutation probability parameters. As such, to get the best results out of the genetic algorithm, you may have to dabble with it first to gain some experience and inside knowledge about such parameters. By doing so, you could better understand how to fine-tune these parameters as your initial guesses and parameter selection strategies become more educated. The pseudocode for the genetic algorithm is shown in Figure 3.3.

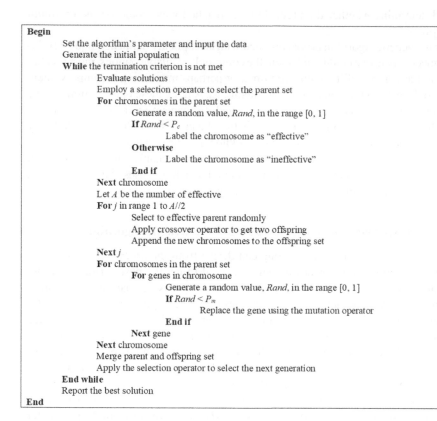

Figure 3.3 Pseudocode for the genetic algorithm.

3.4 Python Codes

The code to implement the genetic algorithm can be found below:

```python
import numpy as np

def init_generator(num_variable, pop_size, min_val, max_val):
    return np.random.uniform(low=min_val,
                             high=max_val,
                             size=(pop_size,num_variable))

def selection(pop, obj_func, R, minimizing=True,
              method='ranking selection'):
    if method == 'ranking selection':
        of = np.apply_along_axis(func1d=obj_func,axis=
1,arr=pop)
        index = np.argsort(of)
        if minimizing:
```

```
                    return pop[index][:R]
            else:
                    return pop[index[::-1]][:R]
    elif method == 'roulette wheel':
        try:
            if not minimizing:
                of = np.apply_along_axis(func1d=obj_func,axis=1,arr=
                pop)
                 selection_prob = of/np.sum(of)
                 pop_index = np.random.choice
                 (a=np.arange(len(pop)),
                  size=R,replace=True,
                  p=selection_prob)
                 return pop[pop_index]
            else:
                    raise TypeError
        except:
            raise TypeError('Roulette wheel cannot be used for
                                minimization, nor negative or null

fitnesses value.')
    elif method == 'tournament selection':
        index=np.random.choice(a=np.arange(len(pop)),size=
        R,replace=True)
        return pop[index]

    else:
        raise NameError(f'method {method} is not defined.')

def cross_over(pop,pc,method='uniform'):
    index = np.random.uniform(size=len(pop))
    selected_parents = pop[index<=pc]
    parents_indeces = np.random.randint(0,len(selected_parents),
                                    size=(2,
                                        len(selected_parents)//2))
    parents_1 = selected_parents[parents_indeces[0]]
    parents_2 = selected_parents[parents_indeces[1]]

    if method == 'one_point':
        C = np.random.randint(pop.shape[1]-1)
        crossover_index = np.where(np.arange(pop.shape[1])<=C,0,1)
        childs_1 = np.where(crossover_index,parents_1,parents_2)
        childs_2 = np.where(crossover_index,parents_2,parents_1)
        childs = np.concatenate((childs_1,childs_2),axis=0)
        return childs
    elif method == 'two_points':
        if pop.shape[1]>=4:
            C1 = np.random.randint(1,pop.shape[1]/2)
            C2 = np.random.randint(pop.shape[1]/2,pop.shape[1]-1)
            crossover_index = np.where(
                                    (np.arange(pop.shape[1])<C1)
                                    |(np.arange(pop.
                                    shape[1])>C2),0,1)
            childs_1 = np.where(crossover_index,parents_1,parents_2)
            ch  ilds_2 = np.where(crossover_index,parents_2,parents_1)
```

```
                    childs = np.concatenate((childs_1,childs_2),axis=0)
                    return childs
            else:
                raise ValueError('To use two points methods
                                 you have to have at least four
                                 variables.')

        elif method == 'uniform':
            crossover_index = np.random.randint(0,2,pop.shape[1])
            childs_1 = np.where(crossover_index,parents_1,parents_2)
            childs_2 = np.where(crossover_index,parents_2,parents_1)
            childs = np.concatenate((childs_1,childs_2),axis=0)
            return childs

        else:
            raise NameError(f'method {method} is not defined.')

def mutation(pop,pm,min_val,max_val,current_iteration,
             iteration,d0,method='uniform'):
    indeces = np.random.uniform(size=pop.shape)<pm
    if method == 'uniform':
        mutated_values = init_generator(pop.shape[1],pop.shape[0],
                                        min_val,max_val)
        mutated_pop = np.where(indeces,mutated_values,pop)
        return mutated_pop

    elif method == 'nonuniform':
        mutated_values = init_generator(pop.shape[1],pop.shape[0],
                                        min_val,max_val)
        d=d0*((iteration-current_iteration)/iteration)
        mutated_pop = np.where(indeces,
                             np.random.uniform(pop-d,pop+d),pop)
        return mutated_pop

    else:
        raise NameError(f'method {method} is not defined.')

def GA_algorithem(pop_size,num_variable,obj_func,R,
                  min_val,max_val,pm=.2,pc=.7, d0=2,
                  iteration=1000,minimizing=True,
                  full_result=False,selection_method='ranking
                  selection',
                  crossover_method='uniform',mutation_method=
                  'uniform'):
    results=np.zeros(iteration)
    NFE=np.zeros(iteration)
    NFE_value=0
    pop=init_generator(num_variable,pop_size,min_val,max_val)
    for i in range(iteration):
        selected_parents=selection(pop,obj_func,R,minimizing=
        minimizing,
                                    method=selection_method)
```

```
    if selection_method=='ranking selection':
        NFE_value+=pop_size
    childs=cross_over(selected_parents,pc,method=crossover_method)
    mutated_childs=mutation(pop,pm,min_val,max_val,
                                  current_iteration=i,
                                  iteration=iteration,d0=d0,
                                  method=mutation_method)
    all_results=np.concatenate((pop,childs,mutated_
    childs),axis=0)
    pop=selection(all_results,obj_func,pop_size,minimizing)
    NFE_value+=len(all_results)
    results[i]=obj_func(pop[0])
    NFE[i]=NFE_value

if not full_result:
    return pop[0], obj_func(pop[0])

else:
    return pop[0], obj_func(pop[0]), results, NFE
```

3.5 Concluding Remarks

One can safely state that genetic algorithm is among the most crucial milestones in the field of meta-heuristic optimization. The novelties used in this algorithm paved the way for the next generation of meta-heuristic optimization algorithms. This was the first notable attempt to implement a parallelized searching mechanism to enumerate through the search space. This would give this algorithm an edge over its predecessors as it can better comb through the search space and avoid being trapped in local optima. And with recent development in parallel computation in modern programming languages, this procedure is not as computationally taxing as one might think. The other notable novelty in this algorithm is that it mimics a natural phenomenon, in this case, the evolutionary process, to form the computational architecture of the algorithm. Looking at the surrounding environment as a source of inspiration would pan out to be a great way to create efficient, general-purpose meta-heuristic optimization algorithms.

In the case of the genetic algorithm, it starts with a set of randomly generated populations. This set would be manipulated using three main operators of the genetic algorithm, namely, crossover, mutation, and selection operators, in an iterative process until a certain termination criterion is met, at which point the computation would come to a halt, and the best solution would be returned as the optimum solution.

Upon closer investigation of the architecture of the genetic algorithm, one could see that it is riddled with a number of parameters that can be fine-tuned to control the performance of this algorithm. While this presumably could provide a lot of flexibility for the user to get the best performance out of this algorithm by fine-tuning it to match the characteristics of a specific optimization problem, this could potentially task the user with an overwhelming process of fine-tuning a lot of parameters, some of which with no intuitively clear impact on the outcome. That said, genetic algorithm is to this very day considered one of the most viable and cited options when it comes to meta-heuristic optimization algorithms.

References

Abualigah, L. & Dulaimi, A.J. (2021). A novel feature selection method for data mining tasks using hybrid sine cosine algorithm and genetic algorithm. *Cluster Computing*, 24, 2161–2176.

Adams, D.B., Watson, L.T., Gürdal, Z., & Anderson-Cook, C.M. (2004). Genetic algorithm optimization and blending of composite laminates by locally reducing laminate thickness. *Advances in Engineering Software*, 35(1), 35–43.

Adeli, H. & Cheng, N.T. (1993). Integrated genetic algorithm for optimization of space structures. *Journal of Aerospace Engineering*, 6(4), 315–328.

Baricelli, N.A. (1962). Numerical testing of evolution theories, part II preliminary tests of performance. Symbiogenesis and terrestrial life. *Acta Biotheoretica*, 16, 99–126.

Bayley, D.J., Hartfield, R.J., Burkhalter, J.E., & Jenkins, R.M. (2008). Design optimization of a space launch vehicle using a genetic algorithm. *Journal of Spacecraft and Rockets*, 45(4), 733–740.

Chen, Z., Jiang, Y., Tong, Z., & Tong, S. (2021). Residual stress distribution design for gear surfaces based on genetic algorithm optimization. *Materials*, 14(2), 366.

De Jong, K.A. (1975). "An analysis of the behavior of a class of genetic adaptive systems." Doctoral Dissertation, University of Michigan.

Du, K.L. & Swamy, M.N.S. (2016). *Search and optimization by metaheuristics: Techniques and algorithms inspired by nature*. Springer International Publishing Switzerland. ISBN: 9783319411910

Fallah-Mehdipour, E., Bozorg-Haddad, O., & Marino, M.A. (2014). Genetic programming in groundwater modeling. *Journal of Hydrologic Engineering*, 19(12), 04014031.

Fogel, L.J., Owens, A.J., & Walsh, M.J. (1966). *Artificial intelligence through simulated evolution*. John Wiley.

Gazonas, G.A., Weile, D.S., Wildman, R., & Mohan, A. (2006). Genetic algorithm optimization of phononic bandgap structures. *International Journal of Solids and Structures*, 43(18–19), 5851–5866.

Goldberg, D.E. (1989). *Genetic algorithms in search, optimization and machine learning*. Addison-Wesley.

Goldberg, D.E., Deb, K., & Korb, B. (1990). Messy genetic algorithms revisited: Studies in mixed size and scale. *Complex Systems*, 4, 415–444.

Hamdia, K.M., Zhuang, X., & Rabczuk, T. (2021). An efficient optimization approach for designing machine learning models based on genetic algorithm. *Neural Computing and Applications*, 33(6), 1923–1933.

Hınçal, O., Altan-Sakarya, A.B., & Ger, A.M. (2011). Optimization of multireservoir systems by genetic algorithm. *Water Resources Management*, 25(5), 1465–1487.

Holland, J.H. (1975). *Adaptation in natural and artificial systems: An introductory analysis with applications to biology, control and artificial intelligence*. University of Michigan Press.

Holland, J.H. (2000). Building blocks, cohort genetic algorithms, and hyperplane-defined functions. *Evolutionary Computation*, 8(4), 373–391.

Iba, K. (1994). Reactive power optimization by genetic algorithm. *IEEE Transactions on Power Systems*, 9(2), 685–692.

Ines, A.V., Honda, K., Gupta, A.D., Droogers, P., & Clemente, R.S. (2006). Combining remote sensing-simulation modeling and genetic algorithm optimization to explore water management options in irrigated agriculture. *Agricultural Water Management*, 83(3), 221–232.

Jahandideh-Tehrani, M., Jenkins, G., & Helfer, F. (2021). A comparison of particle swarm optimization and genetic algorithm for daily rainfall-runoff modelling: A case study for Southeast Queensland, Australia. *Optimization and Engineering*, 22(1), 29–50.

Jarvis, R.M. & Goodacre, R. (2005). Genetic algorithm optimization for pre-processing and variable selection of spectroscopic data. *Bioinformatics*, 21(7), 860–868.

Kleedtke, N., Hua, M., & Pozzi, S. (2021). Genetic algorithm optimization of tin – copper graded shielding for improved plutonium safeguards measurements. *Nuclear Instruments and Methods in Physics Research Section A: Accelerators, Spectrometers, Detectors and Associated Equipment*, 988, 164877.

Michalewicz, Z. (1996). *Genetic algorithms+ data structures= evolution programs*. Springer-Verlag.

Mühlenbein, H. & Schlierkamp-Voosen, D. (1993). Predictive models for the breeder genetic algorithm I. Continuous parameter optimization. *Evolutionary Computation*, 1(1), 25–49.

Turing, A.M. (1980). Computing machinery and intelligence. *Creative Computing*, 6(1), 44–53.

Wager, T.D., & Nichols, T.E. (2003). Optimization of experimental design in fMRI: A general framework using a genetic algorithm. *Neuroimage*, 18(2), 293–309.

Wang, Y. & Wei, C. (2021). Design optimization of office building envelope based on quantum genetic algorithm for energy conservation. *Journal of Building Engineering*, 35, 102048.

Yaghoubzadeh-Bavandpour, A., Bozorg-Haddad, O., Rajabi, M., Zolghadr-Asli, B., & Chu, X. (2022). Application of swarm intelligence and evolutionary computation algorithms for optimal reservoir operation. *Water Resources Management*, 36(7), 2275–2292.

Yang, X.S. (2010). *Nature-inspired metaheuristic algorithms*. Luniver Press. ISBN: 9781905986286

4 Simulated Annealing Algorithm

Summary

Though it has been quite some time since the introduction of the simulated annealing algorithm, it is still considered a viable and relevant meta-heuristic optimization method. The main idea of this algorithm is to mimic the annealing process to form a search strategy that leads to identifying the optimum solution in a search space. In this chapter, we will dig deep and explore the mechanisms used in this algorithm. We get familiar with the simulated annealing algorithm's terminology and see how one can implement this algorithm in the Python programming language. Finally, we will explore the potential merits and drawbacks of this algorithm.

4.1 Introduction

Over the years, the superior performance of the acclaimed genetic algorithm became an indisputable fact that meta-heuristic algorithms could look at the underlying structure of naturally occurring phenomena as a source of inspiration to create an efficient computational algorithm. This was a turning point in the history of meta-heuristic optimization algorithms, as from this point onward, this became the new standard practice to create the computational architecture of meta-heuristic algorithms. Thus, most meta-heuristic algorithms are inspired by a phenomenon that can be seen worldwide. Simulated annealing is another well-regarded optimization algorithm that follows this same principle.

Inspired by the stochastic simulation model of Metropolis et al. (1953), Kirkpatrick et al. (1983) proposed a novel general-purpose, single-solution-based optimization method based on solid materials' annealing process called the simulated annealing algorithm. Like other meta-heuristic optimization methods, this method circumvents the challenges often associated with calculus-based methods, as the algorithm does not rely on gradient information of the objective function. In fact, simulated annealing is considered a direct search method, which utilizes a guided or reinforced search mechanism that directly looks at the objective function values to navigate the search space. Note that this search is conducted in standard simulated annealing by following the trajectory of a single searching agent. While this is not necessarily an indication of the inadequacy of the algorithm's search,

DOI: 10.1201/9781003424765-4

the algorithm is not fully utilizing the potential of parallel computation. That said, alternative computational architectures have been proposed for this algorithm that implements the idea of parallelism (e.g., Czech, 2001).

As for the inspiration behind the simulated annealing algorithm, it tries to imitate the annealing process of a solid matter. *Annealing* is a technical term in metallurgy that refers to sort a heat treatment process that, in effect, alters the microstructure of a solid material. Through this process, the physical and, often, even the chemical properties of solid material would be altered to the point that the end product shows more elasticity and less hardness, making it a more workable substance. In annealing, a solid matter, say a particular alloy of metal or glass, would be heated above its melting point temperature. This temperature would be maintained for a specific duration of time and then gradually reduced with a controlled pace until the melted material would solidify and recrystallize into a defect-free crystalline structure. In the annealing process, heating the materials causes the diffusion of atoms within these substances, and their gradual recrystallization helps the material progress toward the equilibrium state. This would reshape the material in a defect-free structure that, in turn, enhances the qualities of the end product material.

Metropolis et al. (1953) proposed a Markov chain Monte Carlo model that could have simulated the annealing process of a material through a simple stochastic process. Basically, in this model, the atom particles would undergo a small stochastic move that follows the principles of a Markov chain model. Roughly speaking, this simply means that the current position of an atom particle is dependent on its position in the previous stage. If such a movement can reduce the system's energy level, the movement would be deemed successful and, as such, would be accepted. Otherwise, a Monte Carlo-based model is used to test the acceptability of the potential move. The idea is that in this structure, even a non-improving move that increases the overall energy level of the system can be potentially accepted with a probability that follows the *Boltzmann distribution*. This process would be repeated until a certain thermal equilibrium is reached.

The simulation process described above inspired Kirkpatrick et al. (1983) to create a new meta-heuristic algorithm that mimicked the same procedure. In this analogy, the state of the system can be seen as a search agent that would enumerate through feasible solutions. The overall energy level of the system is the objective function or the cost function in case there are constraints in the optimization problem, and reaching the thermal equilibrium is equivalent to converging to the optimum solution. Note that, like other meta-heuristic algorithms, the simulated annealing algorithm is also equipped with a set of parameters through which one can control and fine-tune the optimization process.

Over the years, scholars and researchers have demonstrated the potential of the simulated annealing algorithm in many disciplinary, including but not limited to chemical engineering (Hanke & Li, 2000; Kong et al., 2020), civil engineering (Wu et al., 2020), data science (Tiwari & Roy, 2003; Swarnkar & Tiwari, 2004), energy industry (Jeon & Kim, 2004), geohydrology (Wang & Zheng, 1998; Cunha, 1999), hydrology (Orouji et al., 2013), natural hazard management (Hosseini et al., 2020), nuclear engineering (Tang et al., 2020), pattern recognition (Liu & Huang, 1998), production management (Sridhar & Rajendran, 1993; Meller & Bozer, 1996), project management (Cho & Kim, 1997), and water resources management (Cunha & Sousa, 1999; McCormick & Powell, 2004; Tospornsampan et al., 2005), to name a few. In the upcoming section, we will explore the computational structure of the simulated annealing algorithm.

4.2 Algorithmic Structure of Simulated Annealing Algorithm

As stated, simulated annealing tends to imitate the annealing process of solid matter as a search strategy. To that end, it uses the moving trajectory of a single search agent to enumerate the feasible part of the search space. At each iteration, using a Markov Chain Monte Carlo model, the position of the search agent, here referred to as the *state*, would be updated until a thermal equilibrium is reached, at which point the search would be terminated, and the last state would be returned as the solution to the optimization problem.

The simulated annealing algorithm's flowchart is depicted in Figure 4.1. A closer look would reveal that this algorithm actually consists of three main stages that are the *initiation*, *searching*, and *termination* stages. The search stage comprises two main components. The first component allows the algorithm to generate new states using a special variation of a first-order Markov Chain model. The second component is a strategy by which the acceptance of these newly generated states is tested using a Monte Carlo-based model. The following subsection will discuss each of these stages and their mathematical structures.

4.2.1 Initiation Stage

As we have said, simulated annealing is a single-solution optimization algorithm, which indicates that it relies on a singular search agent. In the simulated annealing algorithm's terminology, a search agent is referred to as a *state*. From a mathematical standpoint, in an optimization problem with N decision variables, an N-dimension coordination system could be used to represent the search space. In this case, any point within the search space, say X, can be represented mathematically as a $1{\times}N$ array as follows:

$$X = \left(x_1, x_2, x_3, \ldots, x_j, \ldots, x_N\right) \tag{4.1}$$

where X is a point in the search space of an optimization problem with N decision variables, and x_j represents the value associated with the jth decision variable.

The algorithm initiates by generating a state randomly within the feasible boundaries of the search space. The algorithm would continually update the position of the search agent and tracks its trajectory during this iterative process until a termination criterion is met, at which point the last state would be returned as the optimum solution.

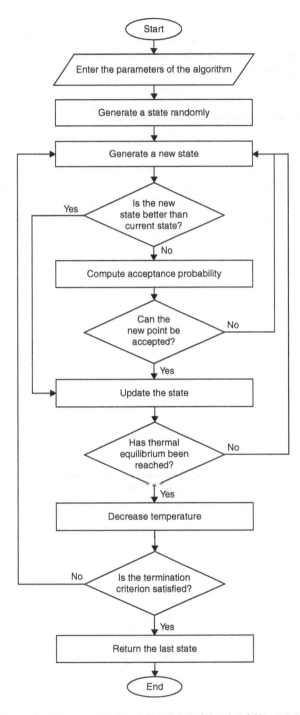

Figure 4.1 The computational flowchart of the simulating annealing algorithm.

4.2.2 Searching Stage

The idea behind the search stage is to enumerate through the search space and locate what could be the optimum solution. In this algorithm, the search is conducted by tracing the trajectory of a single search agent and moving through the search space. The searching stage in the simulated annealing algorithm consists of two main components, one that is responsible for generating a new state and another for testing the acceptability of these stochastically generated states.

The first stage would be to generate potential candidates for a new state to be considered a replacement for the current state. The idea is to conduct a local search in the vicinity of the current state to locate the trajectory that could potentially lead to the optimum result. Various viable stochastic and deterministic schemes are available to conduct a local search. A notable option would be a *random walk*. The random walk can be interpreted as a first-order Markov Chain model, where the values of the current step depend on the variables' values in the previous step. A simple variation of a random walk as a Markov Chain model can be mathematically expressed as follows:

$$V_{t+1} = V_t + \tau \tag{4.2}$$

where V_{t+1} and V_t are the variables at the steps $t+1$ and t, respectively; and τ is a randomly generated value. The general scheme of a random walk in a two-dimensional search space is depicted in Figure 4.2.

In the context of the simulated annealing, each element of the search agent would be subjected to a random walk that can be formulated as follows:

$$x_{i,j}^{new} = x_{i,j} - \varepsilon + \left(2 \times Rand \times \varepsilon\right) \tag{4.3}$$

in which $x_{i,j}^{new}$ represents the altered value for the jth decision variable of the ith state, $x_{i,j}$ is the current value for the jth decision variable of the ith state, *Rand* denotes a randomly generated value ranging from 0 to 1, and ε is the random walk's

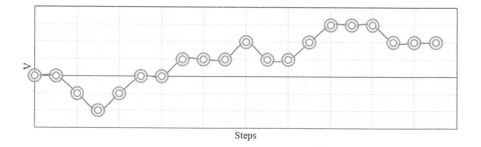

Figure 4.2 A random walk in a two-dimensional search space.

step size, which happens to be a user-defined parameter of the simulated annealing. The value assigned to this parameter should be proportional to the search space size and the characteristics of the objective function. While smaller values would result in a more thorough local search, larger values could be opted for this parameter in larger search spaces or if we deal with smoother objective functions.

Note that the standard simulated annealing algorithm encouraged to conduct of these local searches a specified number of times, denoted by β. This ensures that the local search has reached a temporary thermal equilibrium before moving with the rest of the search. This requires embedding a *nested loop* within the general outer loop of the algorithm. The idea is that through the searching process, we tend to update the state to, in turn, identify a trajectory that leads to what could be the optimum solution. But, the distinctive characteristic of the local search used in this algorithm is that, in simulated annealing, we could accept non-improving moves with certain probability rates. These probabilities are dynamically adjusted through the search process to create a more efficient search. The algorithm is more forgiving when the search is initiated, yet the odds of accepting a non-improving move would decrease as we progress with the search. This ensures that the algorithm would conduct a thorough enough search without being stuck on the local optima. To that end, the algorithm would adjust the probability rate each time it reaches thermal equilibrium. In other words, each time a thermal equilibrium has been reached, the algorithm adjusts the acceptance procedure to be more demanding than the previous stage.

As for the acceptance procedure, the simulated annealing uses Metropolis et al. (1953) as an inspiration to create a Monte Carlo-based model to either accept or reject the newly generated states. The general theme here is that in this model, while all the improving moves are by default accepted, a non-improving move can also be accepted with a certain level of probability that is proportional to its performance. In technical terms, this probability is referred to as the *transition probability*, and Kirkpatrick et al. (1983) proposed using a Boltzmann distribution to compose this *Monte Carlo*-based model. Here the transition probability, denoted by *Pr*, can be computed as follows:

$$\Pr\left(X, X^{new}, \lambda\right) = \begin{cases} 1 & \text{if } f\left(X^{new}\right) \text{ is better than } f\left(x\right) \\ e^{\frac{-\Delta f}{\lambda}} & \text{Otherwise} \end{cases} \tag{4.4}$$

$$\Delta f = \left| f\left(X^{new}\right) - f\left(x\right) \right| \tag{4.5}$$

in which X is the current state, X^{new} is the new state under consideration, $f()$ represents the objective function or the cost function if there are any constraints in the optimization problem, Δf represents the absolute changes in the objective function values, and λ is the acceptance rate control parameter. This parameter, as stated, would be dynamically adjusted through the course of optimization to make the transition from the exploration phase to the exploitation phase. Note that if the new state can outperform the current state, the transition probability is equal to one, meaning that the transition would definitely occur.

To implement this idea, the algorithm would compute the transition probability each time a new state is generated. A random number that ranges between 0 and 1, denoted by *Rand*, is then generated on the spot. Suppose the completed transition probability is greater than the randomly generated number. In that case, the new state will replace the current state, even if the new point is not objectively better than the current state. Bear in mind that the transition probability of a new state that is objectively better than the current state is one, as such, in such cases, the state would continuously be updated.

As for the acceptance rate control parameter (λ), as we have stated, this is something that is dynamically adjusted by the algorithm through the course of optimization. This procedure in simulated annealing lingo is referred to as *temperature reduction*. Every time the algorithm generates a specified number of new states, denoted by β, it is said that the search has reached thermal equilibrium, at which point the algorithm conducts a temperature reduction. That is to say, at each thermal equilibrium, the value associated with the acceptance rate control parameter would be adjusted.

There are different ways to conduct a temperature reduction. The most notable approaches are *linear* and *geometric* temperature reduction (Bozorg-Haddad et al., 2017). As usual, opting for one of these methods as a temperature reduction mechanism can actually constitute as one of the parameters of this algorithm.

In the linear temperature reduction method, we have the following:

$$\lambda_t = \lambda_0 - \alpha \times t \tag{4.6}$$

$$\alpha = \frac{\lambda_0 - \lambda_T}{T} \tag{4.7}$$

where λ_t is the acceptance rate control parameter at the iteration t; λ_0 and λ_T are the values of the acceptance rate control parameter at the start and end of the iteration; T is the maximum number of iterations; and α is the cooling factor. Note that to implement this method, you need to specify the number of iterations explicitly. Here all λ_0, λ_T, and T are user-defined parameters.

The geometric method for temperature reduction is formulated as follows:

$$\lambda_t = \lambda_0 \times \alpha^t \qquad 0 < \alpha < 1 \tag{4.8}$$

Here, the cooling factor can be seen as a user-defined parameter that should be selected from the range [0, 1]. Note that the closer the value is to the upper boundary, the more gradual the temperature reduction. It can also be seen that in this method, you are not limited to any predefined number of iterations. As such, this method is more flexible when selecting the termination criterion method.

4.2.3 Termination Stage

As seen in Figure 4.1, the simulated annealing algorithm would conduct a local search until it reaches thermal equilibrium, executing a temperature reduction procedure. This would gradually make the search strategy to shift from an exploratory phase to the exploitation phase. It has been proven that with enough randomness in the searching process and a gradual thermal reduction, the simulating annealing could eventually converge to the optimum solution (Yang, 2010; Du & Swamy, 2016).

But in order for the algorithm to converge to a solution, like other meta-heuristic algorithms, the sequence of operational structures of this algorithm needs to be executed iteratively until a certain termination criterion is met, at which point the execution of the algorithm would be terminated and the search agents last location, or depending on the way the algorithm is coded, perhaps the best position observed thus far, would be reported as the solution to the optimization problem. Note that without such a termination stage, the algorithm could be executed in an infinite loop. In effect, the termination stage determines whether the algorithm has reached what could be the optimum solution.

As the simulated annealing algorithm is not equipped with an explicitly defined, unique termination mechanism, one could implement the commonly available options, most notably limiting the number of iterations, run time, or perhaps monitoring the improvement made to the best solution in consecutive iterations. Among these options, limiting the number of iterations is arguably the most cited mechanism to create a termination stage for the simulated annealing algorithm. The idea being the process would be executed only for a specified number of times, a parameter known as the maximum iteration. In any case, it should be noted that selecting the termination mechanism is also considered one of the parameters of the algorithm. Bear in mind that these termination criteria may require setting up additional parameters in most cases.

4.3 Parameter Selection and Fine-Tuning the Simulated Annealing Algorithm

Based on the no-free-lunch theorem, fine-tuning an algorithm is quite essential to get the best performance out of any meta-heuristic algorithm, such as the simulated annealing algorithm. This procedure would basically ensure that an algorithm is calibrated to handle the unique characteristics of a given optimization problem. Of course, it is possible to use our intuition, experience, and default values suggested for an algorithm's parameters as a good starting point, one should bear in mind that fine-tuning these parameters is, more than anything, a trial-and-error process. Thus, while it is possible to get a good enough result by having an educated guess for setting the parameters of these algorithms, to get the best possible performance, it is necessary to go through this fine-tuning process. In the case of the simulated annealing algorithm, these parameters are the random walk's step size (ε), the specified number of iterations before reaching thermal equilibrium (β), and opting for a temperature reduction and termination mechanism and, of course, all the

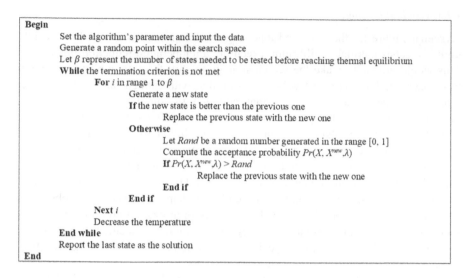

Figure 4.3 Pseudocode for simulating annealing algorithm.

parameters that are embedded in these methods. For instance, if the linear temperature reduction method is selected, the user would also have to fine-tune λ_0, λ_T, and T, or in the case of the geometric method, both λ_0 and cooling factor α are the parameters that need to be defined by the user. A good parameter selection would ensure a smooth transition from the exploratory to the exploitation phase and reach the optimum solution. The pseudocode for the simulated annealing algorithm is shown in Figure 4.3.

4.4 Python Codes

The code to implement the simulated annealing algorithm can be found below:

```
import numpy as np

def init_generator(num_variable, min_val, max_val):
    return np.random.uniform(low=min_val,high=max_val,size=num_
    variable)

def random_walk(x, epsilon):
    return x + np.random.uniform(-epsilon,epsilon,size=len(x))

def acceptance_function(state,new_state,obj_func,
                        lambda_param,minimizing=True):
    of_old = obj_func(state)
    of_new = obj_func(new_state)
    if minimizing:
```

```
        if of_new<=of_old:
            return new_state
        else:
            prob = np.exp(-np.abs(of_old-of_new)/lambda_param)
            rand = np.random.uniform()
            if prob>=rand:
                return new_state
            else:
                return state
    else:
        if of_new>=of_old:
            return new_state
        else:
            prob = np.exp(-np.abs(of_old-of_new)/lambda_param)
            rand = np.random.uniform()
            if prob>=rand:
                return new_state
            else:
                return state

def temperature_reduction(initial_lambda,current_iteration,
                          iteration,final_lambda=None,
                          cooling_factor=None,method='linear'):
    '''method = {'geometric', 'linear'}'''
    if method=='geometric':
        if 0<cooling_factor<1:
            return initial_lambda*(cooling_factor**current_
    iteration)
        else:
            raise ValueError('cooling factor must be between 0
    and 1')
    elif method=='linear':
        cooling_factor=(initial_lambda-final_lambda)/iteration
        return initial_lambda-(cooling_factor*current_iteration)
    else:
        raise NameError(f'method {method} is not defined ')
def SA_algorithem(obj_func,initial_lambda,num_new_state,
                  num_variable,min_val,max_val,iteration=1000,
                  cooling_factor=.5,epsilon=.1,final_lambda=None,
                  temp_reduc_method='geometric',
                  minimizing=True,full_result=False):
    '''method = {'geometric', 'linear'}'''
    results=np.zeros(iteration)
    NFE=np.zeros(iteration)
    NFE_value=0
    state=init_generator(num_variable,min_val,max_val)
    NFE_value+=1
    lambda_value=initial_lambda
    for i in range(iteration):
        for j in range(num_new_state):
            new_state = random_walk(state,epsilon)
```

```
    state   = acceptance_function(state,new_state,obj_func,
                                    lambda_value,
                                    minimizing=minimizing)
        NFE_value+=1
    lambda_value=temperature_reduction(initial_lambda,i,
                                        iteration,final_lambda,
                                        cooling_factor,
                                        method=temp_reduc_method)
        NFE[i]=NFE_value
        results[i]=obj_func(state)
    if not full_result:
        return state, obj_func(state)
    else:
        return state, obj_func(state), results, NFE
```

4.5 Concluding Remarks

The simulated annealing algorithm is a general-purpose, single-solution meta-heuristic optimization method that mimics the annealing process of solid material to create a search strategy to locate the optimum solution with the search space. As we have seen, the computation process starts with positioning the search agent, here called a state, in a randomly generated point within the feasible stage for the search space. The algorithm follows the trajectory of the search agent as it enumerates through the search space. To that end, it generates alternative states using a random walk. This constitutes as a local search to locate possible better solutions. Note that here, unlike what we have seen thus far, the algorithm may even accept non-improving moves using a Monte Carlo-based model with certain odds called transition probability. This improves the possibility that the algorithm would not trap in local optima. After a certain specified number of attempts, the algorithm is said to reach thermal equilibrium. At this point, the algorithm would adjust the transition probability so that accepting non-improving moves is less likely. As such, the algorithm would transition from the exploration phase to the exploitation phase. After the termination criterion is met, the algorithm would return the last state, or possibly the best-observed state, as the best solution during the searching process.

Simulated annealing is a well-regarded algorithm with a simple-enough, straightforward computational algorithm. As such, this is an easy enough algorithm to understand and, more importantly, to execute. More importantly, this is an efficient, robust algorithm, as it was shown that in the fine-tuned algorithm, the final solution is independent of the initial positioning of the search agent (Du & Swamy, 2016). In fact, it is proven that a gradual enough temperature reduction in this method would ensure identifying the optimum solution (Yang, 2010). However, the problem is that this algorithm does not use the full potential of parallelized computation. As such, the convergence speed might not be ideal sometimes, primarily if we deal with a large enough search space, many decision variables, or even a ragged objective function. That said, simulated annealing is still a viable and relevant meta-heuristic optimization method.

References

Bozorg-Haddad, O., Solgi, M., & Loáiciga, H.A. (2017). *Meta-heuristic and evolutionary algorithms for engineering optimization.* John Wiley & Sons. ISBN: 9781119386995

Cho, J.H. & Kim, Y.D. (1997). A simulated annealing algorithm for resource constrained project scheduling problems. *Journal of the Operational Research Society,* 48(7), 736–744.

Cunha, M. (1999). On solving aquifer management problems with simulated annealing algorithms. *Water Resources Management,* 13(3), 153–170.

Cunha, M. & Sousa, J. (1999). Water distribution network design optimization: Simulated annealing approach. *Journal of Water Resources Management,* 125(4), 215–221.

Czech, Z. (2001). Three parallel algorithms for simulated annealing. In *International conference on parallel processing and applied mathematics.* Springer, 210–217.

Du, K.L. & Swamy, M.N.S. (2016). *Search and optimization by metaheuristics: Techniques and algorithms inspired by nature.* Springer International Publishing Switzerland. ISBN: 9783319411910

Hanke, M. & Li, P. (2000). Simulated annealing for the optimization of batch distillation processes. *Computers & Chemical Engineering,* 24(1), 1–8.

Hosseini, F.S., Choubin, B., Mosavi, A., Nabipour, N., Shamshirband, S., Darabi, H., & Haghighi, A.T. (2020). Flash-flood hazard assessment using ensembles and Bayesian-based machine learning models: Application of the simulated annealing feature selection method. *Science of the Total Environment,* 711, 135161.

Jeon, Y.J. & Kim, J. C. (2004). Application of simulated annealing and tabu search for loss minimization in distribution systems. *International Journal of Electrical Power & Energy Systems,* 26(1), 9–18.

Kirkpatrick, S., Gelatt, C.D., & Vecchi, M.P. (1983). Optimization by simulated annealing. *Science,* 220(4598), 671–680.

Kong, W., Tu, X., Huang, W., Yang, Y., Xie, Z., & Huang, Z. (2020). Prediction and optimization of NaV1.7 sodium channel inhibitors based on machine learning and simulated annealing. *Journal of Chemical Information and Modeling,* 60(6), 2739–2753.

Liu, H.C. & Huang, J.S. (1998). Pattern recognition using evolution algorithms with fast simulated annealing. *Pattern Recognition Letters,* 19(5–6), 403–413.

McCormick, G. & Powell, R.S. (2004). Derivation of near-optimal pump schedules for water distribution by simulated annealing. *Journal of the Operational Research Society,* 55(7), 728–736.

Meller, R.D. & Bozer, Y.A. (1996). A new simulated annealing algorithm for the facility layout problem. *International Journal of Production Research,* 34(6), 1675–1692.

Metropolis, N., Rosenbluth, A.W., Rosenbluth, M.N., Teller, A.H., & Teller, E. (1953). Equation of state calculations by fast computing machines. *The Journal of Chemical Physics,* 21(6), 1087–1092.

Orouji, H., Bozorg-Haddad, O., Fallah-Mehdipour, E., & Mariño, M.A. (2013). Estimation of Muskingum parameter by meta-heuristic algorithms. *Proceedings of the Institution of Civil Engineers: Water Management,* 165(1), 1–10.

Sridhar, J. & Rajendran, C. (1993). Scheduling in a cellular manufacturing system: A simulated annealing approach. *International Journal of Production Research,* 31(12), 2927–2945.

Swarnkar, R. & Tiwari, M.K. (2004). Modeling machine loading problem of FMSs and its solution methodology using a hybrid tabu search and simulated annealing-based heuristic approach. *Robotics and Computer-Integrated Manufacturing,* 20(3), 199–209.

Tang, S., Peng, M., Xia, G., Wang, G., & Zhou, C. (2020). Optimization design for super-critical carbon dioxide compressor based on simulated annealing algorithm. *Annals of Nuclear Energy*, 140, 107107.

Tiwari, M.K. & Roy, D. (2003). Solving a part classification problem using simulated annealing-like hybrid algorithm. *Robotics and Computer-Integrated Manufacturing*, 19(5), 415–424.

Tospornsampan, J., Kita, I., Ishii, M., & Kitamura, Y. (2005). Optimization of a multiple reservoir system using a simulated annealing: A case study in the Mae Klong system, Thailand. *Paddy and Water Environment*, 3(3), 137–147.

Wang, M. & Zheng, C. (1998). Ground water management optimization using genetic algorithms and simulated annealing: Formulation and comparison. *Journal of the American Water Resources Association*, 34(3), 519–530.

Wu, K., de Soto, B.G., & Zhang, F. (2020). Spatio-temporal planning for tower cranes in construction projects with simulated annealing. *Automation in Construction*, 111, 103060.

Yang, X.S. (2010). *Nature-inspired metaheuristic algorithms*. Luniver Press. ISBN: 9781905986286

5   Tabu Search Algorithm

Summary

The tabu search algorithm is arguably one of the earliest attempts of computational intelligence-based optimization methods to utilize a memory-based feature to enhance the capabilities of its local search engine. Even though the computational architecture of this algorithm is not particularly complex, thanks to this very feature, the tabu search algorithm still holds itself as an efficient and viable option to handle real-world optimization problems. In this chapter, we will dig deep and explore the mechanisms used in this algorithm. We get familiar with the tabu search algorithm's terminology and see how one can implement this algorithm in the Python programming language. Finally, we will explore the potential merits and drawbacks of this algorithm.

5.1 Introduction

The general idea behind local search-based meta-heuristic algorithms, such as the pattern search algorithm, is to evaluate the points in the vicinity of the search agent and locate the best possible solution in that set. At this point, the search agent would be relocated to the newly identified position. This procedure continues to follow such a trajectory until it reaches a point that could be considered the optimum solution, as no improving move can be identified at that stage. The fundamental flaw here is that such search strategies could identify a trajectory that leads the search agent to a local optimum solution rather than the global one. Avoiding such traps requires precision, experience, delicate calibration of the search engine, and, more importantly, often a lot of computation time. Even with all that, there are no guarantees that the algorithm can always converge to the global optimum solution. This problem is even more pronounced in single-solution-based meta-heuristic algorithms such as the pattern search, where the algorithm relies solely on a single search agent to enumerate through the search space. But as we have seen in the simulated annealing algorithm, if the algorithm adapts a strategy that could also accept non-improving moves, it could potentially avoid being stuck in local optima. And this is the basic idea behind the tabu search algorithm.

DOI: 10.1201/9781003424765-5

Theorized by Glover (1986), the tabu search is a single-solution, stochastic meta-heuristic algorithm that was first designed to handle complex *combinatorial optimization problems* (i.e., optimization problems with finite solution sets) (Glover, 1986, 1989, 1990) but was later modified to handle continues optimization problems as well (Cvijović & Klinowski, 1995; Siarry & Berthiau; 1997; Du & Swamy, 2016). The algorithmic architecture of the tabu search closely resembles the pattern search algorithm. Like the pattern search, here, the algorithm also follows the trajectory of a single search agent that enumerates through the search space using a local search engine. Also, similar to the patterns search algorithm, the algorithm uses the direct search strategy as a guide to finding the right trajectory for the search agent. But what distinguished this algorithm from its predecessors was that it was the first meta-heuristic algorithm with what would become known as an algorithm with memory (Yang, 2010). In fact, it could be argued that, in that sense, the algorithm mimics the human brain, as when we face a problem, we tend to draw from our memoirs or previous experiences to tackle and, in turn, solve the problem. In other words, we test, learn, adapt, and re-try this process until we arrive at the desirable conclusion, which basically happens to be the scheme of the tabu search algorithm.

The word *tabu* or *taboo* simply refers to something that is forbidden. In the core, the tabu search algorithm follows the same simple architecture of the pattern search that is a direct local search but with a notable difference; the tabu search algorithm is equipped with a memory-like feature. This feature gives the algorithm a better sense of the search space and the points previously encountered during the search process. The memory embedded in the tabu algorithm allows it to create a twisted version of the *greedy search* strategy. The point is that, as the algorithm can memorize the previously encountered points, it can avoid revisiting them during the search. In other words, the algorithm would label such points as *tabu points*, and given that it is equipped with memory, it would store them in what is referred to as the *tabu list*. The algorithm would thus prohibit the search agent from returning to these listed points. As such, there may come a time during the search when the search agent is obliged to a non-improving point, as the improving alternatives are listed as tabu points. This would, in turn, decrease the odds of being trapped in local optima. Though it may be a subtle enough change in the overall structure of a local search, it can enhance the algorithm's efficiency. That said, hatching a more efficient memory-based feature for meta-heuristic algorithms, such as the tabu search, is still a topical research field (Yang, 2010).

From a conceptual standpoint, the tabu search has a simple-enough computational structure, making it easy to understand and implement for optimization purposes. Being equipped with a memory-based feature, the algorithm can potentially avoid being trapped in local optima. This would have subtle impacts on this algorithm's efficiency and computational capacities compared to its predecessor local search-based methods or even gradient-based optimization algorithms. In fact, it has been argued that in large-enough iteration, such a feature could have a tangible and meaningful contribution to the computation time and the overall efficiency of the algorithm (Yang, 2010). That said, this algorithm is not without its drawbacks. A notable potential downside is that, in some cases, the algorithm's final solution showed a considerable correlation to the initial point in which the search agent was located (Du & Swamy, 2016). As this is a randomly selected point within the feasible area of the search space, the quality of the final solution and the algorithm's convergence rate could potentially be tied to the placement of the search agent in the first place. The other drawback is that this algorithm utilizes a single search agent to enumerate through the search space. While this is not necessarily problematic in and of itself, it should be noted that the algorithm is not utilizing the full potential of parallelized computation, which can improve search quality and enhance the convergence rate. In conjunction with a proper parameter fine-tuning process, this could drastically reduce the odds of independence between the final solution and the initial point at which the search begins (Du & Swamy, 2016).

That said, when it comes to complex real-world optimization problems, the tabu search is still a viable option for optimization (Bozorg-Haddad et al., 2017). In fact, this algorithm has been successfully implemented for biology (e.g., Zhang et al., 2010), chemical engineering (e.g., Wang et al., 1999), computer science (e.g., Díaz et al., 2008; Hou et al., 2020), data engineering (e.g., Sexton et al., 1998), energy industry (e.g., Gallego et al., 2000; Karamichailidou et al., 2021), hydrology (e.g., Martínez et al., 2010), project management (e.g., de Soto et al., 2017), railway engineering (e.g., Corman et al., 2010), telecommunication (e.g., Costamagna et al., 1998), and water resources management (e.g., da Conceicao Cunha & Ribeiro, 2004), to name a few. It should be noted that, over the years, many variations of the tabu search algorithm have been proposed in the literature, such as reactive tabu search (Battiti & Tecchiolli, 1994), parallel tabu search (Kalinli & Karaboga, 2004), or diversification-driven tabu search (Glover et al., 2010), to name a few. That said, in the following sections, we will explore the computational structure of the standard tabu search algorithm.

5.2 Algorithmic Structure of the Tabu Search Algorithm

As we have seen, the tabu search algorithm is a single-solution localized search with a memory-based feature introduced to the mix. The main task of the embedded memory is to keep track of the previously encountered points in the search space, here called tabu points, and more importantly, to devise a mechanism that prohibits the search agent from revisiting the members of the tabu list. It could be argued that this closely resembles what humans do when faced with a new problem. In these situations, we tend to test, learn, adapt, and re-try this process until we can crack the problem. With that spirit in mind, the algorithm would start the search with a randomly placed point in the search space and follow this agent's trajectory until it reaches what could be the optimum solution to the problem at hand.

The tabu search algorithm's flowchart is shown in Figure 5.1. A closer look would reveal that this algorithm actually consists of three main stages that are the *initiation*, *searching*, and *termination* stages. The mechanism of the searching stage is similar to that of the exploratory move of the pattern search algorithm,

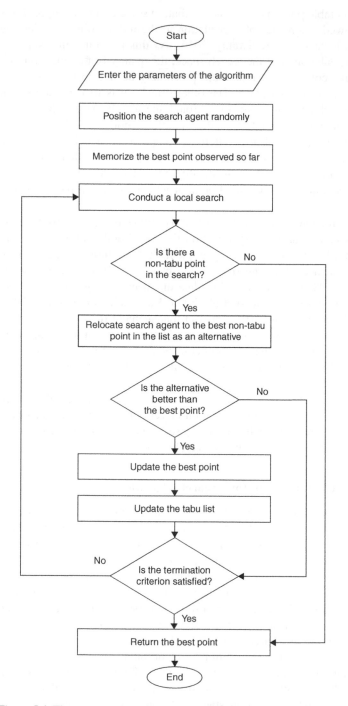

Figure 5.1 The computational flowchart of the tabu search algorithm.

with an additional feature that prevents the search agent from revisiting previously encountered points in the search space, which is technically referred to as tabu points. The following subsection will discuss each of these stages and their mathematical structures.

5.2.1 Initiation Stage

The tabu search algorithm is a single-solution-based meta-heuristic optimization algorithm. As such, it relies only on a single search agent to enumerate through and explore the search space. As we have seen, in an optimization problem with N decision variables, an N-dimension coordination system could be used to represent the search space. In this case, any point within the search space, say X, can be represented mathematically as a $1 \times N$ array as follows:

$$X = \left(x_1, x_2, x_3, \ldots, x_j, \ldots, x_N \right) \tag{5.1}$$

where X is a point in the search space of an optimization problem with N decision variables, and x_j represents the value associated with the jth decision variable.

As shown in Figure 5.1, the tabu search algorithm initiates with locating the single search agent at a randomly generated point within the feasible area of the search space. Given that the algorithm has no previous encounter regarding the search space at the initiation stage, the randomly generated point is temporarily set as the base point. Naturally, this is a placeholder until the algorithm stumbles upon a point with more desirable properties, in which case the base point would be replaced with the new location's coordination. Note that tentative or trial points that would be subsequently generated and tested through the computational process of the tabu search algorithm have the same mathematical structure.

As we have stated, the whole point behind embedding a memory-based feature to the tabu search was to keep track of the previously encountered points in the search space. Thus, in addition to placing the search agent, we need to create a container of a sort, say a *list*, *set*, or *tuple* object, as a vessel to keep a log of the previously encountered points. This container is basically recording the trajectory of the search agent. As such, we ought to initiate the process with an empty container. As we progress and the search agent makes any move, the said container should be updated by adding the properties of the move to the said list.

5.2.2 Searching Stage

The search process for the tabu search algorithm in a continuous search space resembles closely to what we have seen in exploratory moves in the pattern search algorithm. The idea is that the algorithm would actively look at the points in the vicinity of the search agent as tentative or trial points. The most desirable point that is not in the tabu list would be selected as the next destination for the search agent. Bear in mind that this strategy may lead to a non-improving move.

To that end, a mesh grid would be formed around the search agent location. This creates a number of tentative solutions. As we have seen for the patterns search algorithm, there are a few viable options to conduct this task. While each of these methods suggests a different number of tentative or trial points to be evaluated at this stage, the said number is proportional to the number of decision variables of the optimization problem. Naturally, the more decision variables we have, the more tentative points must be evaluated. One of the most common ways to create these mesh grids is the generalized pattern search (GPS) method (Bozorg-Haddad et al., 2017). In the GPS method, only one element of the base point would be either increased or decreased proportionately to the mesh size grid parameter. As such, in the GPS method, the number of tentative points created for each move equals 2×N. This can be mathematically expressed as follows:

$$X_1^{new} = \mu.\left[1,0,0,\ldots,0\right]+X \tag{5.2}$$

$$X_2^{new} = \mu.\left[0,1,0,\ldots,0\right]+X \tag{5.3}$$

$$\vdots$$

$$X_N^{new} = \mu.\left[0,0,0,\ldots,1\right]+X \tag{5.4}$$

$$X_{N+1}^{new} = \mu.\left[-1,0,0,\ldots,0\right]+X \tag{5.5}$$

$$X_{N+2}^{new} = \mu.\left[0,-1,0,\ldots,0\right]+X \tag{5.6}$$

$$\vdots$$

$$X_{2N}^{new} = \mu.\left[,0,0,\ldots,-1\right]+X \tag{5.7}$$

where X_i^{new} is the ith tentative solution, and μ denotes the size of this mesh grid, which happens to be one of the algorithm's parameters. For this parameter, it is crucial to opt for a value that is proportionate to the search space's size and the objective function's characteristics. Failing to do so may lead to skipping the optimum point if the mesh grid size is too large for the search space or facing massive computation time if the mesh grid size is too small. The mesh grid network of tentative points generated by the GPS method in a two-dimensional search space is depicted in Figure 5.2.

As we have stated, all the non-tabu tentative points would be tested after the formation of the mesh grid. Note that if all the tentative points are found in the tabu list, the algorithm will be terminated, and the best solution found thus far will be returned as the optimum solution. Otherwise, the search point would be moved

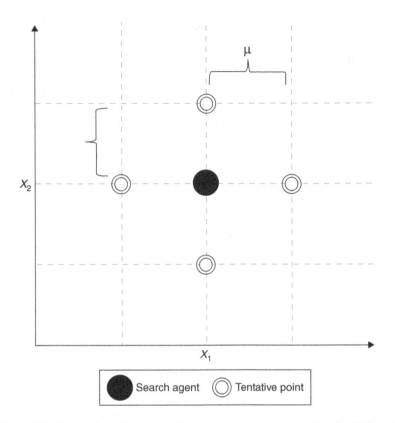

Figure 5.2 The mesh grid network of tentative points generated by the GPS method in a two-dimensional search space.

to the best tentative point, even if this move is deemed non-improving, that is, the current position of the search agent is more desirable than the said point. As we have discussed, this increases the possibility that our search agent, which basic-ally follows the principles of a local search, would not be trapped in local optima. Here, the new position assumed by the search agent would be added to the tabu list. As can be seen, the tabu search basically records the location in which the search agent has been placed. From a practical standpoint, it is not a pragmatic approach to keep track of all the previously encountered points, as this could be quite com-putationally taxing. Bear in mind that you have to check any tentative point against this tabu list, and as we progress through the optimization, this could amount to a lot of computation, especially in real-world problems where we are dealing with a lot of decision variables. So the pragmatic approach would be to limit the length of the tabu list to a specified number. This is a parameter of the tabu search algo-rithm called *tabu tenure*, which is denoted by δ. So when the list is said to be full, that is, it has reached its limit to record tabu points, any time a new point is added, the oldest tabu point encountered by the agent would be dropped from the list.

Greater values for this parameter indicated a *long-term memory* for the algorithm, while passing smaller values would create an algorithm with *short-term memory*. Naturally, the more robust the memory, the more computationally taxing it would be for the algorithm to converge to the solution. But the challenge for short-term memory algorithms is that they may be trapped in a *cycle*. A cycle is basically the algorithm relocating between the same finite number of points because the memory is not strong enough to memorize all the points in the cycle. As such, the algorithm would be unable to break this cycle and keep iterating on these points without exploring other options until the termination criterion is met.

Note that there are different ways to identify a point as a tabu. Of course, the most straightforward way is to label any encountered point as a tabu point and avoid revisiting these points at any cost until they are still on the tabu list. While this is an acceptable way to set up the search engine of the tabu search algorithm, this method, as we have seen, is not without any drawbacks. Such problems would be more pronounced in short-term memories or perhaps when the objective functions are too ragged. The alternative approach would be to opt for a softer touch when it comes to selecting these tabu points. The idea here is to circumvent facing the abovementioned problems; the algorithm can override the general rule of not being able to visit points in tabu lists if an aspiration criterion is met. Depending on the structure of the method, different aspiration criteria could be defined.

That said, there are two well-known approaches in this category *frequency-based memories* and *recently based memories* (Bozorg-Haddad et al., 2017). In the former method, we have to penalize objective functions of the points with an attribute in the tabu list. The more frequently these attributes are visited in previous encounters, the more severe these penalties would be. The penalized objective function would then be used to select the trajectory of the search agent. In the latter approach, the method keeps track of attributes that have changed in the last recorded moves and does not permit repetition. Here, the aspiration criterion could be set up to allow overruling the tabu list if the said move would lead to a better solution than the best previously encountered point thus far (Bozorg-Haddad et al., 2017).

5.2.3 Termination Stage

Based on the instruction depicted in Figure 5.1, the tabu search algorithm would keep on following the trajectory of the search agent as it moves through non-tabu tentative points that were identified in the local search process. It should be noted that the algorithm may have to be forced to terminate the search if all the tentative points are listed as tabu points. Otherwise, like other meta-heuristic algorithms, the sequence of operational structures of this algorithm needs to be executed iteratively until a certain termination criterion is met, at which point the execution of the algorithm would be terminated, and the best point observed thus far would be returned as the solution to the optimization problem at hand. Note that without such a termination stage, the algorithm could be executed in an infinite loop. In effect, the termination stage determines whether the algorithm has reached what could be the optimum solution.

As the tabu search algorithm is not equipped with an explicitly defined, unique termination mechanism, one could implement the commonly available options, most notably limiting the number of iterations, run time, or perhaps monitoring the improvement made to the best solution in consecutive iterations. Among these options, limiting the number of iterations is arguably the most cited mechanism to create a termination stage for the tabu search algorithm. The idea is that the process would be executed only for a specified number of times, a parameter known as the maximum iteration. In any case, it should be noted that selecting the termination mechanism is also considered one of the algorithm's parameters. Bear in mind that in most cases, these terminations may require setting up additional parameters.

5.3 Parameter Selection and Fine-Tuning the Tabu Search Algorithm

Based on the *no-free-lunch theorem*, fine-tuning an algorithm is a critical stage of implementing a meta-heuristic algorithm to ensure that the said algorithm is equipped to handle a given optimization problem. While it is possible to use our intuition, experience, and default values suggested for an algorithm's parameters as a good starting point, this is more than anything a trial-and-error process. Indeed, it is possible to get a good enough result by having an educated guess for setting the parameters of these algorithms, to get the best possible performance, it is necessary to go through this fine-tuning process. As for the tabu search algorithm, these parameters are mesh grid size (μ), tabu tenure (δ), mesh grid generation methods, algorithm's memory type, the termination strategies, and all the possible parameters that are associated with each of these strategies can be added to this list. For instance, if limiting the number of iterations is selected as a termination criterion, the maximum number of iterations would be added to the list of user-defined parameters that need to be fine-tuned by the user. The pseudocode for the tabu search algorithm is shown in Figure 5.3.

```
Begin
        Set the algorithm's parameter and input the data
        Generate a random point within the feasible search space
        Let X* be the best solution observed thus far
        Set the current point as X*
        While the termination criterion is not met
                Conduct a local search in the vicinity of the current search point
                If all neighbors are in the tabu list
                        Stop the algorithm and report X*
                End if
                Relocate the search point to the best non-tabu solution in the neighboring point
                If the search point is better than X*
                        Update X*
                End if
                Update the tabu list
        End while
        Report X*
End
```

Figure 5.3 Pseudocode for the tabu search algorithm.

5.4 Python Codes

The code to implement the standard tabu search algorithm can be found below:

```python
import numpy as np

def init_generator(num_variables, min_val, max_val):
    return np.random.randint(min_val, max_val,num_variables)

def neighboring_points(point, num_variables, step_lenght):

    pos_array = point + step_lenght*np.eye(num_variables,dtype=np.int)
    neg_array = point - step_lenght*np.eye(num_variables,dtype=np.int)
    mesh_grid = np.concatenate((pos_array, neg_array), axis=0)
    return mesh_grid def sorting_mesh_grid(mesh_grid, obj_func,
  minimizing):
    ofs = np.apply_along_axis(obj_func, 1, mesh_grid)
    indeces = np.argsort(ofs)
    if not minimizing:
        indeces = indeces[::-1]
    return mesh_grid[indeces]
def tabu_tester(point, tabu_list):
    return np.any(np.all(point==tabu_list, axis=1))
def evaluator(a, b, obj_func, minimizing):
    if minimizing:
        if obj_func(a)<obj_func(b):
            return True
        else:
            return False
    else:
        if obj_func(a)>obj_func(b):
            return True
        else:
            return False
def tabu_search(num_variables, min_val, max_val, step_lenght, obj_func,
                tabu_tebure, iteration, minimizing=True,
                full_result=False):
    NFE_value = 0
    NFE = np.zeros(iteration)
    results = np.zeros(iteration)
    point = init_generator(num_variables, min_val, max_val)
    best_point = point.copy()
    tabu_list = point.copy()
    tabu_list = point.reshape(1,-1)
    NFE_value += 1
    for i in range(iteration):
      mesh_grid = neighboring_points(point, num_variables, step_lenght)
        mesh_grid = sorting_mesh_grid(mesh_grid, obj_func, minimizing)
        counter = 0
        for j in mesh_grid:
            if not tabu_tester(j, tabu_list):
                point = j
```

```
        tabu_list = np.concatenate((tabu_list,
                                    point.reshape(1,-
                                    1)),axis=0)
            break
        else:
            counter += 1
    if evaluator(point, best_point, obj_func, minimizing):
        best_point = point.copy()
    NFE_value += (counter+1)
    NFE[i] = NFE_value
    results[i] = obj_func(best_point)
    if counter == len(tabu_list):
        break
    if len(tabu_list)>tabu_tebure:
        tabu_list = tabu_list[1:]
if not full_result:
    return best_point, obj_func(best_point)
else:
    return best_point, obj_func(best_point), results, NFE
```

5.5 Concluding Remarks

Tabu search algorithm is a general-purpose, single-solution meta-heuristic optimization method that follows the trajectory of a single search agent that enumerates through the search space using a local search engine. As can be seen, the tabu search algorithm is simply implementing the basic principles of a direct, local search. But, what distinguished this algorithm from its predecessors is that the tabu search algorithm is arguably the first one equipped with a memory-based feature to enhance the capabilities of the local search engine. The newly embedded memory-based feature enables the algorithm to track previously encountered points, referred to as tabu points. It prevents the search agent from visiting them so long as they are still stored in the memory. As one can expect, algorithms with longer term memories would store these positions for more extended periods, while algorithms with short-term memory drop the older positions as they reach their maximum storage capacity to make room for recording the newly encountered tabu points. Note that based on this search strategy, the algorithm may opt for non-improving moves during the search, which decreases the odds of being stuck in local optima.

Introducing a memory-based local search may strike as a subtle improvement at first. Still, if set correctly, it can have a considerable impact on the searching capabilities of the algorithm, the most notable of which is avoiding being trapped in local optima and arguably saving a lot of computational time in large-enough problems. On the other hand, the main shortcoming of this algorithm is that it does not utilize the full potential of parallelized computation, as it relies solely on the trajectory of a single search agent. However, some modified versions of this algorithm tried to address this issue. Other possible drawbacks of the tabu search algorithm are that, in some cases where the algorithm may not have been tuned properly, it was reported that the final solution could be dependent on the initial point. Such problems could be avoided by equipping the algorithm with a more robust, long-term memory and creating a much finer mesh grid for the local search engine. However, this means that the computation may be a bit taxing. Thus, the main challenge in implementing this algorithm is to balance its speed and accuracy by fine-tuning its parameters. That said, the tabu search algorithm is to this day considered an efficient and valid option to tackle optimization problems. More importantly, the benefits of having a memory-based feature, such as the one that was proposed in the tabu search algorithm, were found to be far too lucrative to the point that, nowadays, having a memory-based operator is a standard feature in any descent modern computational intelligence-based method that we know today.

References

Battiti, R. & Tecchiolli, G. (1994). The reactive tabu search. *ORSA Journal on Computing*, 6(2), 126–140.

Bozorg-Haddad, O., Solgi, M., & Loáiciga, H.A. (2017). *Meta-heuristic and evolutionary algorithms for engineering optimization*. John Wiley & Sons. ISBN: 9781119386995

Corman, F., D'Ariano, A., Pacciarelli, D., & Pranzo, M. (2010). A tabu search algorithm for rerouting trains during rail operations. *Transportation Research Part B: Methodological*, 44(1), 175–192.

Costamagna, E., Fanni, A., & Giacinto, G. (1998). A tabu search algorithm for the optimisation of telecommunication networks. *European Journal of Operational Research*, 106(2–3), 357–372.

Cvijović, D. & Klinowski, J. (1995). Taboo search: An approach to the multiple minima problem. *Science*, 267(5198), 664–666.

da Conceicao Cunha, M. & Ribeiro, L. (2004). Tabu search algorithms for water network optimization. *European Journal of Operational Research*, 157(3), 746–758.

de Soto, B.G., Rosarius, A., Rieger, J., Chen, Q., & Adey, B.T. (2017). Using a tabu-search algorithm and 4D models to improve construction project schedules. *Procedia Engineering*, 196, 698–705.

Díaz, E., Tuya, J., Blanco, R., & Dolado, J.J. (2008). A tabu search algorithm for structural software testing. *Computers & Operations Research*, 35(10), 3052–3072.

Du, K.L. & Swamy, M.N.S. (2016). *Search and optimization by metaheuristics: Techniques and algorithms inspired by nature*. Springer International Publishing Switzerland. ISBN: 9783319411910

Gallego, R.A., Romero, R., & Monticelli, A.J. (2000). Tabu search algorithm for network synthesis. *IEEE Transactions on Power Systems*, 15(2), 490–495.

Glover, F. (1986). Future paths for integer programming and links to artificial intelligence. *Computers & Operations Research*, 13(5), 533–549.

Glover, F. (1989). Tabu search—Part I. *ORSA Journal on Computing*, 1(3), 190–206.

Glover, F. (1990). Tabu search—Part II. *ORSA Journal on Computing*, 2(1), 4–32.

Glover, F., Lü, Z., & Hao, J.K. (2010). Diversification-driven tabu search for unconstrained binary quadratic problems. *4OR*, 8(3), 239–253.

Hou, N., He, F., Zhou, Y., & Chen, Y. (2020). An efficient GPU-based parallel tabu search algorithm for hardware/software co-design. *Frontiers of Computer Science*, 14(5), 1–18.

Kalinli, A. & Karaboga, D. (2004). Training recurrent neural networks by using parallel tabu search algorithm based on crossover operation. *Engineering Applications of Artificial Intelligence*, 17(5), 529–542.

Karamichailidou, D., Kaloutsa, V., & Alexandridis, A. (2021). Wind turbine power curve modeling using radial basis function neural networks and tabu search. *Renewable Energy*, 163, 2137–2152.

Martínez, S.I., Merwade, V., & Maidment, D. (2010). Linking GIS, hydraulic modeling, and tabu search for optimizing a water level-monitoring network in South Florida. *Journal of Water Resources Planning and Management*, 136(2), 167–176.

Sexton, R.S., Alidaee, B., Dorsey, R.E., & Johnson, J.D. (1998). Global optimization for artificial neural networks: A tabu search application. *European Journal of Operational Research*, 106(2–3), 570–584.

Siarry, P. & Berthiau, G. (1997). Fitting of tabu search to optimize functions of continuous variables. *International Journal for Numerical Methods in Engineering*, 40(13), 2449–2457.

Wang, C., Quan, H., & Xu, X. (1999). Optimal design of multiproduct batch chemical processes using tabu search. *Computers & Chemical Engineering*, 23(3), 427–437.

Yang, X.S. (2010). *Nature-inspired metaheuristic algorithms*. Luniver Press. ISBN: 9781905986286

Zhang, X., Wang, T., Luo, H., Yang, J.Y., Deng, Y., Tang, J., & Yang, M.Q. (2010). 3D protein structure prediction with genetic tabu search algorithm. *BMC Systems Biology*, 4(1), 1–9.

6 Ant Colony Optimization Algorithm

Summary

The ant colony optimization algorithm is arguably the first notable attempt to introduce swarm intelligence to the field of meta-heuristic optimization. Nowadays, looking up the idea of swarm intelligence as a source of inspiration has become a topical approach to theorizing novel meta-heuristic optimization algorithms. The ant colony optimization algorithm is a stochastic, population-based optimization algorithm that attempts to mimic ants' foraging behavior to create an efficient search engine for optimization problems with discrete search space. In this chapter, we will dig deep and explore the mechanisms used in this algorithm. We get familiar with the ant colony optimization algorithm's terminology and see how one can implement this algorithm in the Python programming language. Finally, we will explore the potential merits and drawbacks of this algorithm.

6.1 Introduction

The main idea behind meta-heuristic optimization algorithms is to provide a stepwise instruction for an iterative computational process that, in turn, helps ensemble a search engine that can thoroughly enumerate through the search space and locate what could potentially be passed as the optimum solution. What gave this branch of optimization methods a general edge over more traditional alternatives were that, while the searching process is being guided here, such guidance is not built upon the derivative of the objective function. As such, these are generally highly efficient and computationally fast methods that are not bound by common issues associated with real-world optimization problems such as high dimensionality, multimodality, epistasis, non-differentiability, and discontinuous search space imposed by constraints (Yang, 2010; Du & Swamy, 2016).

It should be noted that underneath the hood of an efficient meta-heuristic algorithm's search engine, there must be a mechanic that enables the algorithm to switch between what is known as the exploration and exploitation phases. The idea is that while an algorithm should be able to explore the search space fully, it should also be able to identify and, in turn, exploit the most promising areas of the search space and coverage to what could be the optimum solution. This feature

DOI: 10.1201/9781003424765-6

ensures that the algorithm would not prematurely converge to local optima. To that end, it became a common practice to base the computational architecture of meta-heuristic algorithms on a highly efficient phenomenon that can be seen in the real world. Whether it was a physical-based procedure, such as the annealing process of solid matters (Kirkpatrick et al., 1983), or a natural-based phenomenon, such as the gradual evolution of living creatures (Holland, 1975), these inspirations seemed to help create surprisingly highly efficient optimization algorithms with a searching mechanism that resembled what could be passed as computational intelligence. In fact, this is why meta-heuristic algorithms are also sometimes referred to as computational intelligence-based optimization algorithms.

Up until the 1980s, the majority of proposed meta-heuristic algorithms were inspired by phenomena that required a single search agent to enumerate through the search space. In these algorithms, the search agent position would be updated iteratively to create a trajectory that would eventually point to the optimum solution. The one notable exception here would be the genetic algorithm (Holland, 1975). What was interesting about this algorithm was that it pioneered the idea of using multiple search agents rather than a single agent. The phrase *population* was used to refer to this bundle of agents. Resorting to a population-based search rather than a single-solution-based one allowed this meta-heuristic algorithm to take advantage of parallelized computation, which helped create a much faster and more efficient search engine. Here the population would be seen as a single unit, which would be updated iteratively throughout the search process via a set of operators inspired by the natural selection process or what is simply known as evolution. In this simplified, simulated evolution, agents with better properties are more likely to participate in creating the next generation of search agents and pass these superior properties to their offspring. As such, the initial population would keep on evolving until it reaches a point where the best agent in the last population can be passed as the optimum solution to the problem at hand.

It is important to note that what is passed here as the evolution is an iterative stochastic process. The idea is that an agent that is considered to be doing well relative to its population would be given a better chance to be preserved via merging with other well-performing agents. The end product of this margin would be more likely to inherit the suitable properties of both parent agents. Without a random component, this process, however, could easily be trapped in local optima. As such, a mutation process would reintroduce new values back to the mix every now and then. This process would gradually turn the randomly generated initial population into a population with more suiting properties. It is essential to see that the described evolutionary process is not adjusting the properties of individual agents but rather creating a selective environment that basically mixes and matches these agents' properties.

While it is hard to argue with the end result of this procedure, evolution is by no means the only population-based intelligence behavior that can be seen in nature. The other most notable population-based intelligence is what became known as *swarm intelligence*. Swarm intelligence studies the collective intelligence of a group of simple agents that would come as a result of the local interactions of the group's individual members with one another or with their environment (Bonabeau et al., 1999). From the computational science perspective, the social behaviors that can be observed in nature bear the essential characteristics of intelligence behaviors. These behaviors that can be seen for collecting, foraging, hunting, or perhaps a mating ritual were actually formed and perfected over generations to give a species an extra edge to adapt, survive, or perhaps thrive in a given habitat. In other cases, these group behaviors help create a stable equilibrium between several specified that inhabits certain ecosystems. As it turned out, these groups and social behavior can be used as a source of inspiration to create the architecture of an efficient meta-heuristic algorithm. Unlike evolutionary-based search engines, each agent would adapt and modify its properties from the group's collective information. In other words, each individual would *learn* to adjust their performance based on the mechanics of these group and social behaviors in an iterative process until they reach a termination point, at which the best members of the group bear the properties of the optimum solution.

Based on this notion, Dorigo et al. (1991, 1996) proposed arguably the first documented meta-heuristic optimization algorithm, namely, the ant colony optimization, which utilizes the intelligence that can be detected in the context of group behavior of ant colonies. In a nutshell, the ant colony optimization algorithm tries to mimic the foraging behavior of ants and the mechanism by which the colony locates the most suitable path from the nest to the food source to create a search engine that thoroughly enumerates the search space. The ant colony optimization algorithm is a stochastic population-based meta-heuristic algorithm that is based on the notion of a direct search. This means that, like the previously introduced meta-heuristic algorithm, implementing the ant colony optimization algorithm to solve an optimization problem would not require any information regarding the objective function's gradient. Note that ant colony optimization was first theorized to handle *combinatorial optimization problems* with discrete decision variables, such as the well-known benchmark optimization problems of the *traveling salesman* (Dorigo et al., 1991).

The mathematical foundation for the ant colony optimization was laid out by the works of Goss et al. (1989) and Deneubourg et al. (1990). These works proposed a revolutionary and thought-provoking test called the *double-bridge experiment* to study the mechanism behind the ants foraging behavior. Basically, to conduct the test, the ant's nest and a source of food would be connected through two different pathways, such as the one depicted in Figure 6.1. The idea is that among the possible routes from the nest to the food source, one path is deliberately designed to be shorter than the other. The double-bridge experiment is merely a simple optimization problem with a discrete decision variable from a computational science standpoint. Each decision variable denotes a choice between possible paths at each node. As such, if multiple nodes are presented in the routes from the nest to the food source, the problem can be seen as a combinatorial optimization problem, where the string of selected paths at each node would combine together as a route, and the ultimate goal would be to identify a route with minimum distance between the nest and the food sources. In Figure 6.1, for instance, we are dealing with a single decision variable that is selecting between paths 1 and 2. Here, path 1 is the shorter path and, thus, the optimal choice to be selected as a route for the ants to forage their food. What is interesting here is that the double-bridge experiment reveals the mechanism by which an ant colony would collectively help identify this optimal route.

As the experiment initiated, the path selection procedure by the ants seemed more or less arbitrary, as the odds of opting for one path over the other by the first group of foragers seemed utterly random. From a mathematical standpoint, this means at the beginning of the experiment, the probability distribution associated with selecting a path seems to resemble the uniform distribution, as the likelihood of selecting any path is similar to one another (Figure 6.1a). Amazingly, once the first group reaches the food source, the behavior of the ants is altered, as the subsequent waves of foragers seem to be opting for the shorter path without direct communication with the other colony member who has traveled through the route (Figure 6.1b). What is happening here is that the ants are constantly communicating with each other, even if they do not directly contact one another along the way. As the colony attempts to find a way to source food, each ant marks its expedition path via a chemical compound called *pheromone*. Like some other insects, such as honeybees and termites, ants are equipped with a mechanism to release pheromones and detect this substance's concentration in the environment. So, as the first wave of foraging ants selects a route, they mark their path with the pheromone substance. The subsequent waves seem to gravitate toward the path with a higher pheromone concentration. But what makes this mechanism as efficient as it is is that this substance tends to evaporate over time. This means that as time goes by, the pheromone concentration would decrease gradually if not replenished by the foraging ants. Supposedly the amount of pheromone used to mark the shorter path has less time to be evaporated as the ants would keep

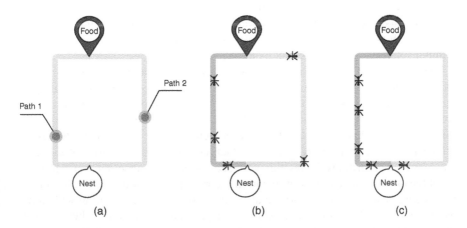

Figure 6.1 The general scheme of a double-bridge experiment with a single node at (a) the beginning of the experiment, (b) mid-experiment, (c) the end of the experiment.

reinforcing the pheromone concentration as they mark the path over and over as they go by. As such, gradually, the colony gravitates toward the shortest route. Note that this still seems to be a stochastic process, meaning it is unlikely but not impossible for an individual ant to opt for a path with a lower pheromone concentration (Figure 6.1c). Goss et al. (1989) captured the governing dynamics of ant's path selection in a double-bridge experiment with a single node through a stochastic model that is as follows:

$$\phi_j = \frac{\left(\mu_j + z\right)^h}{\sum_{j=1}^{2}\left(\mu_j + z\right)^h} \qquad j \in \{1,2\} \tag{6.1}$$

in which ϕ_j is the probability of selecting the jth path, μ_j represents the number of foraging ants that have previously opted for the jth path, and z and h are the model parameters that need to be calibrated with the experimental data.

Inspired by the mechanisms mentioned above, Dorigo et al. (1991, 1996) proposed a novel meta-heuristic algorithm that mimics the group behavior of ant colonies as they attempt to forage for food. As stated, this population-based stochastic algorithm was developed initially to solve combinatorial optimization problems with discrete decision variables. In the context of the ant colony optimization analogy, the population of the search agent can be seen as the ant colony, and as such, each individual search agent would then be interpreted as the route opted by an ant in this colony. The idea is that through an iterative process, these ants would enumerate through the discrete search space. In a stochastic process similar to what we have discussed earlier, in each iteration, they tend to mark the routes using a virtual pheromone distribution mechanism, where the amount of pheromone concentration would be proportional to the objective function of the said route. This pheromone consternation would then be transformed into a probability distribution value by which the algorithm attempts to regenerate a new colony of ants. This process would be continued until a termination criterion is met, at which point the ant with the most suitable route would be returned as the optimum solution to the problem at hand.

Arguably, one of the most notable things that could be associated with the ant colony optimization algorithm is that it was able to elevate the idea of creating a searching strategy by simply taking inspiration from nature by reinterpreting and capturing the intelligence that can be observed in group and social behavior of living creatures. This notion would create a tectonic shift in how meta-heuristic optimization algorithms are perceived. In fact, nowadays, the majority of modern meta-heuristic algorithms follow this very same idea. Of course, the main limitation of the original ant colony optimization algorithm was that it could only handle discrete search spaces. The algorithm also runs into issues when dealing with non-positive objective functions or maximization problems in general. These issues were usually attributed to the way the pheromone distribution mechanism was mathematically presented in the original variation of the ant colony optimization algorithm. Of course, as we have seen in earlier chapters, there are simple mathematical tricks to circumvent some of these issues, but the original idea behind this algorithm has been used to create a more general and modified version of the algorithm that could even handle continuous decision spaces. Some notable variations are the elitist ant system (Dorigo et al., 1996), ant-Q (Gambardella & Dorigo, 1995), ant colony system (Gambardella & Dorigo, 1996; Dorigo & Gambardella, 1997), hypercube ant system (Blum & Dorigo, 2004), virtual ant algorithm (Yang et al., 2006), and *Pachycondyla apicalis* ants algorithm (Monmarché et al., 2000), to name a few.

That said, the standard ant colony optimization algorithm is still to this day a reasonable and viable option for optimization, as it has been successfully implemented for chemical engineering (e.g., Babanezhad et al., 2020), computer science (e.g., Yu et al., 2015), data science (e.g., Parpinelli et al., 2002), economy (e.g., Uthayakumar et al., 2020), energy industry (e.g., Toksarı, 2007), geohydrology (e.g., Abbaspour et al., 2001), healthcare (e.g., Liu et al., 2021), marine engineering (e.g., Liang & Wang, 2020), mining engineering (e.g., Zhang et al., 2020), pharmaceutical sciences (e.g., Korb et al., 2007), project management (e.g., Merkle et al., 2002), structural engineering (e.g., Camp et al., 2005), telecommunication (e.g., Sim & Sun, 2003), and water resources management (e.g., Maier et al., 2003), to name a few. In the following sections, we will explore the computational structure of a standard ant colony optimization algorithm.

6.2 Algorithmic Structure of the Ant Colony Optimization Algorithm

As we have seen, the ant colony optimization algorithm is based on mimicking the group behavior that governs the foraging patterns of an ant colony. As a population-based meta-heuristic algorithm, the algorithmic architecture of the ant colony optimization is based upon a parallelized computation, that is, the algorithm would operate on a set of bundled search agents rather than a single one. Each search agent represents a virtual ant that enumerates through the discrete search space and marks it with what could be interpreted as virtual pheromone concentrations. These values would then be transposed to probabilistic values by which the algorithm attempts to stochastically update the positions of the ant colony's population set. This process would be repeated until it reaches a specified termination point, where the best position of the last population, or perhaps even the best position observed thus far, would be returned as the optimum solution.

The ant colony optimization algorithm's flowchart is depicted in Figure 6.2. A closer look at the architecture of the ant colony optimization algorithm would reveal that it actually consists of three main stages that are the *initiation, foraging*, and *termination* stages. To conduct the foraging stage, the ant colony optimization algorithm is equipped with a pheromone-laying and -following mechanism, which is the algorithm's core computational mechanic to stochastically update the ants' position in a way that more suitable routes are more likely to be selected. The following subsection will discuss each of these stages and their mathematical structures.

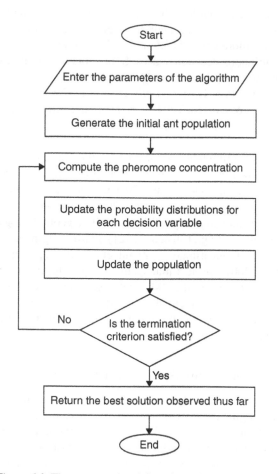

Figure 6.2 The computational flowchart of the ant colony optimization algorithm.

6.2.1 Initiation Stage

Ant colony optimization algorithm is a population-based meta-heuristic algorithm, and as such, it is designed to handle multiple search agents that would enumerate through a discrete search space. As we have seen, in an optimization problem with N decision variables, an N-dimension coordination system could be used to represent the search space. In this case, any point within the search space, say X, can be represented mathematically as a $1{\times}N$ array as follows:

$$X = \left(x_1, x_2, x_3, \ldots, x_j, \ldots, x_N\right) \tag{6.2}$$

where X represents a route selected by an ant in the search space of an optimization problem with N decision variables, and x_j represents the value associated with the *j*th decision variable. It should be noted that, here, we are dealing with a discrete search space. This means that the set variable could assume a set of finite acceptable values for each decision variable. These sets could be *monotonic*, as all the decision variables have the same set of acceptable values. The optimization algorithm could be *non-monotonic*, where the values for each decision variable vary. This can be mathematically represented as follows:

$$x_j \in \{v_{j,1}, v_{j,2}, v_{j,3}, \ldots, v_{j,k}, \ldots, v_{j,P}\} \qquad \forall j \tag{6.3}$$

where $v_{j,k}$ represents the *k*th possible value for the *j*th decision variable, and *P* is the number of possible values that can be assumed by the *j*th decision variable.

The ant colony optimization algorithm would start by randomly generating a series of routes within the feasible boundaries of the search space. This bundle of arrays can be mathematically expressed as $M \times N$ matrix, where *M* denotes what is technically referred to as the population size. Each row represents a route in the search space in such a structure. Note that the population size is one of the parameters of the ant colony optimization algorithm. A population, denoted by *pop*, can be represented as follows:

$$pop = \begin{bmatrix} X_1 \\ X_2 \\ \vdots \\ X_i \\ \vdots \\ X_M \end{bmatrix} = \begin{bmatrix} x_{1,1} & x_{1,2} & \cdots & x_{1,j} & \cdots & x_{1,N} \\ x_{2,1} & x_{1,2} & \cdots & x_{2,j} & \cdots & x_{2,N} \\ & & & \vdots & & \\ x_{i,1} & x_{i,2} & \cdots & x_{i,j} & \cdots & x_{i,N} \\ & & & \vdots & & \\ x_{M,1} & x_{M,2} & \cdots & x_{M,j} & \cdots & x_{M,N} \end{bmatrix} \tag{6.4}$$

where X_i represents the *i*th route in the population, and $x_{i,j}$ denotes the *j*th decision variable of the *i*th route. It should be noted that for each column, the values would be drawn from the set associated with the possible values for the said decision variable.

According to the ant colony optimization algorithm's analogy, each route is, in fact, a combination of possible values that are drawn from the sets associated with each decision variable to create a string of discrete values. To put this in the context of the double-bridge experiment, you can assume that each decision variable is a node, where a series of paths, or what are the possible values that can be assumed by that decision variable, would link this variable to the next one. A scaled version of this process can be captured by a web-like graph that accounts for all the potentially viable routes in the search space. The idea is we want to locate the most desirable route among all the possible candidates. In the context of the double-bridge experiment, the desirability of a route was interpreted as the shortest pathway from the colony's nest to the food source. For an optimization problem, the desirability of a route would be interpreted as the value of the objective function, or perhaps the cost function if there are constraints, for the said string. The better the objective function value computed for an array, the more desirable the route. Thus, when the algorithm evaluates a route, it calculates the objective function of the said route. This would help the algorithm mark the search space with pheromone concentration, where the concentration of this virtual compound should be proportionate to the routes' desirability.

6.2.2 Foraging Stage

Apart from what we have discussed thus far, there is another notable way the ant colony optimization algorithm distinguishes itself from other meta-heuristic optimization algorithms we have explored to this point. Thus far, the strategy to update the search agent's position of the meta-heuristic optimization methods we have studied, whether they were single-solution-based algorithms like simulated annealing or population-based ones such as the genetic algorithm, was to operate on the position of the search agents in the previous iteration. In some cases, this meant that the algorithm followed the trajectory of the search agents as it tweaked the component of these arrays, such as what we have seen in pattern search (Hooke & Jeeves, 1961), tabu search (Glover, 1989, 1990), and simulated annealing (Kirkpatrick et al., 1983); or the algorithm would mix and match a portion of these arrays to generate a new string, similar to what we have in the genetic algorithm (Holland, 1975). But thanks to its algorithmic architecture, which is based on parallelized computation, the ant colony optimization algorithm offers a fresh perspective on how the updating procedure can be handled. The idea is rather than operating on the entirety of the string, the algorithm could focus on individual components and the way they are distributed.

To do that, the algorithm would need a mechanism to monitor the frequency in which values would be selected for the routes and link these distributions to how good these routes actually are. This is where the pheromone-laying and -following mechanism would come into the picture. The good routes need to be marked with a virtual substance to indicate how desirable these paths are to other ants. The more desirable a path, the more the concentration of this compound would be. Note that pheromone markings are taking place on the set of possible values for each decision variable. This can be represented as follows:

$$C_j = \left\{ c_{j,1}, c_{j,2}, c_{j,3}, \ldots, c_{j,k}, \ldots, c_{j,P} \right\} \qquad \forall j \tag{6.5}$$

in which C_j is the set that stores the pheromone concentration for the jth decision variable, and $c_{j,k}$ denoted the pheromone concentration for the kth value of the jth decision variable.

The algorithm needs to start with a blank canvas, and as such, the pheromone concentration for all the possible values of all decision, variables would be initially assumed to be 0. As the algorithm proceeds with its computation, these values would be updated in each iteration based on the distribution of the values for each decision variable and the overall performance of their corresponding routes. Basically, a pheromone distribution mechanism would allocate the virtual substance to the possible values for each decision value, where values that are associated with better routes would have more pheromone concentration than others. In such a structure, it is crucial to embed a mechanism that would gradually decrease these concentrations to ensure that the algorithm would not stick to a single path. This ensures that the algorithm would not be trapped in local optima, as new values would always have a chance, as slim as it may be, to be stochastically selected, while it is still likely for more desirable paths to be selected in this process. This can be mathematically modeled as follows:

$$c_{j,k}^{new} = (1-\rho) \times c_{j,k} + \sum_{i=1}^{M} c_{j,k}^{i} \qquad \forall k,j \tag{6.6}$$

in which $c_{j,k}^{new}$ is the updated pheromone concentration for the kth possible value of the jth decision variable; $c_{j,k}$ is the current pheromone concentration for the kth possible value of the jth decision variable; $c_{j,k}^{i}$ denotes the pheromone intensity of the kth possible value of the jth decision variable that is deposited by the ith route; and finally, ρ represents the evaporation rate or pheromone decay rate, which happens to be a parameter of the ant colony optimization method that ranges between 0 and 1.

The pheromone deposited by each route here, denoted by F, also needs to be calculated in a way that is proportional to the objective function value. Thus, we need a transformation mechanism to interpret the objective function values into pheromone deposits associated with each route. The general theme here is that the better the objective function, the more pheromone deposits are associated with that route. Later down the road, pheromone concentrations themselves would need to undergo another transformation to be changed into probabilistic values, and they need to be non-negative values in the first place. And this is where the practical challenges to implement the ant colony optimization become more apparent. As it turns out, defining a general transformation function is rather challenging, if not impossible. Obviously, the first thing to note here is that different transformation functions are needed for maximization and minimization problems. If, for instance, the values of the objective function in the maximization problem are non-negative, these values could be directly used as pheromone deposit values. If we are dealing with the minimization of a problem with non-zero positive objective function values, you can implement the following as the transformation function (Bozorg-Haddad et al., 2017):

$$c_{j,k}^i = \begin{cases} \dfrac{\varnothing}{F(X_i)} & if \ x_{i,j} = v_{j,k} \\ 0 & otherwise \end{cases} \qquad \forall k,j,i \tag{6.7}$$

where ϕ is a constant value that can be defined by the user. You can also adjust the formula to be able to handle 0 values for the objective function by the below-modified transformation function:

$$c_{j,k}^i = \begin{cases} \left| \dfrac{\varnothing}{F(X_i) + |F(X_i)| \times \varepsilon} \right| & if \ x_{i,j} = v_{j,k} \\ 0 & otherwise \end{cases} \qquad \forall k,j,i \tag{6.8}$$

in which ε is a small, non-negative value that should be selected by the user. In any case, bear in mind that selecting a transformation function can often be a challenging task that may require tweaking or even defining a customized function. Again, the general rule of thumb is that the function must assign higher pheromone deposits to better objective functions, and the pheromone deposits must be a non-negative value.

Next, we need to convert these pheromone concentrations into probabilities by which certain values would be selected for each of the decision variables. As we have discussed, the general theme is the more pheromone concentration, the more likely it is for a value to be selected for the next set of routes. Note that, in essence, we assign a discrete probability distribution (i.e., mass distribution function) for each decision variable. This can be mathematically formulated as follows:

$$pr_{j,k} = \frac{\left(c_{j,k}\right)^\alpha \times \left(\vartheta_{j,k}\right)^\beta}{\sum_{d=1}^{P}\left[\left(c_{j,d}\right)^\alpha \times \left(\vartheta_{j,d}\right)^\beta\right]} \qquad \forall k,j \tag{6.9}$$

where $pr_{j,k}$ denotes the probability of selecting the kth value for the jth decision variable; $\vartheta_{j,k}$ represents the prior empirical information available for the kth value for the jth decision variable; and finally, both α and β are influence parameters, which are two of the parameters for the ant colony optimization algorithm. Both these values must be non-negative, which needs to be fine-tuned to match the optimization problem. It should be noted that typically the values associated with these parameters are assumed to be $\alpha \approx \beta \approx 2$ (Yang, 2010). As for the prior empirical information, this is any prior information about the desirability of each given value that may exist before doing any optimization. In some practical cases, through sheer experiment or perhaps even purely based on some sense of intuition, the user may feel like some values are more likely to produce better routes than others. Thus, one can see that these values serve as some weight assignment-like mechanism. As such, they need to be positive values, where the greater these so-called weights, the more likely that value will be selected in the process. While good prior empirical information could speed up the computational process and help the algorithm converge to the optimal solution, inaccurate input data could jeopardize the integrity of the emerging solutions. Thus, it is crucial to use this component cautiously, and if no prior empirical information is available, they can all be assumed to be equal to 1, which, mathematically speaking, neutralizes the effect of this component. As the final note on the generated probabilistic values, it should be noted that given their nature, the sum of the computed probabilities for all the possible values of a given decision variable should always be equal to 1. This means that mathematically speaking, for each of the decision variables, the following condition should always hold:

$$\sum_{d=1}^{P} pr_{j,d} = 1 \qquad \forall j \tag{6.10}$$

After computing these mass distribution functions, the algorithm would attempt to generate the next iteration of the routes in a stochastic manner. The point is that while we want more desirable values to have better odds of being selected to compose the new routes, we do not want to exclude any given value from the process. Eliminating values from this selection procedure could increase the odds of being trapped in local optima. Thus, as unlikely as it is, even the worst values can be possibly selected based on this procedure. Even though many of the computer programming languages these days, including Python, offer easy-to-use built-in functions or libraries that can select a value based on the mass distribution function, what happens in the backend of these functions is usually a simple mathematical trick. What happens is that they first convert the mass distribution function into the cumulative distribution function. The cumulative probability for the kth value of the jth decision variable, denoted by $Pr_{j,k}$, can be calculated as follows:

$$Pr_{j,k} = \sum_{d=1}^{k} pr_{j,d} \qquad \forall k, j \tag{6.11}$$

Then, a random value denoted by *Rand* for each decision variable in the range of [0, 1] would be generated. For each decision variable, the first value with a cumulative probability greater than the randomly generated number would be returned as the randomly selected value for that decision variable. This procedure would be repeated until M new routes that form the new population set are generated.

6.2.3 Termination Stage

Based on the foraging stage described above, the ant colony optimization algorithm would attempt to update the population in a stochastic manner in each iteration. This means that after each update, the pheromone deposits and their concentration would be computed for the values available for each decision variable. These concentrations would then be transformed into probabilistic values and using this mass distribution, the algorithm would regenerate a new set of routers. As we have said, the stochastic nature of this process ensures that while better values are more likely to appear in the regenerated routes, we are not excluding any particular values from the process.

Like other meta-heuristic algorithms, the sequence of operational structures of this algorithm needs to be executed iteratively until a certain termination criterion is met, at which point the execution of the algorithm would be terminated and the best route in the last iteration, or depending on the way the algorithm is coded, perhaps the best route observed thus far, would be reported as the solution to the optimization problem. Note that without such a termination stage, the algorithm could be executed in an infinite loop. The termination stage would, in effect, determine whether the algorithm has reached what could be the optimum solution.

As the ant colony optimization algorithm is not equipped with an explicitly defined, unique termination mechanism, one could implement the commonly available options, most notably limiting the number of iterations, run time, or perhaps monitoring the improvement made to the best solution in consecutive iterations. Among these options, limiting the number of iterations is arguably the most cited mechanism to create a termination stage for the ant colony optimization algorithm. The idea is the process would be executed only for a specified number of times, a parameter known as the maximum iteration. In any case, it should be noted that the selection of the termination mechanism is also considered one of the algorithm's parameters. Bear in mind that in most cases, these termination mechanisms may require setting up additional parameters.

6.3 Parameter Selection and Fine-Tuning the Ant Colony Optimization Algorithm

From the *no-free-lunch theorem*, one can conclude that fine-tuning an algorithm is essential to get the best performance out of a meta-heuristic algorithm. This would basically ensure that an algorithm is equipped to handle the unique characteristics of a given optimization problem. Of course, it is possible to use our intuition, experience, and default values suggested for an algorithm's parameters as a good starting point, one should bear in mind that fine-tuning these parameters is, more than anything, a trial-and-error process. Thus, while it is possible to get a good enough result by having an educated guess for setting the parameters of these algorithms, to get the best possible performance, it is necessary to go through this fine-tuning process.

```
Begin
    Set the algorithm's parameter and input the data
    Generate the initial population
    While the termination criterion is not met
            Evaluate solutions
            Update pheromone concentration for all the variables
            Update the probability distribution for all the variables
            Update the population set using the new distributions
    End while
    Report the best solution observed thus far
End
```

Figure 6.3 Pseudocode for the ant colony optimization algorithm.

In the case of the ant colony optimization algorithm, these parameters are population size (M), the evaporation rate or pheromone decay rate (ρ), influence parameters (α and β), and of course, opting for a pheromone to probability transformation function, termination criterion, and all the parameters that are associated with these methods. For instance, if limiting the number of iterations has been selected as a termination criterion, the maximum iteration (T) is another parameter that needs to be defined by the user. As can be seen here, the ant colony optimization algorithm is packed with many parameters, which makes the fine-tuning process a bit more challenging than other population-based optimization algorithms we have seen thus far. Furthermore, the impact of some of these parameters on the final outcome can still be a bit vague for first-time users. In other words, while some parameters have an obvious effect on the emerged solution that one could intuitively understand, say the population size or the maximum number of iterations, the effects of others are not so intuitively clear, say evaporation rate or influence parameters. As such, to get the best results out of the ant colony optimization algorithm, you may have to dabble with it first to gain some experience and inside knowledge about such parameters. By doing so, you could better understand how to fine-tune these parameters as your initial guesses and parameter selection strategies become more educated. The pseudocode for the ant colony optimization algorithm is shown in Figure 6.3.

6.4 Python Codes

The code to implement the standard ant colony optimization algorithm can be found below:

```python
import numpy as np

def init_generator(pop_size, num_variables, min_val, max_val):
    return np.random.randint(min_val,max_val+1,
                             size=(pop_size,num_variables))

def sgn(x):
    return np.where(x>=0, 1, -1)
```

```python
def pheremon_concentration_value(x, obj_func, minimizing, epsilon):
    f = obj_func(x)
    if minimizing:
        return np.abs(1/(f+sgn(f)*epsilon))
    else:
        return f

def pheremon_calculator(index_set, pop, obj_func, minimizing,
    epsilon):
    value = 0
    for i in index_set:
        value += pheremon_concentration_value(pop[i], obj_func,
                                              minimizing, epsilon)
    return value

def pheremon_values(pop, num_variables, min_val, max_val,
                    obj_func, minimizing, epsilon):
    outcome = np.zeros((num_variables, max_val-min_val+1))
    options = np.arange(min_val, max_val+1)
    for i in range(num_variables):
        for j in range(len(options)):
            index = np.array(np.argwhere(pop[:,i]==options[j]).flat)
            outcome[i, j] = pheremon_calculator(index, pop,
                                                obj_func, minimizing,
                                                epsilon)
    return outcome

def pheremon_update(pheremon_val, evaporation_rate, pop,
                    num_variables, min_val, max_val,
                    obj_func, minimizing, epsilon):
    a = (1-evaporation_rate)*pheremon_val
    b = pheremon_values(pop, num_variables, min_val,
                        max_val, obj_func, minimizing, epsilon)
    return a+b

def probabilty_calculator(pheremon_val, alpha, beta,
                          epsilon, desirability=None):
    x = pheremon_val+epsilon
    if desirability==None:
        desirability=np.ones_like(pheremon_val)
    denomerator = np.sum((x**alpha)*(desirability**beta),
                         axis=1).reshape(-1,1)
    return ((x**alpha)*(desirability**beta))/denomerator

def update_pop(pop, pop_size, num_variables, prob, min_val, max_val):
    new_pop = np.zeros_like(pop)
    options = np.arange(min_val, max_val+1)
    for i in range(num_variables):
        new_pop[:,i] = np.random.choice(options,
                                        size=(pop_size),p=prob[i])
    return new_pop
```

```python
def best_solution(pop, obj_func, minimizing):
    results = np.apply_along_axis(obj_func, 1, pop)
    if minimizing:
        index = np.argmin(results)
    else:
        index = np.argmax(results)
    return pop[index]


def evaluate(a, b, minimizing, obj_func):
    of_a = obj_func(a)
    of_b = obj_func(b)
    if minimizing:
        if of_a<of_b:
            return a
        else:
            return b
    else:
        if of_a>of_b:
            return a
        else:
            return b


def ant_colony_optimization(min_val, max_val, pop_size, num_variables,
                            evaporation_rate, alpha, beta, iteration,
                            obj_func, minimizing, desirability=None,
                            epsilon=1e-5, full_results=False):
    results = np.zeros(iteration)
    NFE = np.zeros(iteration)
    NFE_value = 0
    pop = init_generator(pop_size, num_variables, min_val, max_val)
    pheremon_val = np.zeros((num_variables, max_val-min_val+1))
    xbest = best_solution(pop, obj_func, minimizing)
    for i in range(iteration):
        pheremon_val = pheremon_update(pheremon_val, evaporation_rate,
                                       pop, num_variables, min_val,
                                       max_val, obj_func, minimizing,
                                       epsilon)
        prob = probabilty_calculator(pheremon_val, alpha, beta,
                                     0, desirability)
        pop = update_pop(pop, pop_size, num_variables, prob,
                         min_val, max_val)
        NFE_value += pop_size
        localbest = best_solution(pop, obj_func, minimizing)
        xbest = evaluate(xbest, localbest, minimizing, obj_func)
        results[i] = obj_func(xbest)
        NFE[i] = NFE_value
    if not full_results:
        return xbest, obj_func(xbest)
    else:
        return xbest, obj_func(xbest), results, NFE
```

6.5 Concluding Remarks

The ant colony optimization algorithm is a stochastic population-based meta-heuristic optimization method based on mimicking an ant colony's foraging behavior. The algorithm starts the search process by generating a set of randomly composed feasible routes. The desirability of these routes would then be calculated, and the values would then be used in a mechanism that basically marks the finite possible values available for each decision variable with a virtual pheromone deposit. The more frequently these values appear in better-performing routes, the more pheromone concentration they would receive. The pheromone concentration would then be transformed into probabilistic values by which the algorithm attempts to regenerate a stochastically new set of routes. The process would be repeated until the algorithm hits a termination stage, at which point the best route observed thus far would be returned as the optimal solution.

Although the ant colony optimization is definitely not the first meta-heuristic algorithm that tries to imitate a natural phenomenon per se, it pioneered a new branch of these optimization algorithms where the search engine is based on the social and group behavior of a specific species, what is, later on, became known as the swarm intelligence. These days, most modern meta-heuristic optimization algorithms try to follow in the footsteps of the ant colony optimization algorithm, as they typically attempt to mimic the swarm intelligence-based behavior that can be seen in nature. The other novelty in the algorithmic structure of the ant colony optimization is the way by which it utilized parallelized computation. Most meta-heuristic algorithms would attempt to see the search agents as a whole unit, and as such, they either follow their trajectory as they enumerate through the search space or they mix and match the components of previously encountered points to generate a new position for the search space. As we have seen. This is not the case for the ant colony optimization, as this algorithm would analyze the frequency in which the component would appear in good routes. As such, the components of more desirable routes are more likely to reappear in the newly generated routes, while the algorithm is designed in a way that even components of the bad routes would not be excluded from the search. This ensures a more thorough search that is less likely to be trapped in local optima. However, this makes things a bit more advanced and computationally taxing from a pure computational stand-point than previous methods. Furthermore, the pheromone-laying and -following mechanisms that enable the algorithm to execute this idea restrict the standard version of the algorithm to discrete search spaces. That said, there are modified variations of this algorithm that utilize the basic principles of this algorithm even to handle search spaces composed of continuous decision variables. More importantly, even the standard version of the ant colony optimization algorithm is a reasonably powerful method to handle real-world combinatorial optimization problems, and as such, it is still considered a viable and topical option to this very day.

References

Abbaspour, K.C., Schulin, R., & Van Genuchten, M.T. (2001). Estimating unsaturated soil hydraulic parameters using ant colony optimization. *Advances in Water Resources*, 24(8), 827–841.

Babanezhad, M., Behroyan, I., Nakhjiri, A.T., Marjani, A., Heydarinasab, A., & Shirazian, S. (2020). Liquid temperature prediction in bubbly flow using ant colony optimization algorithm in the fuzzy inference system as a trainer. *Scientific Reports*, 10(1), 1–14.

Blum, C. & Dorigo, M. (2004). The hyper-cube framework for ant colony optimization. *IEEE Transactions on Systems, Man, and Cybernetics–Part B*, 34(2), 1161–1172.

Bonabeau, E., Dorigo, M., Théraulaz, G., & Theraulaz, G. (1999). *Swarm intelligence: From natural to artificial systems*. Oxford University Press.

Bozorg-Haddad, O., Solgi, M., & Loáiciga, H.A. (2017). *Meta-heuristic and evolutionary algorithms for engineering optimization*. John Wiley & Sons. ISBN: 9781119386995

Camp, C.V., Bichon, B.J., & Stovall, S.P. (2005). Design of steel frames using ant colony optimization. *Journal of Structural Engineering*, 131(3), 369–379.

Deneubourg, J.L., Aron, S., Goss, S., & Pasteels, J.M. (1990). The self-organizing exploratory pattern of the argentine ant. *Journal of Insect Behavior*, 3(2), 159–168.

Dorigo, M. & Gambardella, L.M. (1997). Ant colonies for the traveling salesman problem. *BioSystems*, 43(2), 73–81.

Dorigo, M., Maniezzo, V., & Colorni, A. (1991). Positive feedback as a search strategy. Dipartimento di Elettronica, Politecnico di Milano, Milan, Italy, Technical Report, 91-016.

Dorigo, M., Maniezzo, V., & Colorni, A. (1996). The ant system: Optimization by a colony of cooperating ants. *IEEE Transactions on Systems Man and Cybernetics–Part B*, 26(1), 29–42.

Du, K.L. & Swamy, M.N.S. (2016). *Search and optimization by metaheuristics: Techniques and algorithms inspired by nature*. Springer International Publishing Switzerland. ISBN: 9783319411910

Gambardella L.M. & Dorigo M. (1995). Ant-Q: A reinforcement learning approach to the traveling salesman problem. In *Proceedings of the 12th international conference on machine learning*, Tahoe City, CA, USA.

Gambardella, L.M. & Dorigo, M. (1996). Solving symmetric and asymmetric TSPs by ant colonies. In *Proceedings of the International Conference on Evolutionary Computation*, Nagoya University, Japan.

Glover, F. (1989). Tabu search—Part I. *ORSA Journal on Computing*, 1(3), 190–206.

Glover, F. (1990). Tabu search—Part II. *ORSA Journal on Computing*, 2(1), 4–32.

Goss, S., Aron, S., Deneubourg, J.L., & Pasteels, J.M. (1989). Self-organized shortcuts in the Argentine ant. *Naturwissenschaften*, 76(12), 579–581.

Holland, J.H. (1975). *Adaptation in natural and artificial systems: An introductory analysis with applications to biology, control and artificial intelligence*. University of Michigan Press.

Hooke, R. & Jeeves, T.A. (1961). "Direct Search" solution of numerical and statistical problems. *Journal of the ACM*, 8(2), 212–229.

Kirkpatrick, S., Gelatt, C.D., & Vecchi, M.P. (1983). Optimization by simulated annealing. *Science*, 220(4598), 671–680.

Korb, O., Stützle, T., & Exner, T.E. (2007). An ant colony optimization approach to flexible protein–ligand docking. *Swarm Intelligence*, 1(2), 115–134.

Liang, Y. & Wang, L. (2020). Applying genetic algorithm and ant colony optimization algorithm into marine investigation path planning model. *Soft Computing*, 24(11), 8199–8210.

Liu, L., Zhao, D., Yu, F., Heidari, A.A., Li, C., Ouyang, J., ... & Pan, J. (2021). Ant colony optimization with Cauchy and greedy Levy mutations for multilevel COVID 19 X-ray image segmentation. *Computers in Biology and Medicine*, 136, 104609.

Maier, H.R., Simpson, A.R., Zecchin, A.C., Foong, W.K., Phang, K.Y., Seah, H.Y., & Tan, C.L. (2003). Ant colony optimization for design of water distribution systems. *Journal of Water Resources Planning and Management*, 129(3), 200–209.

Merkle, D., Middendorf, M., & Schmeck, H. (2002). Ant colony optimization for resource-constrained project scheduling. *IEEE Transactions on Evolutionary Computation*, 6(4), 333–346.

Monmarché, N., Venturini, G., & Slimane, M. (2000). On how *Pachycondyla apicalis* ants suggest a new search algorithm. *Future Generation Computer Systems*, 16(8), 937–946.

Parpinelli, R.S., Lopes, H.S., & Freitas, A.A. (2002). Data mining with an ant colony optimization algorithm. *IEEE Transactions on Evolutionary Computation*, 6(4), 321–332.

Sim, K.M. & Sun, W.H. (2003). Ant colony optimization for routing and load-balancing: survey and new directions. *IEEE Transactions on Systems, Man, and Cybernetics – Part A: Systems and Humans*, 33(5), 560–572.

Toksarı, M.D. (2007). Ant colony optimization approach to estimate energy demand of Turkey. *Energy Policy*, 35(8), 3984–3990.

Uthayakumar, J., Metawa, N., Shankar, K., & Lakshmanaprabu, S.K. (2020). Financial crisis prediction model using ant colony optimization. *International Journal of Information Management*, 50, 538–556.

Yang, X.S. (2010). *Nature-inspired metaheuristic algorithms*. Luniver Press. ISBN: 9781905986286

Yang, X.S., Lees, J.M., & Morley, C.T. (2006). Application of virtual ant algorithms in the optimization of CFRP shear strengthened precracked structures. In *Proceedings of the International Conference on Computational Science*, Berlin, Heidelberg.

Yu, Q., Chen, L., & Li, B. (2015). Ant colony optimization applied to web service compositions in cloud computing. *Computers & Electrical Engineering*, 41, 18–27.

Zhang, S., Bui, X.N., Trung, N.T., Nguyen, H., & Bui, H.B. (2020). Prediction of rock size distribution in mine bench blasting using a novel ant colony optimization-based boosted regression tree technique. *Natural Resources Research*, 29(2), 867–886.

7 Particle Swarm Optimization Algorithm

Summary

The particle swarm optimization algorithm is arguably one of the most revered meta-heretic algorithms. Its abstract and straightforward computational structure allowed it to take parallelized computation to the next level. Though it was not the first swarm intelligence-based meta-heuristic algorithm per se, it absolutely played a critical role in making this branch into the ever-expanding and topical subject it is today. In this chapter, we will dig deep and explore the mechanisms used in this algorithm. We would get familiar with the particle swarm optimization algorithm's terminology and see how one can implement this algorithm in the Python programming language. Finally, we will explore the potential merits and drawbacks of this algorithm.

7.1 Introduction

Swarm intelligence is the technical term to identify the collective intelligence of simple agents in the context of social or group behavior (Du & Swamy, 2016). Often, this intelligent-like behavior presents itself in the collective behaviors of decentralized and self-organizing agents, where the detected patterns in the actions of these individuals are derived from the local interactions of individual components with one another or their surrounding environment (Bonabeau et al., 1999). The perception of intelligence in the actions of these swarms stems from the fact that, while the isolated behavior of individual components may often appear self-centered, when these components come together, the same behavior serves the group as a whole. Thus, these seemingly simple-enough behaviors would enable the swarm to achieve something much more complicated than any individual member could have achieved. Thus, simply put, the general theme of swarm intelligence is to break down a challenging and complicated matter by assigning simple-enough manageable tasks to a number of agents. As these agents tend to carry out their individually defined missions through collaboration, cooperation, or even competition, their collective efforts would be coordinated toward the group's primary goal.

In the early 1990s, the emerging field of meta-heuristic optimization seemed interested in the enticing idea of swarm intelligence, as several scholars attempted

DOI: 10.1201/9781003424765-7

to incorporate and merge these two ideas. Back then, meta-heuristic optimization algorithms showed a lot of potentials and were on the verge of establishing themselves as a reliable alternative to conventional optimization methods, as it was claiming that it is, unlike their rivals, not bound by high dimensionality, multimodality, epistasis, non-differentiability, and discontinuous search space imposed by constraints (Yang, 2010; Du & Swamy, 2016). To its credit, the majority of proposed meta-heuristic optimization algorithms available at that point in time were, in fact, descent methods that could handle real-world optimization problems. The issue, however, with these methods was that given the nature of these algorithms, there could be no absolute guarantee that the algorithm could converge to the global optimum. This problem was even more pronounced with single solution-based algorithms as they tangled with more complicated real-world optimization problems. On the other hand, population-based algorithms such as the genetic algorithm (Holland, 1975) and the ant colony optimization algorithm (Dorigo et al., 1991, 1996) seemed to handle this issue better. While there was still no assurance of convergence to the global optimum solution, these algorithms proved more robust and dependable because they could conduct a more exhaustive search by employing multiple search agents instead of a single agent.

While these population-based meta-heuristic algorithms with parallelized computational capabilities showed great potential regarding real-world complex optimization problems, not many options were available back in the day if you tend to opt for this branch of algorithms. As efficient as it is, the genetic algorithm is riddled with many parameters, which makes the fine-tuning process of the algorithm rather challenging, especially given that the role of many of these parameters on the final outcome of the algorithm is not easy to understand for non-experts users intuitively. The available version of the ant colony optimization algorithm, the other viable population-based option at the time, was limited to handling optimization problems with discrete decision variables. This void presented an excellent research potential in this field as it seemed that there was still room for improvement to utilize every ounce of the computational power of the emerging, more robust, and more accessible computers of that era. This is where the particle swarm optimization algorithm enters the picture.

By mixing the general idea of swarm intelligence and meta-heuristic optimization, Kennedy and Eberhart (1995) were able to theorize a new algorithm called the particle swarm optimization. The particle swarm optimization algorithm is a population-based stochastic approach based on the basic principles of direct search; that is, it looks directly at the values of the objective function rather than resorting to gradient-based search. This algorithm takes inspiration from the generic social and group behavior that can be observed in a school of fish, a flock of birds, or a bee colony. Note that, unlike other algorithms we have studied thus far, the algorithm does not attempt to capture in detail a specific phenomenon but rather presents an elegant abstract representation of a simple social behavior that can be seen in nature. The point here is that even this abstract description of the group behavior could cultivate an algorithmic structure that governs the autonomous search agents toward what could be the optimum solution.

As we stated, the particle swarm optimization algorithm is basically a series of simple instructions that tries to model the generic group behavior of animals such as fish, birds, or bees as they tend to move from one place to another in a pack. As a population-based algorithm, particle swarm optimization would utilize a set of autonomously defined search agents that would enumerate through the search space. Based on the particle swarm optimization analogy, each search agent is referred to as a *particle*, and the set of these particles is known as the population or the *swarm*. The general theme here is that these particles would attempt to move within the search space and locate what could be the most desirable point or what we know as the optimum solution. What constitutes this group behavior as intelligent is that the particles are not only allowed to communicate with one another and influence one another's behavior, each component is equipped with a memory-like feature enabling it to keep track of the best position it encountered thus far. As such, while each search agent is enumerating through the search space on its own, as a unit, these particles are effectively enhancing the search capabilities of each member of the swarm.

The particle swarm optimization is also considered a trajectory-based meta-heuristic algorithm that would track each search agent's trajectory as it enumerates through the search space. What distinguishes the particle swarm optimization algorithm from other trajectory-based optimization algorithms we have seen thus far is that it proposes a more sophisticated mechanism to update the position of the search agents than the local search-based engine we have seen in the pattern search algorithm (Hooke & Jeeves, 1961) or the tabu search algorithm (Glover, 1989, 1990). In the trajectory-based optimization algorithm we have studied thus far, the algorithm would conduct a local search, which evaluates the tentative points in the vicinity of the search agent to update its position. Here, on the other hand, the algorithm would resort to vector-based mathematics to create a trajectory for each given search agent. To do that, the algorithm would identify two aspiration points in the search space for each particle. The algorithm would then adjust the motion of each particle to align them with these aspiration points. For each given particle, these aspiration points are the best position visited by the swarm thus far and the best point previously encountered by the said particle. As can be seen, this is not just a blind local search like we have seen previously in trajectory-based meta-heuristic algorithms, as this searching mechanism can be seen as an enhanced vector computation to update the position of the search agent. In fact, this feature enables the particle swarm optimization algorithm to locate the region of the optimum solution much quicker than non-swarm intelligence-based, population-based algorithms such as the genetic algorithm, though it may take relatively longer for the algorithm to converge to the optimum solution (Du & Swamy, 2016).

The architecture of the particle swarm optimization algorithm is actually fairly straightforward. It starts by randomly placing a set of particles in the feasible portion of the search space. The algorithm would then identify the aspiration points for all the particles. This requires to establishing a memory-based feature dedicated to keeping track of all the best positions encountered by all the particles and also recording the best point visited by the entire swarm. These aspiration points would then be used to update the position of the particles, as they would be guided toward their aspiration points individually. This process would be repeated until a termination point is reached, at which point the algorithm would return the best point observed thus far as the optimum solution.

As we have seen, the particle swarm optimization's computational algorithm is based on simple and primitive vector-based mathematics. Coupling this with utilizing parallelized computation and less bookkeeping when it comes to tapping into the computer's memory are some of the most notable merits of the particle swarm optimization algorithm. This does not mean that the particle swarm optimization algorithm has no drawbacks. The issue often attributed to this algorithm is the *stagnation* of particles (Du & Swamy, 2016). This means that there are times that the particles would stay idle as they reach certain points in the search space, and the algorithm is unable the upgrade their position, meaning that the said particle would be left out from the search for some time until it can possibly accept newly imposed motions. The algorithm's performance could also suffer drastically with a smaller swarm, so there can be no guarantee to reach local optima in these cases (Du & Swamy, 2016). The other generic problem that is associated with trajectory-based meta-heuristic algorithms is that if the algorithm is not equipped with a mechanism to keep the motion vectors in check, the particles may sometimes bounce out of the feasible part of the search space, causing the algorithm even to return infeasible results in some cases. This is not the end of the world, of course, as it can be easily addressed by either mounting the algorithm with a boundary-checking mechanism or alternatively treating the feasible range of decision variables as simple constraints. This means that violating these conditions would cause penalizing the objective function, forcing the algorithm to stay within the feasible bounds of the search space.

Over the years, many variants of the particle swarm optimization algorithm were proposed in the literature, some of which are comprehensive learning particle swarm optimization (Liang et al., 2006), adaptive particle swarm optimization (Zhan et al., 2009), and chaotic particle swarm optimization (Alatas et al., 2009), to name a few. That said, the standard particle swarm optimization algorithm is still considered a viable option to handle real-world optimization problems. In fact, the particle swarm optimization algorithm has been successfully used for agricultural engineering (e.g., Noory et al., 2012; Fallah-Mehdipour et al., 2013), data science (e.g., Salerno, 1997), energy sector (e.g., Abido, 2002), environmental studies (e.g., Chuanwen & Bompard, 2005), computational science (e.g., De Falco et al., 2007; Kuo et al., 2011), financial management (e.g., Zhu et al., 2011), geohydrology (e.g., Bozorg-Haddad et al., 2013), geotechnical engineering (e.g., Hajihassani et al., 2018), mechanical engineering (e.g., Chen et al., 2020), medical science (e.g., Sharif et al., 2020), petroleum engineering (e.g., Onwunalu & Durlofsky, 2010), and water resources planning and management (e.g., Montalvo et al., 2008; Yaghoubzadeh-Bavandpour et al., 2022). In the following sections, we will explore the computational structure of the standard particle swarm optimization algorithm.

7.2 Algorithmic Structure of the Particle Swarm Optimization Algorithm

As we have seen, the particle swarm optimization algorithm is based on mimicking a simplified generic group behavior of species such as birds, bees, and fish as they tend to move from one location to another. As a population-based meta-heuristic algorithm, the algorithmic architecture of the particle swarm optimization algorithm is based on parallelized computation, that is, the algorithm would operate on a set of bundled search agents rather than a single one. As each search agent, referred to as particles, would enumerate through the search space, the algorithm will try to coordinate its position based on two aspiration points, the best point observed thus far by the entire swarm and the best point previously encountered by the particle. This is why the particle swarm optimization algorithm is considered a trajectory-based meta-heuristic optimization algorithm, as the position updating mechanism is based on vector-based mathematics. As such, the algorithm would use these aspiration points to update the position of the particles in the swarm. This process would be repeated until it reaches a specified termination stage, at which point the best position observed thus far would be returned as the optimum solution to the problem at hand.

The swarm particle optimization algorithm's flowchart is depicted in Figure 7.1. A closer look at the architecture of the particle swarm optimization algorithm

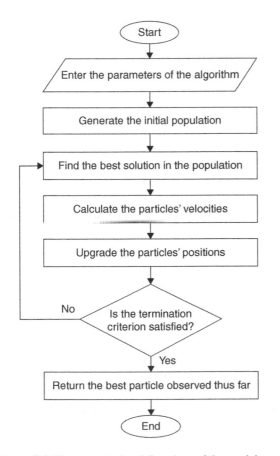

Figure 7.1 The computational flowchart of the particle swarm optimization algorithm.

would reveal that it actually consists of three main stages that are the *initiation*, *searching*, and *termination* stages. To conduct the searching stage, the particle swarm optimization algorithm would resort to vector-based mathematics to stochastically upgrade the position of the particles. Using this structure, the algorithm would conduct a thorough search and locate what could be the optimum solution to the problem at hand. The following subsection will discuss each of these stages and their mathematical structures.

7.2.1 Initiation Stage

The particle swarm optimization algorithm is a population-based meta-heuristic algorithm, and as such, it works with multiple search agents, here called particles, that would enumerate through the search space. As we have seen, in an optimization problem with N decision variables, an N-dimension coordination system could be used to represent the search space. In this case, any point within the search space, say X, can be represented mathematically as a $1 \times N$ array as follows:

$$X = \left(x_1, x_2, x_3, \ldots, x_j, \ldots, x_N\right) \tag{7.1}$$

where X represents a particle in the search space of an optimization problem with N decision variables, and x_j represents the value associated with the jth decision variable.

The particle swarm optimization algorithm initiates its searching process by randomly placing a series of particles within the feasible boundaries of the search space. This bundle of arrays, which in the particle swarm optimization algorithm's terminology is referred to as a swarm, can be mathematically expressed as $M \times N$ matrix, where M denotes the number of particles or what is technically referred to as the population size. In such a structure, each row represents a single particle. Note that population size is one of the many parameters of the particle swarm optimization algorithm. A population, denoted by *pop*, can be represented as follows:

$$pop = \begin{bmatrix} X_1 \\ X_2 \\ \vdots \\ X_i \\ \vdots \\ X_M \end{bmatrix} = \begin{bmatrix} x_{1,1} & x_{1,2} & \cdots & x_{1,j} & \cdots & x_{1,N} \\ x_{2,1} & x_{1,2} & \cdots & x_{2,j} & \cdots & x_{2,N} \\ & & \vdots & & & \\ x_{i,1} & x_{i,2} & \cdots & x_{i,j} & \cdots & x_{i,N} \\ & & \vdots & & & \\ x_{M,1} & x_{M,2} & \cdots & x_{M,j} & \cdots & x_{M,N} \end{bmatrix} \tag{7.2}$$

where X_i represents the ith particle in the population, and $x_{i,j}$ denotes the jth decision variable of the ith particle.

The randomly generated swarm contains a set of feasible positions in the search space. These particles would then keep moving within the search space, but their motion would be dictated by two main aspiration points, which are the best point observed thus far and the best position encountered by each particle through the search. This means that, in addition to the structure we have created thus far, we need a memory-based feature that records these encountered points in the space. To keep track of the best points observed by the individual particles, we would also need a matrix structure such as the one we have used to create the swarm in Equation (7.2). In such a matrix, each row can be represented as follows:

$$Pbest_i = \left(p_{i,1}, p_{i,2}, p_{i,3}, \ldots, p_{i,j}, \ldots, p_{i,N} \right) \qquad \forall i \tag{7.3}$$

in which $Pbest_i$ represents the best-observed position by the ith particle, and $p_{i,j}$ denotes the value in the jth dimension for the best position encountered by the ith particle.

As the algorithm initiates, the said matrix would be identical to the swarm. But after each upgrade in the position of the particles in each iteration, the objective function for the new position would be evaluated. If the new position is deemed more desirable, it will replace the best-recorded value for the said particle.

The algorithm also needs to keep track of the best point observed thus far by the entire swarm. This can be represented as follows:

$$Gbest = \left(g_1, g_2, g_3, \ldots, g_j, \ldots, g_N \right) \tag{7.4}$$

in which $Gbest$ denotes the best position encountered by the swarm, and g_j is the value of the best position in the jth dimension. Mathematically speaking, this is the best position in the memory matrix introduced earlier. As such, each time any best-encountered position in this matrix gets updated, the algorithm checks that against the best-recorded point. If the new position happens to outperform the best position observed thus far, the new value will replace the record stored in the memory. The algorithm would then use these aspiration points to help coordinate the particles' motions.

7.2.2 Searching Stage

The idea behind the particle swarm optimization algorithm is to coordinate the motions of the particles as they enumerate the search space. To that end, we need to incorporate vector-based mathematics in our computation. As we have seen, every particle's position in the search space can be mathematically represented as an array. Any motion in the particles can be captured via a *vector*. A vector has a similar structure to an array, with a notable distinction that, unlike an array, the vector conveys movement in the space. In this context, if an array specifies a given point in the search space, a vector can be seen as an arrow to depict a moving motion. Adding or subtracting a vector to an array would convey that motion to the said array, causing it to relocate within the search space from its starting point to a new position. The mathematics behind this is rather simple, as it is basic arithmetic carried out component-wise, that is, the components with the same dimension would be matched together and follow the arithmetic procedure.

The velocity in the particle swarm optimization algorithm captures the movement notion. In physics, you may recall that the velocity denoted the change in the position of an object with respect to a frame of reference that is a function of time. This would not only determine the *speed* of the motion but also convey information about the *direction* of the move as well. But, in the context of trajectory-based meta-heuristic optimization algorithms, velocity needs a slightly revised interpretation; unlike what we have seen in physics, the frame of reference for these movements cannot be defined as a function of time. In a meta-heuristic optimization algorithm, the movement of the search agent occurs each time the algorithm iterates. Of course, the frequency of these iterations can be seen as a function of time, but the amount of time needed here would rely on external matters such as the processing power of the system, the quality of the code, or the compiler that execute them, which is not relevant to what we are looking for. As such, in this context, the frame of reference for the velocity vector can be defined as a function of iteration. As we are tracking these motions one iteration at a time, the velocity vector would be identical to the repositioning vector of the search agent, which can be mathematically expressed as follows:

$$V_i = \left(v_{i,1}, v_{i,2}, v_{i,3}, \ldots, v_{i,j}, \ldots, v_{i,N}\right) \qquad \forall i \tag{7.5}$$

in which V_i represents the velocity vector of the *i*th particle, and $v_{i,j}$ denotes the velocity value of the *i*th particle in the *j*th dimension.

As we have stated, the particle swarm optimization algorithm is based on constantly imposing a new motion on the particles to upgrade their position based on two aspiration points, which are the best point observed by each individual particle and the best position encountered by the swarm thus far. In the context of vector mathematics we have seen above, this means that in each iteration, a velocity vector would be computed for each particle to guide the said particle toward its new position. In the first iteration, these values are randomly generated as the particles have no previous encounter with the search space. But from that point onward, in any iteration, the algorithm would need to compute these guiding motions for each particle in any given dimension. This can be mathematically expressed as follows:

$$v_{i,j}^{new} = \omega \times v_{i,j} + C_1 \times Rand_1 \times \left(p_{i,j} - x_{i,j}\right) + C_2 \times Rand_2 \times \left(g_j - x_{i,j}\right) \qquad \forall i,j \tag{7.6}$$

in which $v_{i,j}^{new}$ denotes the upgraded velocity value of the ith particle in the jth dimension; ω represents the inertia weight parameter, which would be dynamically adjusted in each given iteration; C_1 and C_2 are, respectively, cognitive and social parameters, which happen to be two of the parameters of the particle swarm optimization algorithm; and $Rand_1$ and $Rand_2$ are two randomly generated values within the [0, 1] range. Note that having these random components would give this process a stochastic nature. Their purpose is to impose some randomness to influence the particle's motion obtained from this stage. The idea behind this randomness is the exact same idea behind mutation operators in the genetic algorithm, as, without them, the algorithms would not have a good sense for exploration. In this case, without this component, the particles are not enticed to explore beyond the area occupied by the swarm. Failing to add these components would seriously jeopardize the integrity of the emerging solutions from these algorithms. As for the cognitive and social parameters, these can be seen as a simple weight to determine the amount of influence the motion can receive from the aspiration point. By tampering with these parameters, you can prioritize the motions of the particles to be more influenced by the best point encountered thus far or perhaps the best previously encountered position by each particle. As one can assume, tilting the weight balance toward the cognitive parameter makes the algorithm to base the motion more on the particles' memory, while using a more pronounced social parameter would ensure that the swarm memory mainly drives the particle's motions. Given their weight nature, both these values must be non-negative and fine-tuned to match the optimization problem. It should be noted that typically the values associated with these parameters are assumed to be $\alpha \approx \beta \approx 2$ (Bozorg-Haddad et al., 2017).

It should be noted that in some cases, the motions of the particles might be too swift, or perhaps the particles may come to a halt. In the former case, the particles could not fully explore the search space and may even bounce out of the feasible space, while in the latter case, the algorithm would come to something that is referred to as stagnation. In the context of a trajectory-based meta-heuristic optimization algorithm, stagnation means that the positions of one or perhaps more searching agents are not being updated through the iterations. Besides being a waste of computational power, this means that the algorithm is not utilizing its full potential to enumerate through the search space, which could, in turn, jeopardize the final solution returned by the algorithms. As such, one suggestion to remedy to handle these problems is to put a cap on both the upper and lower limits of the velocity vector, which can be expressed as follows:

$$V_j^L < v_{i,j} < V_j^U \qquad \forall i, j \tag{7.7}$$

in which V_j^L and V_j^U represent the lower and upper bounds of the velocity vector along the jth dimension, respectively. Accordingly, the upgraded values for the velocity vector should be able to hold the above-described condition. Failing to do so would mean that the value would be replaced by one of these boundary values, depending on which limit has been violated by the generated velocity. While this is a legitimate attempt to keep the particles' motion in check, in practice, you need to have a solid idea about the search space and the movement of particles, which requires a tone of experiment and intuition. Even with that, there is no guarantee the particles would not bounce out of the feasible space. A more reasonable alternative would be to impose the constraints on the updated position rather than the velocity vectors. After the algorithm's position of the search agents is updated, the new value would be tested against the lower and upper feasible bonds in each dimension. If the said values violate these boundary conditions, the new value would be replaced by one of the boundary values, depending on which limit has been violated by the new position. As such, we can ensure that the generated values always remain within the feasible boundaries of the search space.

As for the inertia weight, this is a non-negative parameter that is dynamically adjusted by the algorithm in each iteration to decrease gradually over time. From a mathematical standpoint, this parameter determines the particles' inertia at each given time, that is, how much of an influence the previous velocity vector has on the upgraded version. The larger values would mean the particle could mostly keep its momentum, and the new trajectory of the particle would be influenced mainly by the same trajectory that led the particle to its current position. There are two known methods to compute the intertie weights, namely, *linear* and *geometric*. As usual, opting for one of these methods can actually constitute as one of the parameters of this algorithm.

In the linear method, the inertia weight values would gradually decrease from two constant values in each iteration in a linear fashion, which can be formulated as follows:

$$\omega_t = \omega_0 - \left[(\omega_T - \omega_T) \times \frac{t}{T} \right] \qquad \forall t \tag{7.8}$$

where ω_t denotes the value of intertie at the tth iteration, ω_0 is the initial inertia weight, ω_T is the inertia weight at the last iteration, and T represents the total number of iterations. Note that to implement this method, you need to specify the number of iterations explicitly. In this case, ω_0, ω_T, and T are all algorithm user-defined parameters.

In the geometric method, the inertia weight at the tth iteration can be calculated as follows:

$$\omega_t = \omega_0 \times \gamma^t \qquad \forall t \tag{7.9}$$

where γ is a reduction rate parameter that should be selected from the range $(0, 1)$. Note that the closer the value to the upper boundary, the more gradual the reduction of the inertia weight parameter in each iteration would be. It can also be seen that in this method, you are not limited to any predefined number of iterations. As such, this method is more flexible when selecting the termination criterion method.

After the velocity vector gets computed for each of the particles in the swarm, the algorithm attempts to upgrade their positions by applying the moving motion of the vector to the particles' current positions. The position upgrading procedure can be mathematically expressed as follows:

$$X_i^{new} = \left(x'_{i,1}, x'_{i,2}, x'_{i,3}, \ldots, x'_{i,j}, \ldots, x'_{i,N} \right) \qquad \forall i \tag{7.10}$$

$$x'_{i,j} = x_{i,j} + v_{i,j} \qquad \forall i, j \tag{7.11}$$

where X_i^{new} represents the upgraded ith particle, and $x'_{i,j}$ denotes the new value for the jth decision variable of the ith particle.

7.2.3 Termination Stage

Based on the searching stage described above, the particle swarm optimization algorithm would attempt to update the position of all the particles in the swarm in each iteration. As stated, this would mean that after each upgrade, the new positions would be tested against the best-encountered positions recorded for each particle. If the new position happens to be the best position assumed by the particle, the record will be updated to reflect this fact. Upon any upgrade in the memory matrix, the algorithm would also check the new position against the best-recorded value by the entire swarm, and if need be, this value would be replaced as well. These values would be used as aspiration points to resume computing the trajectory of the particles.

Like other meta-heuristic algorithms, the sequence of operational structures of this algorithm needs to be executed iteratively until a certain termination criterion is met, at which point the execution of the algorithm would be terminated, and the best position observed by the swarm would be reported as the solution to the optimization problem. Note that without such a termination stage, the algorithm would potentially be executed in an infinite loop. The termination stage would, in effect, determines whether the algorithm has reached what could be the optimum solution.

As the particle swarm optimization algorithm is not equipped with an explicitly defined, unique termination mechanism, one could implement the commonly available options, most notably limiting the number of iterations, run time, or perhaps monitoring the improvement made to the best solution in consecutive iterations. Among these options, limiting the number of iterations is arguably the most cited mechanism to create a termination stage for the particle swarm optimization algorithm. The idea is that the process would be executed only for a specified number of times, a parameter known as the maximum iteration. In any case, it should be noted that selecting the termination mechanism is also considered one of the algorithm's parameters. Bear in mind that in most cases, these termination mechanisms may require setting up additional parameters.

7.3 Parameter Selection and Fine-Tuning the Particle Swarm Optimization Algorithm

From the *no-free-lunch theorem*, one can conclude that fine-tuning an algorithm is essential to get the best performance out of a meta-heuristic algorithm. This would basically ensure that an algorithm is equipped to handle the unique characteristics of a given optimization problem. Of course, it is possible to use our intuition, experience, and default values suggested for an algorithm's parameters as a good starting point, one should bear in mind that fine-tuning these parameters is, more than anything, a trial-and-error process. Thus, while it is possible to get a good enough result by having an educated guess for setting the parameters of these algorithms, to get the best possible performance, it is necessary to go through this fine-tuning process.

In the case of the particle swarm optimization algorithm, these parameters are population size (M), cognitive parameter (C_1), social parameter (C_2), and of course, opting for inertia weigh computation mechanism, termination criterion, and all the parameters that are associated with these methods. For instance, if limiting the number of iterations has been selected as a termination criterion, the maximum iteration (T) is another parameter that needs to be defined by the user. Or in case the linear inertial weight assignment method is selected, the user would need to pass the initial interim weight (ω_0), inertia weight at the last iteration (ω_T), and the maximum iteration (T). As seen here, the particle swarm optimization algorithm is packed with many parameters, making the fine-tuning process a bit more challenging. That said, the impact of some of these parameters on the final outcome can still be a bit vague for first-time users. In other words, while some parameters have an obvious effect on the emerged solution that one could intuitively understand, say the population size or the maximum number of iterations, the effects of others are not so intuitively clear, say cognitive and social parameters. As such, to get the best results out of the particle swarm optimization algorithm, you may have to dabble with it first to gain some experience and inside knowledge about such parameters. By doing so, you could better understand how to fine-tune these parameters as your initial guesses and parameter selection strategies become more educated. The pseudocode for the particle swarm optimization algorithm is shown in Figure 7.2.

```
Begin
        Set the algorithm's parameter and input the data
        Generate the initial population
        Generate the initial velocity values randomly
        Record the best solution observed thus far, denoted by Gbest
        While the termination criterion is not met
                For a particle in the population
                        Record the best position observed by the particle, denoted by Pbest_particle
                Next particle
                If the new best position observed in the process
                        Update Gbest
                End if
                For a particle in the population
                        Compute the particle's velocity
                        Update the particle's position
                Next particle
                Update inertia weight parameter
        End while
        Report Gbest as the solution
End
```

Figure 7.2 Pseudocode for the particle swarm optimization algorithm.

7.4 Python Codes

The code to implement the particle swarm optimization algorithm can be found below:

```python
import numpy as np

def init_generation(num_variables, pop_size, min_val, max_val):
    return np.random.uniform(low=min_val, high=max_val,
                             size=(pop_size, num_variables))

def best_solution(pop, obj_func, minimizing):
    result = np.apply_along_axis(func1d=obj_func, axis=1, arr=pop)
    if minimizing:
        index = np.argmin(result)
    else:
        index = np.argmax(result)
    return (pop[index], result[index])

def merge(pop, new_pop, obj_func, minimizing):
    result=np.apply_along_axis(func1d=obj_func, axis=1, arr=pop)
    new_result = np.apply_along_axis(func1d=obj_func,
                                     axis=1, arr=new_pop)
    if minimizing:
        index = np.reshape(result>new_result,newshape=(-1,1))
    else:
        index = np.reshape(result<new_result,newshape=(-1,1))
    merge = np.where(index,new_pop,pop)
    return merge
```

```python
def inertia_computation(inertia_0, inertia_final, iteration):
    a = np.arange(1,iteration+1)/iteration
    b = inertia_0-inertia_final
    return inertia_0-(a*b)

def update_velocity(pop, pbest, inertia, velocity,
                    gbest, cognative_param,
                    social_param, upper_velocity, lower_velocity):
    rand=np.random.uniform(size=(pop.shape[0], pop.shape[1]))
    a = (velocity*inertia)
    b = cognative_param*rand*(pbest-pop)
    c = social_param*rand*(gbest[0]-pop)
    new_velocity= a + b + c
    new_velocity=np.where(new_velocity>upper_velocity,
                          upper_velocity,new_velocity)
    new_velocity=np.where(new_velocity<lower_velocity,
                          lower_velocity,new_velocity)
    return new_velocity

def PSO_algorithem(num_variables, pop_size, min_val,
                   max_val, obj_func, inertia_0=.8,
                   inertia_final=.4, iteration=1000,
                   upper_velocity=np.inf, lower_velocity=-np.inf,
                   cognative_param=2, social_param=2, minimizing=True,
                   full_result=False):
    results=np.zeros(iteration)
    NFE=np.zeros(iteration)
    NFE_value=0
    inertia_values = inertia_computation(inertia_0,
                                         inertia_final, iteration)
    pop=init_generation(num_variables, pop_size, min_val, max_val)
    NFE_value+=pop_size
    pbest=pop.copy()
    gbest=best_solution(pbest, obj_func, minimizing)
    velocity=np.zeros_like(pop)
    for i in range(iteration):
        velocity=update_velocity(pop, pbest, inertia_values[i],
                                 velocity, gbest,
                                 cognative_param, social_param,
                                 upper_velocity, lower_velocity)
        pop+=velocity
        pbest=merge(pop, pbest, obj_func, minimizing)
        gbest=best_solution(pbest, obj_func, minimizing)
        NFE_value+=pop_size
        NFE[i]=NFE_value
        results[i]=gbest[1]
    if not full_result:
        return gbest
    else:
        return gbest[0], gbest[1], results, NFE
```

7.5 Concluding Remarks

To this day, the particle swarm optimization algorithm is considered one of the most revered meta-heuristic algorithms that can handle complex real-world problems rather efficiently. From a technical standpoint, the particle swarm optimization algorithm can be categorized as a population-based algorithm that tracks the trajectory of multiple search agents as they enumerate through the search space. The algorithm attempts to coordinate the movement of individual agents to make a cohesive unit where each component not only attempts to improve its own position but also helps steer the entire swarm toward the more rewarding section of the search space. As such, the algorithm would triangulate the position of what could be the optimum solution. What distinguished the particle swarm optimization algorithm from its population-based predecessors was that it introduced a novel and efficient way to utilize the capacities of parallelized computation. As such, it often performs better when locating the optimum solution's general vicinity compared to other population-based algorithms, such as the genetic algorithm. That said, converging to the optimum solution was still considered a challenge, as the standard version of this algorithm would often require quite some iteration to converge to the optimum solution.

From a mathematical standpoint, the algorithm is based on simple vector-based arithmetic operations. Coupling this with the abstract algorithmic structure of the algorithm and low-memory cost to keep track of the best solution encountered by the particle makes this algorithm a formidable choice from the computation point of view. That said, this algorithm is not without its drawbacks. The lack of any guarantee to converge to the optimum solution, stagnation, and the potential to bounce out of feasible space are some of the most notable issues associated with the particle swarm optimization algorithm. In most cases, these problems can often be addressed by fine-tuning the algorithm or equipping it with external operators that essentially keep the motions of the particles in check. Overall, it can be safely stated that the particle swarm optimization algorithm can still be considered one of the best choices for handling complex real-world optimization problems.

References

Abido, M.A. (2002). Optimal design of power-system stabilizers using particle swarm optimization. *IEEE Transactions on Energy Conversion*, 17(3), 406–413.

Alatas, B., Akin, E., & Ozer, A.B. (2009). Chaos embedded particle swarm optimization algorithms. *Chaos, Solitons & Fractals*, 40(4), 1715–1734.

Bonabeau, E., Dorigo, M., Théraulaz, G., & Theraulaz, G. (1999). *Swarm intelligence: From natural to artificial systems.* Oxford University Press.

Bozorg-Haddad, O., Solgi, M., & Loáiciga, H.A. (2017). *Meta-heuristic and evolutionary algorithms for engineering optimization.* John Wiley & Sons. ISBN: 9781119386995

Bozorg-Haddad, O., Tabari, M.M.R., Fallah-Mehdipour, E., & Mariño, M.A. (2013). Groundwater model calibration by meta-heuristic algorithms. *Water Resources Management*, 27(7), 2515–2529.

Chen, H., Fan, D. L., Fang, L., Huang, W., Huang, J., Cao, C., ... & Zeng, L. (2020). Particle swarm optimization algorithm with mutation operator for particle filter noise reduction in mechanical fault diagnosis. *International Journal of Pattern Recognition and Artificial Intelligence*, 34(10), 2058012.

Chuanwen, J. & Bompard, E. (2005). A self-adaptive chaotic particle swarm algorithm for short term hydroelectric system scheduling in deregulated environment. *Energy Conversion and Management*, 46(17), 2689–2696.

De Falco, I., Della Cioppa, A., & Tarantino, E. (2007). Facing classification problems with particle swarm optimization. *Applied Soft Computing*, 7(3), 652–658.

Dorigo, M., Maniezzo, V., & Colorni, A. (1991). Positive feedback as a search strategy. Dipartimento di Elettronica, Politecnico di Milano, Milan, Italy, Technical Report, 91-016.

Dorigo, M., Maniezzo, V., & Colorni, A. (1996). The ant system: Optimization by a colony of cooperating ants. *IEEE Transactions on Systems Man and Cybernetics–Part B*, 26(1), 29–42.

Du, K.L. & Swamy, M.N.S. (2016). *Search and optimization by metaheuristics: Techniques and algorithms inspired by nature.* Springer International Publishing Switzerland. ISBN: 9783319411910

Fallah-Mehdipour, E., Bozorg-Haddad, O., & Mariño, M.A. (2013). Extraction of multicrop planning rules in a reservoir system: Application of evolutionary algorithms. *Journal of Irrigation and Drainage Engineering*, 139(6), 490–498.

Glover, F. (1989). Tabu search—Part I. *ORSA Journal on Computing*, 1(3), 190–206.

Glover, F. (1990). Tabu search—Part II. *ORSA Journal on Computing*, 2(1), 4–32.

Hajihassani, M., Armaghani, D.J., & Kalatehjari, R. (2018). Applications of particle swarm optimization in geotechnical engineering: A comprehensive review. *Geotechnical and Geological Engineering*, 36(2), 705–722.

Holland, J.H. (1975). *Adaptation in natural and artificial systems: An introductory analysis with applications to biology, control, and artificial intelligence.* University of Michigan Press, Ann Arbor, MI.

Hooke, R. & Jeeves, T.A. (1961). "Direct Search" solution of numerical and statistical problems. *Journal of the ACM*, 8(2), 212–229.

Kennedy, J. & Eberhart, R. (1995). Particle swarm optimization. In *Proceedings of IEEE International Conference on Neural Networks*, Perth, WA, USA.

Kuo, R.J., Chao, C. M., & Chiu, Y.T. (2011). Application of particle swarm optimization to association rule mining. *Applied Soft Computing*, 11(1), 326–336.

Liang, J.J., Qin, A.K., Suganthan, P.N., & Baskar, S. (2006). Comprehensive learning particle swarm optimizer for global optimization of multimodal functions. *IEEE Transactions on Evolutionary Computation*, 10(3), 281–295.

Montalvo, I., Izquierdo, J., Pérez, R., & Tung, M.M. (2008). Particle swarm optimization applied to the design of water supply systems. Computers & Mathematics with *Applications*, 56(3), 769–776.

Noory, H., Liaghat, A.M., Parsinejad, M., & Bozorg-Haddad, O. (2012). Optimizing irrigation water allocation and multicrop planning using discrete PSO algorithm. *Journal of Irrigation and Drainage Engineering*, 138(5), 437–444.

Onwunalu, J.E. & Durlofsky, L.J. (2010). Application of a particle swarm optimization algorithm for determining optimum well location and type. *Computational Geosciences*, 14(1), 183–198.

Salerno, J. (1997). Using the particle swarm optimization technique to train a recurrent neural model. In *Proceedings of 9th IEEE International Conference on Tools with Artificial Intelligence*, Newport Beach, CA, USA.

Sharif, M., Amin, J., Raza, M., Yasmin, M., & Satapathy, S.C. (2020). An integrated design of particle swarm optimization (PSO) with fusion of features for detection of brain tumor. *Pattern Recognition Letters*, 129, 150–157.

Yaghoubzadeh-Bavandpour, A., Bozorg-Haddad, O., Rajabi, M., Zolghadr-Asli, B., & Chu, X. (2022). Application of swarm intelligence and evolutionary computation algorithms for optimal reservoir operation. *Water Resources Management*, 36(7), 2275–2292.

Yang, X.S. (2010). *Nature-inspired metaheuristic algorithms*. Luniver Press. ISBN: 9781905986286

Zhan, Z.H., Zhang, J., Li, Y., & Chung, H.S.H. (2009). Adaptive particle swarm optimization. *IEEE Transactions on Systems, Man, and Cybernetics-Part B*, 39(6), 1362–1381.

Zhu, H., Wang, Y., Wang, K., & Chen, Y. (2011). Particle Swarm Optimization (PSO) for the constrained portfolio optimization problem. *Expert Systems with Applications*, 38(8), 10161–10169.

8 Differential Evolution Algorithm

Summary

With all its nuances, novelties, and abstract computational structure, the differential evolution algorithm takes the idea of evolutionary computation to the next level. As such, the differential evolution algorithm is arguably one of the most formidable meta-heuristic algorithms to handle complex real-world problems. In this chapter, we will dig deep and explore the mechanisms used in this algorithm. We would get familiar with the differential evolution algorithm's terminology and see how one can implement this algorithm in the Python programming language. Finally, we will explore the potential merits and drawbacks of this algorithm.

8.1 Introduction

Despite its great potential and promising preliminary result, it took quite some time for Holland's (1975) work on what became known as the genetic algorithm to become one of the mainstream approaches for optimization. As a meta-heuristic optimization method, the genetic algorithm seemed to have a good grasp on handling problems associated with dimensionality, multimodality, epistasis, non-differentiability, and discontinuous search space imposed by constraints, some of the most notable pitfalls that posed significant challenges for even the best calculus- and numeric-based optimization methods at the time (Yang, 2010; Du & Swamy, 2016). But what disguised the genetic algorithm from its contemporary meta-heuristic optimization algorithms was that it could enumerate through the search space using multiple search agents rather than limiting its computational structure to a single agent, which was the norm back in those days (Hooke & Jeeves, 1961; Glover, 1989, 1990). Unfortunately, the lack of readily available computers that could handle parallelized computers both from a software and hardware standpoint and the intricate algorithmic architecture of the genetic algorithm in comparison to its contemporary rivals hold the public back from seeing the genetic algorithm as a practical solution to tackle real-world optimization problems. In fact, it was not until the mid-1980s that scholars and researchers started to appreciate the genetic algorithm as the technology of the time could finally handle parallelized computation to an acceptable degree (e.g., Goldberg, 1989).

DOI: 10.1201/9781003424765-8

In the early 1990s, the emerging field of meta-heuristic optimization seemed to catch up and embraced the topical subject of swarm intelligence as several scholars attempted to incorporate and merge these two ideas. Algorithms such as the ant colony optimization algorithm (Dorigo et al., 1991, 1996) and the particle swarm intelligence (Kennedy & Eberhart, 1995) showed promising results regarding the complex real-world optimization algorithm. Thanks to their algorithmic structures, which were based on parallelized computation, the algorithms could establish themselves as a more robust alternative with a better convergence rate than their non-population-based meta-heuristic algorithm rivals. Often, these algorithms were presented with simple, easy-to-implement structures with only a few parameters, which made fine-tuning procedure, the inseparable part of using meta-heuristic optimization algorithms, a much more manageable task. However, this was not the case for the genetic algorithm, which is notoriously known for being riddled with too many parameters, most of which do not intuitively clearly impact the final result. As importantly, it seemed that the genetic algorithm was not taking advantage of the full potential of parallelized computation to the same extent as its rivals at the time.

Storn and Price (1997) came up with a brilliant major revision for the genetic algorithm that revolutionized the idea of evolutionary-based meta-heuristic optimization. While this was still based on the same general principles of the genetic algorithm, the nuances of this groundbreaking idea were enough to distinguish it as a stand-alone evolutionary-based meta-heuristic algorithm, namely, the differential evolutionary algorithm. Storn and Price (1997) needed to take things one step further than the genetic algorithm to theorize this algorithm. To do that, they could not have been bound by mimicking a biologically plausible process. As such, while the differential evolutionary algorithm is still reciting the same components of the genetic algorithm, from an evolutionary biology standpoint, the inspiration behind this algorithm is a purely hypothetical abstract representation of the evolutionary process.

Structurally speaking, the differential evolution algorithm is based on the same generic principles of evolutionary algorithms, such as the genetic algorithm. The difference here is that the differential evolution would go out of its way to make the computational procedure as efficient as possible, even if it creates a biologically implausible algorithm. In other words, the differential evolution algorithm justifies its purely hypothetical inspirational source by creating a more efficient algorithmic architecture than traditional evolutionary algorithms such as the genetic algorithm. That said, these differences are not so pronounced that they can relieve the differential evolution algorithm from employing the core operational mechanisms of evolutionary computation. As such, we are still using operators such as *crossover*, *mutation*, and *selection*, though the differential evolution algorithm's take on these operations is a bit different from what we have seen in previously introduced evolutionary algorithms. Note that the common analogy used in evolutionary algorithms also applies to this algorithm. As such, the search agents here are still referred to as *chromosomes*, each component of these chromosomes is called a *gene*, and the bundle of chromosomes is known as a *population* or a *generation*. Like other

evolutionary algorithms, differential evolution is also considered a stochastic, iterative algorithm that attempts to update the population's properties until it reaches
a termination stage. At this point, the best solution in the last generation would be
reported as the optimum solution.

Here, like most population-based meta-heuristic algorithms, the computational
procedures initiate by randomly generating a population within the feasible boundaries of the search space. Next, the algorithm would manipulate the population
by calling the mutation operator. However, here, unlike traditional evolutionary
algorithm, the mutation operation is based on *perturbation*, which in this context
means rather than randomly swapping some of the components of the search agents
with feasible values, here the operator would combine the properties of two other
distinct members of the population to form a new array. What is important here is
that the new population would not immediately replace the old generation; instead,
it would be stored in the memory as an alternative population. What is important to
note here is that while this is still considered a stochastic process, all the members
of the old generation would be subject to mutation. Next, the crossover operation
would be called to manipulate the mutated population. However, while in most
evolutionary computations, the crossover is interpreted as mixing the members
of the current generation to generate new offspring, in this case, the crossover is
responsible for mixing the old and mutated generations with one another. While
this is still a stochastic procedure by nature, it guarantees that at least one element,
or what is here referred to as genes, would be swapped between two populations
for each chromosome. Again, it is essential to note that, unlike most conventional crossover operators where the operation deals with strings of arrays that get
swapped as a whole bundle, here we are resorting to entry-wise or component-
wise mathematics. This means that the operation deals with each individual gene
rather than a string of genes. After applying these operators, the algorithm would
attempt to merge these old and new generations using the *greedy strategy*. The idea
is for each search agent, only the improving changes would take effect, otherwise,
the chromosome would remain as is (Yaghoubzadeh-Bavandpour et al., 2022a).
This ensures that the algorithm would preserve good chromosomes along the way.
These procedures would be repeated iteratively until a certain termination criterion
is met, at which point the best chromosome in the last population would be reported
as the optimum solution.

From a computational standpoint, one can even state that differential evolution
is more in line with parallelized computation ideas than traditional evolutionary-
based meta-heuristic algorithms. This can be mainly attributed to the fact that it
embraces the idea of entry-wise computation, which can be handled via vector-
based mathematics. As evidenced by swarm intelligence-based algorithms, this
sort of calculation can create more efficient algorithmic structures as it permits
implanting parallelized computations in a more practical way (Yaghoubzadeh-
Bavandpour et al., 2022b). As such, the differential evolution algorithm can be
seen as a bridge between the two words of evolutionary computation and swarm
intelligence. It is worth noting that for the accurate and efficient algorithm that

it is, the differential evolution algorithm is based on surprisingly straightforward procedures, which makes it easy to implement for practical optimization problems. The algorithm has far less parameters than conventional evolutionary algorithms, such as the genetic algorithm, but the main problem is that, like its predecessors, the impact of these parameters on the final outcome is still not intuitively obvious for beginners. More importantly, given the perturbation nature of its operators and the implementation of greedy strategy in the backbone of this algorithm, the main problem associated with this algorithm is the often slow rate of convergence or even premature convergence to local optima (Du & Swamy, 2016). That said, these issues are often not something that careful fine-tuning cannot address.

Over the years, many variants of the differential evolution algorithm have been proposed in the literature, some of which are self-adaptive differential evolution (Zhao et al., 2011), sinusoidal differential evolution (Draa et al., 2015), adaptive guided differential evolution (Mohamed & Mohamed, 2019), and multi-population differential evolution (Chen et al., 2020), to name a few. That said, the standard differential evolutionary algorithm is still considered a viable option to handle real-world optimization problems to the point that it has been successfully used for a wide range of optimization problems inducing but not limited to computer science (e.g., Tsai et al., 2013), financial science (e.g., Chauhan et al., 2009), energy industry (e.g., Abou El Ela et al., 2010), geophysics (e.g., Růžek & Kvasnička, 2001), hydrology (e.g., Xu et al., 2012), image processing (e.g., Aslantas, 2009), medical science (e.g., Lei et al., 2014), project management (e.g., Zou et al., 2011), remote sensing (e.g., De Falco et al., 2008), structural engineering (e.g., Tang et al., 2008), and water resources planning and management (e.g., Suribabu, 2010; Hong et al., 2018). In the following sections, we will explore the computational structure of a standard differential evolutionary algorithm.

8.2 Algorithmic Structure of the Differential Evolution Algorithm

The differential algorithm is a stochastic, population-based direct search algorithm with a foot in evolutionary computation and swarm intelligence. The main idea here is to fully utilize the notion of parallelized computation to create a much more efficient algorithm, even if that means that the evolutionary inspiration behind this algorithm is not biologically plausible. To that end, differential evolution's algorithmic architecture was based on perturbation, which, in turn, required the algorithm to resort to vector-based mathematics. As such, the algorithm needed to introduce new interpretations of conventional evolutionary-based operators such as selection, mutation, and crossover. The common feature in these altered operators was that they could handle entry-wise procedures in opposition to operating on two or three chunks of strings. Moreover, by embedding the greedy strategy in its algorithmic structure, the differential evolution algorithm advocated using a more elite-oriented perspective when upgrading the population. The main idea, in a nutshell, was that any update from the old generation to the new one was only permitted if the said move was an improving one.

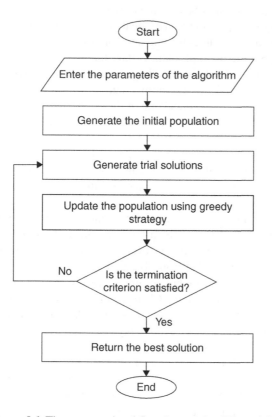

Figure 8.1 The computational flowchart of the differential evolution algorithm.

The differential evolution algorithm's flowchart is depicted in Figure 8.1. A closer look at the architecture of this algorithm would reveal that it actually consists of three main stages that are the initiation, reproduction, and termination stages. The reproduction stage itself consists of three main operators that are *selection, crossover*, and *mutation*. Using this structure, the algorithm would conduct a thorough search and locate what could be the optimum solution to the problem at hand. The following subsection will discuss each of these stages and their mathematical structures.

8.2.1 Initiation Stage

The differential evolution is a population-based meta-heuristic optimization algorithm, and as such, it works with multiple search agents, here called chromosomes or genomes, that would enumerate through the search space. As we have seen, in an optimization problem with N decision variables, an N-dimension coordination system could be used to represent the search space. In this case, any point within the search space, say X, can be represented mathematically as a $1 \times N$ array as follows:

$$X = \left(x_1, x_2, x_3, \ldots, x_j, \ldots, x_N \right) \tag{8.1}$$

where X represents a chromosome in the search space of an optimization problem with N decision variables, and x_j represents the value associated with the jth decision variable, or what is technically known as a gene.

The differential evolution algorithm starts with randomly placing a series of chromosomes within the feasible boundaries of the search space. This bundle of arrays, which is referred to as the population in evolutionary computation's terminology, can be mathematically expressed as $M \times N$ matrix, where M denotes the number of chromosomes or what is technically referred to as the population size. In such a structure, each row represents a single chromosome. Note that population size is one of the many parameters of the differential evolution algorithm. A population, denoted by *pop*, can be represented as follows:

$$pop = \begin{bmatrix} X_1 \\ X_2 \\ \vdots \\ X_i \\ \vdots \\ X_M \end{bmatrix} = \begin{bmatrix} x_{1,1} & x_{1,2} & \cdots & x_{1,j} & \cdots & x_{1,N} \\ x_{2,1} & x_{1,2} & \cdots & x_{2,j} & \cdots & x_{2,N} \\ & & & \vdots & & \\ x_{i,1} & x_{i,2} & \cdots & x_{i,j} & \cdots & x_{i,N} \\ & & & \vdots & & \\ x_{M,1} & x_{M,2} & \cdots & x_{M,j} & \cdots & x_{M,N} \end{bmatrix} \tag{8.2}$$

where X_i represents the ith particle in the population, and $x_{i,j}$ denotes the jth gene of the ith chromosome.

The initially generated population represents the first generation of chromosomes. As we progress, the values stored in the *pop* matrix will be altered according to the computational structure of the differential evolution algorithm's reproduction stage. By the end of this iterative computation process, when the termination criterion is met, one or possibly multiple chromosomes will converge to the optimum solution.

8.2.2 Reproduction Stage

The reproduction stage is a general term used here to refer to a procedure in the differential evolution algorithm by which the older generation would evolve into a new one. The reproduction stage consists of three main pillars that are *selection*, *crossover*, and *mutation* operators. These operators are used as alteration tools to ultimately increase the odds of reproducing a better population from the last generation. The main idea here is to create a stochastic mechanism by which the child population would, in all likelihood, inherit suitable biological properties from the parent population. However, note that the interpretation of these operators in this algorithm differs from what we expect from these operators in other evolutionary computation-based meta-heuristic optimization algorithms. For one, the selection operator, though it is still in effect, is completely dissolved in the structure of the other two operators and is turned into a purely random selection procedure with no prior assumption or tendency to prioritize any chromosome over the others. The other notable difference is that, here, the offspring population that gets reproduced using these operators is not necessarily the new generation. In other words, this newly generated population can be seen as a set of tentative solutions that would be selected as the chromosomes of the new generation if and only if they are deemed an improvement over their counterparts in the previous generations. This is to say that the merging process of the old generation and the tentative solutions to create the new generation of chromosomes are governed by the greedy strategy. Lastly, note that these operators are based on entry-wise computation, which is another distinctive feature of the differential evolution algorithm from the other evolutionary-based meta-heuristic algorithms we have explored thus far. This feature would, in turn, enable the differential evolution algorithm to embrace the idea of parallelized computation. These operators are explored in the following subsections.

8.2.2.1 Mutation Operator

The first step to creating the tentative solutions is through the mutation operators. The primary application of traditional mutation operators was limited to imposing minor tweaks to the population by randomly replacing a few genes with new feasible values. Here, however, the mutation operator more or less creates the foundation of the tentative solution by mixing the components of the old generation. Using perturbation, each old chromosome would be blended with two other distinct randomly selected chromosomes, creating a new tentative solution. As such, the bottom line is that the proposed mutation operator in the differential evolution algorithm would effectively create a tentative population that is effectively the permutated version of the old generation. Here, the selection operator, which dictates the paired distinct chromosomes involved in creating each of these new tentative solutions, is based on pure random selection. This operation can be mathematically expressed as follows:

$$X_i' = \left(x_{i,1}', x_{i,2}', x_{i,3}', \ldots, x_{i,j}', \ldots, x_{i,N}' \right) \qquad \forall i \tag{8.3}$$

$$x_{i,j}' = x_{r,j} + \delta \times \left(x_{r',j} - x_{r'',j} \right) \qquad \forall i,j \,\&\, r \neq r' \neq r'' \neq j \tag{8.4}$$

$$r = IntRand(1, M) \tag{8.5}$$

$$r' = IntRand(1, M) \tag{8.6}$$

$$r'' = IntRand(1, M) \tag{8.7}$$

in which X_i' denotes the ith mutated solution; $x_{i,j}'$ represents the jth gene of the ith mutated chromosome; r, r', and r'' are three randomly selected indexes from the population; *IntRand* is a function that picks random integer value in the range 1–M; and finally, δ denotes the mutation factor, which is a user-defined parameter of the differential evolution algorithm that can range from 0 to 2. For practical cases, it has been suggested that the value of this parameter should be selected from the range [0, 1] to get a more stable performance out of the algorithm (Yang, 2010; Du & Swamy, 2016). However, this may lead to premature convergence in some cases, as it limits the algorithm's capabilities to explore remote areas of the search space.

8.2.2.2 Crossover Operator

The primary application of traditional crossover operators was to regenerate offspring by mixing two existing chromosomes. Here, however, the idea of a crossover is to impose minor adjustments to the old generation by fusing it with the mutated tentative solutions. Thus, the point here is to merge the old generation with the mutated solutions to create a new set of tentative solutions. Using perturbation, some of the genes in the old chromosomes would be swapped with the genes in the chromosomes stored in the tentative solution. In a sense, the crossover would mitigate the randomness of mutation operation, as it only allows some of these new genes to be imposed in the tentative solution. While this is still a stochastic process by nature, it is designed in a way to ensure that at least one gene for each tentative solution would be selected from the mutated population. This operation can be mathematically expressed as follows:

$$X_i'' = \left(x_{i,1}'' , x_{i,2}'' , x_{i,3}'' , \ldots , x_{i,j}'' , \ldots , x_{i,N}'' \right) \qquad \forall i \tag{8.8}$$

$$x_{i,j}'' = \begin{cases} x_{i,j}'' & if \quad Rand \leq C \ or \ j = b_i \\ x_{i,j} & if \quad Rand > C \ and \ j \neq b_i \end{cases} \qquad \forall i, j \tag{8.9}$$

$$b_i = IntRand(1, N) \qquad \forall i \tag{8.10}$$

in which X_i denotes the ith tentative solution; $x_{i,j}$ represents the jth gene of the ith tentative chromosome; *Rand* is a randomly generated number within the range [0, 1]; b_i is a randomly selected index for the ith chromosome to ensure that at least one of the genes with the said index would be replaced with the mutated values; and finally, C is the crossover constant, which is another user-defined parameter of the differential evolution algorithm that can range from 0 to 1.

As the last step of this stage, the algorithm would use the greedy strategy to combine the old and tentative solutions and create a new generation of chromosomes. The gist of this strategy is that it compares the counterpart chromosomes of both sets and opts for the objectively better one. The selected chromosomes would then be used to form the next generation. This procedure can be mathematically expressed as follows:

$$X_i^{new} = \begin{cases} X_i'' & if & X_i' \ is\,better\,than\,X_i \\ X_i & otherwise \end{cases} \quad \forall i \qquad (8.11)$$

where X_i^{new} denotes the ith chromosome for the new generation.

8.2.3 Termination Stage

Based on the instruction in Figure 8.1, the algorithm would generate a set of tentative solutions as a pool of potentially viable chromosomes for the next generation through mutation and crossover. Then using the greedy strategy, the algorithm would create the next generation of child chromosomes. The next generation would then be relabeled as the current generation to go through a similar computation process.

As can be seen, like other meta-heuristic algorithms, the sequence of operational structures of this algorithm needs to be executed iteratively until a certain termination criterion is met, at which point the execution of the algorithm would be terminated, and the best chromosome in the last generation would be reported as the solution to the optimization problem. Note that without such a termination stage, the algorithm would potentially be executed in an infinite loop. The termination stage would, in effect, determine whether the algorithm has reached what could be the optimum solution.

As the differential evolution algorithm is not equipped with an explicitly defined, unique termination mechanism, one could implement the commonly available options, most notably limiting the number of iterations, run time, or perhaps monitoring the improvement made to the best solution in consecutive iterations. Among these options, limiting the number of iterations is arguably the most cited mechanism to create a termination stage for the differential evolution algorithm. The idea is that the process would be executed only for a specified number of times, a parameter known as the maximum iteration. In any case, it should be noted that selecting the termination mechanism is also considered one of the algorithm's parameters. Bear in mind that in most cases, these termination mechanisms may require setting up additional parameters.

8.3 Parameter Selection and Fine-Tuning Differential Evolution Algorithm

One of the main conclusions that one can derive from the *no-free-lunch theorem* is that fine-tuning an algorithm is essential to get the best performance out of a meta-heuristic algorithm. This would basically ensure that an algorithm is equipped to handle the unique characteristics of a given optimization problem. Of course, it is possible to use our intuition, experience, and default values suggested for an algorithm's parameters as a good starting point, one should bear in mind that fine-tuning these parameters is, more than anything, a trial-and-error process. Thus, while it is possible to get a good enough result by having an educated guess for setting the parameters of these algorithms, to get the best possible performance, it is necessary to go through this fine-tuning process.

In the case of the differential evolution algorithm, these parameters are population size (M), mutation factor (δ), crossover constant (C), opting for the termination criterion, and of course, all the parameters that are embedded for this process. For instance, if limiting the number of iterations has been selected as a termination criterion, the maximum iteration (T) is another parameter that needs to be defined by the user.

As can be seen, the differential evolutionary algorithm has a few parameters, which makes the fine-tuning process a bit more manageable than other population-based optimization algorithms we have seen thus far. That said, the main challenge here is that while some parameters have a noticeable effect on the emerged solution that one could intuitively understand, say the population size or the maximum number of iterations, the effects of others are not so intuitively clear, say the crossover constant. As such, to get the best results out of the differential evolution algorithm, you may have to dabble with it first to gain some experience and inside knowledge about such parameters. By doing so, you could better understand how to fine-tune these parameters as your initial guesses and parameter selection strategies become more educated. The pseudocode for the differential evolution algorithm is shown in Figure 8.2.

```
Begin
        Set the algorithm's parameter and input the data
        Generate the initial population
        Let M denote the population size
        Evaluate the population
        While the termination criterion is not met
                For i in range 1 to M
                        Generate a solution ??ᵢ' via mutation operator
                        Generate the tentative solution ??ᵢ'''' using crossover operator
                        If ??ᵢ'''' is better than Xᵢ
                                Replace Xᵢ with ??ᵢ''''
                        End if
                Next i
        End while
        Report the best solution
End
```

Figure 8.2 Pseudocode for the differential evolution algorithm.

8.4 Python Codes

The code to implement the differential evolution algorithm can be found below:

```python
import numpy as np

def init_generator(num_variables, pop_size, min_val, max_val):
    return np.random.uniform(min_val, max_val,
                             size=(pop_size, num_variables))

def mutation_index_system(pop_size):
    index_set = np.zeros((pop_size,pop_size),dtype=np.int) +
                np.arange(pop_size)
    dropable_index = (pop_size+1)*np.arange(pop_size)
    result = np.reshape(np.delete(index_set, dropable_index)
    return result, (pop_size, pop_size-1))

def mutation(pop, index, mutation_factor, min_val ,max_val):
    pop_size = len(pop)
    mut_pop = np.zeros_like(pop)
    for i in range(pop_size):
        a,b,c = np.random.choice(index[i],3,replace=False)
        mut_pop[i]=pop[a]+mutation_factor*(pop[b]-pop[c])
    mut_pop=np.where(mut_pop>max_val, max_val, mut_pop)
    mut_pop=np.where(mut_pop<min_val, min_val, mut_pop)
    return mut_pop

def crossover(pop,mut_pop, p_crossover):
    prob=np.random.uniform(size=(pop.shape[0],pop.shape[1]))
    b_index=np.random.randint(0,pop.shape[1],size=pop.shape[0])
    for i in range(prob.shape[0]):
        prob[i][b_index[i]]=0
    return np.where(prob<p_crossover,mut_pop,pop)

def merge(pop, crossover_pop, obj_func, minimizing):
    result = np.apply_along_axis(func1d=obj_func, axis=1, arr=pop)
    cross_result = np.apply_along_axis(func1d=obj_func,
                                       axis=1, arr=crossover_pop)
    if minimizing:
        replacing_index = np.argwhere(cross_result<result)
    else:
        replacing_index = np.argwhere(cross_result>result)
    pop[replacing_index]=crossover_pop[replacing_index]
    return pop

def evaluation(pop, obj_func, minimizing):
    result = np.apply_along_axis(func1d=obj_func, axis=1, arr=pop)
    if minimizing:
        index = np.argmin(result)
    else:
        index = np.argmax(result)
    return pop[index], result[index]
```

```
def DE_algorithem(num_variables, min_val, max_val, obj_func, pop_
   size=50,
                   mutation_factor=.5, p_crossover=.5, iteration=100,
                   minimizing=True, full_result=False):
   results=np.zeros(iteration)
   NFE=np.zeros(iteration)
   NFE_value=0
   index = mutation_index_system(pop_size)
   pop=init_generator(num_variables, pop_size, min_val, max_val)
   for i in range(iteration):
       mut_pop=mutation(pop, index, mutation_factor, min_val, max_val)
       crossover_pop=crossover(pop, mut_pop, p_crossover)
       pop=merge(pop, crossover_pop, obj_func, minimizing)
       NFE_value+=pop_size
       NFE[i]=NFE_value
       results[i]=evaluation(pop, obj_func, minimizing)[1]
   X, best_of = evaluation(pop, obj_func, minimizing)
   if not full_result:
       return X, best_of
   else:
       return X, best_of, results, NFE
```

8.5 Concluding Remarks

The differential evolution algorithm is still seen as a formidable meta-heuristic algorithm when it comes to tackling complex real-world problems. Although, from a technical standpoint, the differential evolution algorithm is considered an evolutionary computation-based algorithm, it distinguishes itself from the conventional features commonly associated with this branch of meta-heuristic optimization algorithms. The main nuance in the architecture of this algorithm is that it introduces the idea of entry-wise operation to the basic structure of the evolutionary computation. In and of itself, this seemingly subtle change allows the algorithm to take full advantage of the potential behind parallelized computation. By doing so, this algorithm addresses one of the main drawbacks that is commonly associated with evolutionary-based algorithms, that is, the lack of an algorithmic structure that can utilize parallelized computation at its full capacity. Through perturbation, the refined operators of this algorithm would use the search agents' properties to coordinate the components of the potential next generation. This, in a sense, makes the differential evolution algorithm a combination of two schools of thought in meta-heuristic optimization, namely, evolutionary comparison and swarm intelligence. Another essential feature of the differential evolution algorithm is that it implements the greedy strategy to merge the old generation and the tentative solution. By doing so, the algorithm ensures that we are at least doing as well as the previous iteration in each iteration. Any upgrade in any of the chromosomes is only permitted if it improves performance. Of course, if not careful, this could lead to premature convergence or being trapped in local optima. It should be noted that the algorithm is based on simple and straightforward mathematical procedures with limited parameters that need fine-tuning. However, like many other evolutionary-based meta-heuristic algorithms, the main problem here is that the effects of some of these parameters on the final solution are not so intuitively clear, which could make things a bit challenging for those who are not that familiar with the structure of the evolutionary-based meta-heuristic algorithms. That said, the differential evolution algorithm is a viable option for handling practical optimization problems.

References

Abou El Ela, A.A., Abido, M.A., & Spea, S.R. (2010). Optimal power flow using differential evolution algorithm. *Electric Power Systems Research*, 80(7), 878–885.

Aslantas, V. (2009). An optimal robust digital image watermarking based on SVD using differential evolution algorithm. *Optics Communications*, 282(5), 769–777.

Chauhan, N., Ravi, V., & Chandra, D.K. (2009). Differential evolution trained wavelet neural networks: Application to bankruptcy prediction in banks. *Expert Systems with Applications*, 36(4), 7659–7665.

Chen, H., Heidari, A.A., Chen, H., Wang, M., Pan, Z., & Gandomi, A.H. (2020). Multi-population differential evolution-assisted Harris hawks optimization: Framework and case studies. *Future Generation Computer Systems*, 111, 175–198.

De Falco, I., Della Cioppa, A., Maisto, D., & Tarantino, E. (2008). Differential evolution as a viable tool for satellite image registration. *Applied Soft Computing*, 8(4), 1453–1462.

Dorigo, M., Maniezzo, V., & Colorni, A. (1991). Positive feedback as a search strategy. Dipartimento di Elettronica, Politecnico di Milano, Milan, Italy, Technical Report, 91-016.

Dorigo, M., Maniezzo, V., & Colorni, A. (1996). The ant system: Optimization by a colony of cooperating ants. *IEEE Transactions on Systems Man and Cybernetics–Part B*, 26(1), 29–42.

Draa, A., Bouzoubia, S., & Boukhalfa, I. (2015). A sinusoidal differential evolution algorithm for numerical optimisation. *Applied Soft Computing*, 27, 99–126.

Du, K.L. & Swamy, M.N.S. (2016). *Search and optimization by metaheuristics: Techniques and algorithms inspired by nature*. Springer International Publishing Switzerland. ISBN: 9783319411910

Glover, F. (1989). Tabu search—part I. *ORSA Journal on Computing*, 1(3), 190–206.

Glover, F. (1990). Tabu search—part II. *ORSA Journal on Computing*, 2(1), 4–32.

Goldberg, D.E. (1989). *Genetic algorithms in search, optimization and machine learning*. Addison-Wesley.

Holland, J.H. (1975). *Adaptation in natural and artificial systems: An introductory analysis with applications to biology, control, and artificial intelligence*. University of Michigan Press.

Hong, H., Panahi, M., Shirzadi, A., Ma, T., Liu, J., Zhu, A. X., ... & Kazakis, N. (2018). Flood susceptibility assessment in Hengfeng area coupling adaptive neuro-fuzzy inference system with genetic algorithm and differential evolution. *Science of the Total Environment*, 621, 1124–1141.

Hooke, R. & Jeeves, T.A. (1961). "Direct Search" solution of numerical and statistical problems. *Journal of the ACM*, 8(2), 212–229.

Kennedy, J. & Eberhart, R. (1995). Particle swarm optimization. In *Proceedings of IEEE International Conference on Neural Networks*, Perth, WA, USA.

Lei, B., Tan, E.L., Chen, S., Ni, D., Wang, T., & Lei, H. (2014). Reversible watermarking scheme for medical image based on differential evolution. Expert Systems with Applications, 41(7), 3178–3188.

Mohamed, A.W. & Mohamed, A.K. (2019). Adaptive guided differential evolution algorithm with novel mutation for numerical optimization. *International Journal of Machine Learning and Cybernetics*, 10(2), 253–277.

Růžek, B. & Kvasnička, M. (2001). Differential evolution algorithm in the earthquake hypocenter location. *Pure and Applied Geophysics*, 158(4), 667–693.

Storn, R. & Price, K. (1997). Differential evolution – a simple and efficient heuristic for global optimization over continuous spaces. *Journal of Global Optimization*, 11(4), 341–359.

Suribabu, C.R. (2010). Differential evolution algorithm for optimal design of water distribution networks. *Journal of Hydroinformatics*, 12(1), 66–82.

Tang, H., Xue, S., & Fan, C. (2008). Differential evolution strategy for structural system identification. *Computers & Structures*, 86(21–22), 2004–2012.

Tsai, J.T., Fang, J.C., & Chou, J.H. (2013). Optimized task scheduling and resource allocation on cloud computing environment using improved differential evolution algorithm. *Computers & Operations Research*, 40(12), 3045–3055.

Xu, D.M., Qiu, L., & Chen, S.Y. (2012). Estimation of nonlinear Muskingum model parameter using differential evolution. *Journal of Hydrologic Engineering*, 17(2), 348–353.

Yaghoubzadeh-Bavandpour, A., Bozorg-Haddad, O., Rajabi, M., Zolghadr-Asli, B., & Chu, X. (2022b). Application of swarm intelligence and evolutionary computation algorithms for optimal reservoir operation. *Water Resources Management*, 36(7), 2275–2292.

Yaghoubzadeh-Bavandpour, A., Bozorg-Haddad, O., Zolghadr-Asli, B., & Gandomi, A.H. (2022a). Improving approaches for meta-heuristic algorithms: A brief overview. In Bozorg-Haddad, O., Zolghadr-Asli, B. eds. *Computational intelligence for water and environmental sciences*. Springer Singapore, 35–61.

Yang, X.S. (2010). *Nature-inspired metaheuristic algorithms*. Luniver Press. ISBN: 9781905986286

Zhao, S.Z., Suganthan, P.N., & Das, S. (2011). Self-adaptive differential evolution with multi-trajectory search for large-scale optimization. *Soft Computing*, 15(11), 2175–2185.

Zou, D., Liu, H., Gao, L., & Li, S. (2011). An improved differential evolution algorithm for the task assignment problem. *Engineering Applications of Artificial Intelligence*, 24(4), 616–624.

9   Harmony Search Algorithm

Summary

With all its nuances, novelties, and abstract computational structure, the harmony search algorithm is arguably one of the most formidable meta-heretic algorithms to handle discrete complex real-world optimization problems. In this chapter, we will dig deep and explore the mechanisms used in this algorithm. We would get familiar with the harmony search algorithm's terminology and see how one can implement this algorithm in the Python programming language. Finally, we will explore the potential merits and drawbacks of this algorithm.

9.1 Introduction

Whether you are an avid music fan, a casual listener, or perhaps a musician, it is a safe bet to state that music has become an integrated part of our life. It can move us emotionally, bring forth a long-lost memory, pump up a crowd on a dance floor, or help eliminate piled-up anxiety. The amount of music's influence on today's culture is so palpable that often if there is no music around, one cannot help but wonder if there is something missing. The brilliant Canadian-American actor, comedian, writer, and producer Jim Carrey famously joked about this in one of his roles "*you know the trouble with real life? There's no danger music.*"

What makes us revere music the way we do, one might never know with absolute certainty. Like any other art, perhaps this is the piece's authenticity that resonates with us, the audience, on an emotional level. We, the listener, feel contented with something beautiful that often we cannot even put our hand on, but we know deep down that we do enjoy it. Music, like other art forms, is, of course, subjective. We often agree if a piece works, even if we cannot explain why. The artists themselves may seek the reason, for it is the artist's job to make some harmony out of chaos. There is no better place to witness this notion than a Jazz club. Individual artists would improvise around a single theme to create one coherent and harmonious piece on the spot. The composed music is a result of individual contributors' conscious attempts to synchronize their musical efforts with one another. While each instrument would undoubtedly have a clear role, they must work alongside

DOI: 10.1201/9781003424765-9

each other with respect to others' musical range and, most importantly, the band's central musical theme. This is why the musical improvisation of a seasoned band would not turn into a chaos of haphazard and random notes.

Inspired by musical improvisation and music composition in general, Geem et al. (2001) theorized a new meta-heuristic optimization algorithm called the harmony search algorithm. For the most part, the harmony search algorithm resembles the computational structure of a population-based meta-heuristic algorithm. As such, the algorithm would deal with an architecture that is built around multiple search agents which are bundled together in a single set. Unlike other population-based meta-heuristic algorithms we have seen thus far, the harmony search algorithm would only generate, or in this case compose, a single new search agent in each iteration. But even more interestingly, this newly composed entry would not be checked against the best solution, which is the common practice, but rather it would be compared to the worst search agent currently stored in the memory. As such, the whole procedure can be seen as the algorithm constantly raising the bar in each iteration to push the set toward what could be the optimum solution. It should be noted that while the standard version of the harmony search algorithm was theorized to handle combinatorial optimization problems with discrete decision variables, this unique take on the meta-heuristic optimization can be, and, in fact, had been, tweaked with minor computation adjustment to handle search spaces with continuous decision variables as well.

As stated, the harmony search algorithm is a music-inspired population-based meta-heuristic optimization algorithm that attempts to mimic the basic process that goes into music improvisation and music composition in general. As a population-based algorithm, naturally, the algorithm's computation would be initiated by randomly generating a set of feasible search agents, which are collectively referred to as the *harmony memory* or simply *memory*. Each search agent is called a harmony, composed of notes or pitches in the harmony search algorithm terminology. From the mathematical standpoint, these notes or pitches are, in fact, decision variable values and can be seen as the notes played by individual instruments to create harmony collectively. Naturally, the general idea is to find the most pleasing harmony, that is, identifying the harmony with the most desirable objective function. After initiation, the algorithm would evaluate the randomly generated harmonies to mark the best and worst ones stored in the memory. As can be seen, the base of this evolution is the objective function values rather than its gradient-related information, making the harmony search algorithm a direct search-oriented method. Upon initiation, the algorithm would generate a single harmony using stochastic-based procedures. The general theme of these procedures is to tap into the memory to generate a potentially new harmony. Interestingly, the new harmony would then be checked against the worst harmony stored in the memory and replaced in case it is deemed more pleasant or has a better objective function value from the mathematical point of view. This process would be repeated until a certain termination criterion is met, at which point the algorithm would be terminated, and the best harmony stored in the memory would be returned as the solution to the optimization problem.

One of the most notable distinctions of this algorithm from its predecessors is that it generates a new search agent in each iteration despite being a population-based algorithm. And as significantly, as opposed to the norm to evaluate the new solutions against the best-stored values, this search agent would be compared to the worst-stored search agent in the memory. This unique take on the meta-heuristic approach for optimization has merits and drawbacks. Like other meta-heuristic algorithms, the harmony search algorithm can generally circumvent the pitfalls commonly associated with gradient-based methods, most notably the struggle to handle high dimensionality, multimodality, epistasis, non-differentiability, and discontinuous search space imposed by constraints (Yang, 2010). More specific-ally, there is reasonable experimental evidence to suggest that the harmony search algorithm's convergence rate and the final solution are not dramatically affected by the randomness embedded in the initialization stage (Du & Swamy, 2016), making it a potentially robust meta-heuristic algorithm. As importantly, the algorithm has few parameters that, as evidence suggests, are much more forgiving when it comes to parameter selection, making the algorithm fine-tuning, which happens to be one of the main challenges of implementing meta-heuristic algorithms, much more manageable (Yang, 2010). Coupling these with the simple, easy-to-implement instructions of the algorithm makes it a formidable, viable option to handle real-world optimization problems. That said, as the algorithm updates a single solution in each iteration, it is not utilizing the full potential of parallelized computation. As such, the convergence rate may suffer a bit compared to other fully parallelized computation-based algorithms.

Over the years, many variants of the harmony search algorithm were proposed in the literature, some of which are improved harmony search (Mahdavi et al., 2007), global-best harmony search (Omran & Mahdavi, 2008), self-adaptive har-mony search (Wang & Huang, 2010), intelligent tuned harmony search (Yadav et al., 2012), and enhanced harmony search algorithm (Maheri & Narimani, 2014), to name a few. That said, the standard harmony search algorithm is still considered a viable option to handle real-world optimization problems. In fact, the standard harmony search algorithm has been successfully used for architectural engineering (e.g., Fesanghary et al., 2012), diagnostic science (e.g., Karbhari et al., 2021), edu-cation science (e.g., Al-Betar & Khader, 2012), energy industry (e.g., Rao et al., 2012), environmental engineering (e.g., Ayvaz, 2010), geohydrology (e.g., Ayvaz, 2009), hydrology (e.g., Arsenault et al., 2014), project management (e.g., Geem, 2010), structural engineering (e.g., Lee et al., 2005), and water resources planning and management (e.g., Geem, 2006). In the following sections, we will explore the computational structure of a standard harmony search algorithm.

9.2 Algorithmic Structure of the Harmony Search Algorithm

As a population-based meta-heuristic optimization method, the harmony search algorithm is, to some extent, based on parallelized computation, though as we would see in this section, the computational architecture of this algorithm partially differs from conventional population-based methods. Here, unlike other population-based methods we have explored thus far, the intention behind the population is not to keep track of the search agents' positions as they enumerate through the search space, but rather the population serves as a *memory pool* to store some potentially viable solutions. The other notable difference in this algorithm's structure from what we traditionally came to expect from population-based algorithms is that it only creates a single tentative option, here called harmony, in each iteration. While, of course, this means that the so-called composing of the tentative harmony is less computationally taxing than the conventional approach, which is to update the entire population, this also means that the algorithm may not be utilizing the full potential behind parallelized computation. After the new harmony is composed, the algorithm checks this tentative option against the worst recorded harmony in its memory. It replaces it with the new harmony if the tentative point has better properties than the worst harmony. Again, this procedure goes against the traditional approach of meta-heuristic algorithms, as they tend to constantly track the best position in the search space rather than eliminate the worst one. But here, the algorithm would constantly keep eliminating the worst record, pushing the threshold bar for the memory to store better results in each iteration. This process would be repeated until it reaches a specified termination stage, at which point the best harmony record in the memory would be returned as the optimum solution to the problem at hand.

The harmony search algorithm's flowchart is depicted in Figure 9.1. A closer look at the architecture of the harmony search algorithm would reveal that it actually consists of three main stages that are the initiation, composing, and termination stages. The composing stage itself consists of three main strategies that are *memory*, *randomization*, and *pitch adjustment*. Using this structure, the algorithm would conduct a thorough search and locate what could be the optimum solution to the problem at hand. The following subsection will discuss each of these stages and their mathematical structures.

9.2.1 Initiation Stage

The harmony search is a population-based algorithm, though, as we have stated, the application of this population differs from what we came to expect from populations in meta-heuristic algorithms. As we have seen, in an optimization problem with N decision variables, an N-dimension coordination system could be used to represent the search space. In this case, any point within the search space, say X, can be represented mathematically as a $1 \times N$ array as follows:

$$X = \left(x_1, x_2, x_3, \ldots, x_j, \ldots, x_N\right) \tag{9.1}$$

where X represents a harmony composed of individual notes, N denotes the number of notes or decision variables, and x_j represents the value associated with the jth decision variable. It should be noted that, here, we are dealing with a discrete search space. This means that the set variable could assume a set of finite

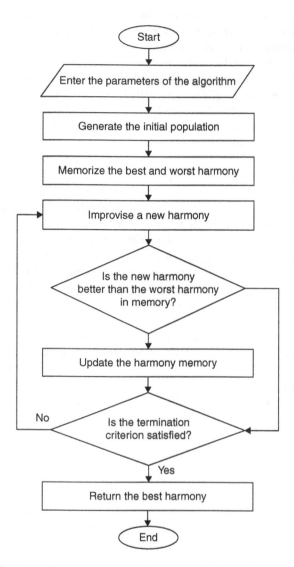

Figure 9.1 The computational flowchart of the harmony search algorithm.

acceptable values for each decision variable. These sets could be *monotonic*, as all the decision variables have the same set of acceptable values, or the optimization algorithm could be non-monotonic, where the values for each decision variable vary. This can be mathematically represented as follows:

$$x_j \in \left\{ v_{j,1}, v_{j,2}, v_{j,3}, \ldots, v_{j,k}, \ldots, v_{j,P} \right\} \qquad \forall j \tag{9.2}$$

where $v_{j,k}$ represents the kth possible value for the jth decision variable, and P is the number of possible values that can be assumed by the jth decision variable.

The harmony search algorithm would start by randomly generating a series of harmonies within the search space's feasible boundaries and storing them in its memory. This bundle of arrays can be mathematically expressed as $M \times N$ matrix called *harmony memory*, where M denotes what is technically referred to as the *population size*. In such a structure, each row represents a harmony. Note that the population size is one of the parameters of the harmony search algorithm. The harmony memory, denoted by HM, can be represented as follows:

$$HM = \begin{bmatrix} X_1 \\ X_2 \\ \vdots \\ X_i \\ \vdots \\ X_M \end{bmatrix} = \begin{bmatrix} x_{1,1} & x_{1,2} & \cdots & x_{1,j} & \cdots & x_{1,N} \\ x_{2,1} & x_{1,2} & \cdots & x_{2,j} & \cdots & x_{2,N} \\ & & \vdots & & & \\ x_{i,1} & x_{i,2} & \cdots & x_{i,j} & \cdots & x_{i,N} \\ & & \vdots & & & \\ x_{M,1} & x_{M,2} & \cdots & x_{M,j} & \cdots & x_{M,N} \end{bmatrix} \tag{9.3}$$

where X_i represents the ith harmony in the harmony memory, and $x_{i,j}$ denotes the jth decision variable of the ith harmony. It should be noted that for each column, the values would be drawn from the set associated with the possible values for the said decision variable.

According to the harmony search algorithm's analogy, each harmony is, in fact, a combination of possible values that is drawn from the sets associated with each decision variable to create a string of discrete values. After the harmony memory is generated, they would get evaluated, and both the best and worst harmonies would be marked. The algorithm would then continue with its procedure by attempting to compose a new tentative harmony.

9.2.2 Composing Stage

Presumably, as you deconstruct the improvisation process of musicians, you could come to the conclusion that the general theme of composing a new tune on the fly is based on three main principles. Firstly, the musician may recite a certain musical number, say a famous song that stuck to one's mind, right from memory. Alternatively, the musician may opt to play something close, but not precisely the same piece of music, by refining and adjusting a few notes on the go. Lastly, the musician may go for something completely fresh by compiling new notes together to create an original piece. Basically, the harmony search algorithm formularized a way to capture these strategies mathematically to compose a new harmony.

As such, the idea behind the composing stage is to generate a tentative harmony. As stated, the harmony search algorithm would only compose a new harmony at each iteration. To do that, the algorithm would utilize three main strategies, namely, memory, randomization, and pitch adjustment strategies, which basically represent the three principles by which a musician would improvise a new musical number. It is important to note that here the memory plays the role of a pool from which the musician could compose a new piece. To create this tentative harmony, the algorithm would go through each of these stochastic-based strategies individually. At the end of this, the algorithm is able to compose a new tentative harmony, ready to be tested against the worst recorded harmony in the memory.

9.2.2.1 Memory Strategy

The memory strategy is the backbone of improvising a new harmony. The idea is that the musicians would tap into their memories to recite a tune they have dabbled with earlier. The general theme is that the new harmony would be a mishmash of notes stored in the memory. The schematic idea behind this strategy is depicted in Figure 9.2.

As expected, this is a stochastic process, as the algorithm randomly selects different notes from the memory to compose a fresh harmony. In the core, it performs the same function as the crossover operator did in the genetic algorithm, as they both recycle the properties of the search agents in previous iterations to generate a new one. Of course, in the genetic algorithm, we would be generating two new tentative offspring chromosomes, and we only selected two parent solutions,

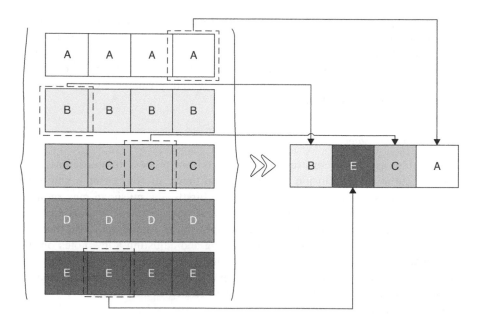

Figure 9.2 The general theme of composing a new harmony using the memory stagey.

while here, we only compose one harmony, and in all likelihood, more than two old harmonies would be cited in the composing process.

From a mathematical standpoint, the memory strategy could be formulated as follows:

$$x'_j = x_{i,j} \qquad \forall j \tag{9.4}$$

$$i = IntRand(1, M) \tag{9.5}$$

where x'_j represents the jth decision variable of the tentative harmony, and *IntRand* denotes a function that would randomly select an integer from the range 1 to M. Bear in mind that the algorithm would use this function until all the decision variables of the tentative harmony are cited from the memory pool.

9.2.2.2 Randomization Strategy

Randomization is based on the pure improvisation principle, where the musician cooks a melody on the spot. The purpose of this strategy is to reintroduce the values outside the memory into the mix. This ensures that the algorithm would do a more thorough search and reduce the chance of being stuck in local optima. The randomization strategy for the harmony search is the equivalent of mutation to the genetic algorithm.

From a mathematical point of view, as one can expect, a purely random process is needed to capture the essence of this strategy. To that, a random value within the range [0, 1] would be generated for each decision variable in the tentative harmony composed via the memory strategy. These randomly generated values would then be checked against a user-defined parameter called harmony memory conceding rate, denoted by *HMCR*. Anytime the random value is above the threshold set by the *HMCR* parameter, the corresponding decision variable would be selected to go through the randomization strategy. This means the selected values would be randomly replaced by a feasible value available for that particular decision variable. This can be mathematically formulated as follows:

$$x'_j = v_{j,k} \tag{9.6}$$

$$k = IntRand(1, P) \tag{9.7}$$

Note that this is a purely random process, and as such, while there is a possibility that more than one value of the tentative harmony would be replaced via randomization strategy, it is also possible that the tentative harmony would not be affected at all. As a note on the *HMCR* parameter, remember that while very high values would compromise the searching capabilities of the algorithm as it becomes highly unlikely to let non-memorized values back into the mix through the random-ization stage, very low values for this parameter could affect the convergence rate of the algorithm. While parameter fine-tuning is needed here, it is suggested that a good general starting point for *HMCAR* could be 0.9 (Du & Swamy, 2016).

9.2.2.3 Pitch Adjustment Strategy

After the tentative harmony was generated through the memory strategy and pos-sibly altered via the randomization strategy, the algorithm would put the said har-mony through one last process, the pitch adjustment strategy. This strategy is also based on pure randomness and resembles the same principles we have seen in ran-domization. The idea here is to mimic the improvisation principle where the musi-cian refines and tweaks a memorized tune to some extent so that while the new melody has a sense of freshness, it is still based on memorized tunes. To capture this, the algorithm may randomly select some of the notes and replace them with one pitch higher or lower than its current note. Again, note that the selection process here is purely random, but what differs from the randomization strategy we explored earlier is that here we conduct a refined local search for the most part. This ensures that we get as close as possible to the optimum solutions once we locate its vicinity.

To conduct the pitch adjustment strategy, the algorithm would generate a random value within the range [0, 1] for each decision variable in the composed tentative harmony. These randomly generated values would then be checked against a user-defined pitch adjustment rate parameter, denoted by *PAR*. Whenever the random value fell beneath the threshold set by the *PAR* parameter, the corresponding deci-sion variable would be selected to go through the pitch adjustment strategy. Then another random value within the range [0, 1] would be generated randomly. If the second randomly generated value is below 0.5, the note would be replaced with one lower pitch; otherwise, one higher pitch would be used to replace the selected decision variable. This can be mathematically captured as follows for a case where the *j*th note with the value $v_{j,k}$ has been selected for pitch adjustment:

$$x_j' = \begin{cases} v_{j,k+1} & \text{if} \quad Rand > 0.5 \\ v_{j,k-1} & \text{otherwise} \end{cases} \tag{9.8}$$

Again, it is essential to bear in mind that this is a purely random process, and as such, while there is a possibility that the pitch adjustment strategy would alter more than one value of the tentative harmony, it is also possible that the tentative harmony would not be affected at all. As a note on the *PAR* parameter, note that while very low values would compromise the refined local searching capabilities of the algorithm as it becomes highly unlikely to do any pitch refinement, very high values for this parameter could affect the convergence rate of the algorithm. While parameter fine-tuning is needed here, it is suggested that a good general starting point for the *PAR* parameter could be 0.8 (Du & Swamy, 2016).

Finally, after the tentative harmony is composed, the algorithm would evaluate it against the worst-stored harmony in its memory. If the freshly composed tentative harmony is deemed better than the worst harmony, their place would be swapped, otherwise, the tentative solution would be tossed aside, and the process would be repeated.

9.2.3 Termination Stage

Based on the composition stage described above, in each iteration, the harmony search algorithm would compose a new harmony, which would be tested against the worst recorded harmony in the algorithm's memory and replaced in case the new harmony is deemed more desirable. As such, in each iteration, the algorithm eliminates the undesirable solutions and pushes the search boundaries toward what could potentially be the most desirable solutions.

Like other meta-heuristic algorithms, the sequence of operational structures of this algorithm needs to be executed iteratively until a certain termination criterion is met, at which point the execution of the algorithm would be terminated, and the best harmony recorded in the memory would be reported as the solution to the optimization problem. Note that without such a termination stage, the algorithm would potentially be executed in an infinite loop. The termination stage would, in effect, determine whether the algorithm has reached what could be the optimum solution.

As the harmony search algorithm is not equipped with an explicitly defined, unique termination mechanism, one could implement the commonly available options, most notably limiting the number of iterations, run time, or perhaps monitoring the improvement made to the best solution in consecutive iterations. Among these options, limiting the number of iterations is arguably the most cited mechanism to create a termination stage for the harmony search optimization algorithm. The idea is that the process would be executed only for a specified number of times, a parameter known as the maximum iteration. In any case, it should be noted that the selection of a termination mechanism is also considered one of the algorithm's parameters. Bear in mind that in most cases, these terminations may require setting up additional parameters.

9.3 Parameter Selection and Fine-Tuning the Harmony Search Algorithm

From the *no-free-lunch theorem*, one can conclude that fine-tuning an algorithm is essential to get the best performance out of a meta-heuristic algorithm. This would basically ensure that an algorithm is equipped to handle the unique characteristics of a given optimization problem. Of course, it is possible to use our intuition, experience, and default values suggested for an algorithm's parameters as a good starting point, one should bear in mind that fine-tuning these parameters is, more than anything, a trial-and-error process. Thus, while it is possible to get a good enough result by having an educated guess for setting the parameters of these algorithms, to get the best possible performance, it is necessary to go through this fine-tuning process.

In the case of the harmony search algorithm, these parameters are population size (M), harmony memory conceding rate ($HMCR$), pitch adjustment rate parameter (PAR), and of course, opting for the termination criterion, and all the parameters that are associated with these methods. For instance, if limiting the number of iterations has been selected as a termination criterion, the maximum iteration (*T*) is another parameter that needs to be defined by the user. As can be seen here, the harmony search optimization algorithm has a few parameters, which makes the fine-tuning process a bit more manageable than other population-based optimization algorithms we have seen thus far. More importantly, studies suggest that the final solutions reported by the harmony search algorithm are not as sensitive to parameter selection as other meta-heuristic algorithms if they are relatively reasonably tuned. That said, to get the absolute best results out of the harmony search algorithm, it is best to dabble with these algorithms first to gain some experience and inside knowledge about such parameters. By doing so, you could better understand how to fine-tune these parameters as your initial guesses and parameter selection strategies become more educated. The pseudocode for the harmony search algorithm is shown in Figure 9.3.

```
Begin
        Set the algorithm's parameter and input the data
        Let M denote the number of harmonies and N represent the number of notes
        Let HMCR denote the harmony memory considering the rate parameter
        Let PAR denote the pitch adjustment rate parameter
        Generate the initial population
        Evaluate the population
        Memorize the worst harmony
        While the termination criterion is not met
                Let X^new denote a new improvised harmony
                For j in range 1 to N
                        Let i denote a randomly selected number in the range 1 to M
                        n^j_n = n^j_nn
                Next j
                For j in range 1 to N
                        Generate a random value, Rand, in the range [0, 1]
                        If Rand > HMCR
                                Assign a randomly selected feasible value to n^j_n
                        End if
                Next j
                For j in range 1 to N
                        Generate a random value, Rand, in the range [0, 1]
                        If Rand < PAR
                                Generate a random value, Rand_2, in the range [0, 1]
                                If Rand_2 > 0.5
                                        Swap n^j_n with one upper pitch
                                Otherwise
                                        Swap n^j_n with one lower pitch
                                End if
                        End if
                Next j
                Compose the new harmony X^new
                If X^new is better than the worst harmony
                        Replace the worst harmony with X^new
                End if
        End while
        Find and report the best harmony
End
```

Figure 9.3 Pseudocode for the harmony search algorithm.

9.4 Python Codes

The code to implement the harmony search algorithm can be found below:

```python
import numpy as np

def init_generator(pop_size, num_variables, min_val, max_val):
    return np.random.randint(min_val, max_val+1,
                            (pop_size, num_variables))

def sorting_pop(pop, obj_func, minimizing):
    results = np.apply_along_axis(obj_func, 1, pop)
    indeces = np.argsort(results)
    if not minimizing:
        indeces = indeces[::-1]
    return pop[indeces]

def evaluator(a, b, obj_func, minimizing):
    of_a = obj_func(a)
    of_b = obj_func(b)
    if minimizing:
        if of_a<of_b:
            return True
        else:
            return False
    else:
        if of_a>of_b:
            return True
        else:
            return False

def memory_strategy(pop, pop_size, num_variables):
    indeces = np.random.randint(0, pop_size, num_variables)
    temp = pop[indeces]*np.eye(num_variables)
    return np.sum(temp, axis = 0).astype(int)

def random_selection(new_x, HMCR, num_variables, min_val, max_val):
    rand = np.random.uniform(0,1,num_variables)
    return np.where(rand>HMCR, np.random.randint(min_val, max_val+1),
                    new_x)

def pitch_adjustment(new_x, PAR, num_variables):
    rand_index = np.random.uniform(0,1, num_variables)
    rand_values = np.random.uniform(0,1, num_variables)
    temp = np.where((rand_index<PAR)&(rand_values>.5), new_x+
    1, new_x)
    temp = np.where((rand_index<PAR)&(rand_values<=.5), new_x-
    1, temp)
    return temp
```

```
def harmony_search(pop_size, num_variables, min_val, max_val,
  HMCR, PAR,
                  obj_func, iteration, minimizing, full_result=False):
    NFE_value = 0
    NFE = np.zeros(iteration)
    results = np.zeros(iteration)
    pop = init_generator(pop_size, num_variables, min_val, max_val)
    NFE_value += pop_size
    pop = sorting_pop(pop, obj_func, minimizing)
    for i in range(iteration):
        worst = pop[-1]
        new_x = memory_strategy(pop, pop_size, num_variables)
        new_x = random_selection(new_x, HMCR, num_variables,
                                min_val, max_val)
        new_x = pitch_adjustment(new_x, PAR, num_variables)
        if evaluator(new_x, worst, obj_func, minimizing):
            pop[-1] = new_x
        pop = sorting_pop(pop, obj_func, minimizing)
        NFE_value += 1
        NFE[i] = NFE_value
        results[i] = obj_func(pop[0])
    if not full_result:
        return pop[0], obj_func(pop[0])
    else:
        return pop[0], obj_func(pop[0]), results, NFE
```

9.5 Concluding Remarks

The harmony search algorithm is still seen as a formidable meta-heuristic algorithm when it comes to tackling complex real-world problems. Although the harmony search algorithm is considered a population-based algorithm from a technical standpoint, it distinguishes itself from the conventional computational structures we expect from this algorithm. For instance, the algorithm does not treat the initial population as a set of search agents. Instead, these are seen as a set of records that would be used as a memory pool to create a single new tentative solution. Generating a single solution rather than updating the entire population makes the algorithm much less computationally taxing. However, it also means that the algorithm is not fully taking advantage of the potential of parallelized computation. The other nuance in this algorithm is that the composed tentative solution is tested against the worst record in the memory in contrast to the conventional method, which checks these tentative options against the best solution. The idea is that in each iteration, the algorithm would eliminate the worst record from memory and, as such, triangulate on the vicinity of the area containing the optimum solution. Thanks to its redefined local search strategies, the algorithm would then return what could be the optimum solution. What is interesting is that the entire computation architecture of this algorithm is based on stochastic-based strategies, yet it shows promising results when it comes to handling complex real-world optimization problems. Coupling these features with a few parameters that need tuning and simple computational procedures makes the algorithm viable for handling practical optimization problems.

References

Al-Betar, M.A. & Khader, A.T. (2012). A harmony search algorithm for university course timetabling. *Annals of Operations Research*, 194(1), 3–31.

Arsenault, R., Poulin, A., Côté, P., & Brissette, F. (2014). Comparison of stochastic optimization algorithms in hydrological model calibration. *Journal of Hydrologic Engineering*, 19(7), 1374–1384.

Ayvaz, M.T. (2009). Application of harmony search algorithm to the solution of groundwater management models. *Advances in Water Resources*, 32(6), 916–924.

Ayvaz, M.T. (2010). A linked simulation – optimization model for solving the unknown groundwater pollution source identification problems. *Journal of Contaminant Hydrology*, 117(1–4), 46–59.

Du, K.L. & Swamy, M.N.S. (2016). *Search and optimization by metaheuristics: Techniques and algorithms by nature*. Springer International Publishing Switzerland. ISBN: 9783319411910

Fesanghary, M., Asadi, S., & Geem, Z.W. (2012). Design of low-emission and energy-efficient residential buildings using a multi-objective optimization algorithm. *Building and Environment*, 49, 245–250.

Geem, Z.W. (2006). Optimal cost design of water distribution networks using harmony search. *Engineering Optimization*, 38(03), 259–277.

Geem, Z.W. (2010). Multiobjective optimization of time-cost trade-off using harmony search. *Journal of Construction Engineering and Management*, 136(6), 711–716.

Geem, Z.W., Kim, J.H., & Loganathan, G.V. (2001). A new heuristic optimization algorithm: Harmony search. *Simulation*, 76(2), 60–68.

Karbhari, Y., Basu, A., Geem, Z.W., Han, G.T., & Sarkar, R. (2021). Generation of synthetic chest X-ray images and detection of COVID-19: A deep learning based approach. *Diagnostics*, 11(5), 895.

Lee, K.S., Geem, Z.W., Lee, S.H., & Bae, K.W. (2005). The harmony search heuristic algorithm for discrete structural optimization. *Engineering Optimization*, 37(7), 663–684.

Mahdavi, M., Fesanghary, M., & Damangir, E. (2007). An improved harmony search algorithm for solving optimization problems. *Applied Mathematics and Computation*, 188(2), 1567–1579.

Maheri, M.R. & Narimani, M.M. (2014). An enhanced harmony search algorithm for optimum design of side sway steel frames. *Computers & Structures*, 136, 78–89.

Omran, M.G. & Mahdavi, M. (2008). Global-best harmony search. *Applied Mathematics and Computation*, 198(2), 643–656.

Rao, R.S., Ravindra, K., Satish, K., & Narasimham, S.V.L. (2012). Power loss minimization in distribution system using network reconfiguration in the presence of distributed generation. *IEEE Transactions on Power Systems*, 28(1), 317–325.

Wang, C.M. & Huang, Y.F. (2010). Self-adaptive harmony search algorithm for optimization. *Expert Systems with Applications*, 37(4), 2826–2837.

Yadav, P., Kumar, R., Panda, S.K., & Chang, C.S. (2012). An intelligent tuned harmony search algorithm for optimisation. *Information Sciences*, 196, 47–72.

Yang, X.S. (2010). *Nature-inspired metaheuristic algorithms*. Luniver Press. ISBN: 9781905986286

10 Shuffled Frog-Leaping Algorithm

Summary

Inspired by how the memetic paradigm would influence the foraging behavior of a group of frogs as they attempt to search for a food source in their natural habitat, the shuffled frog-leaping algorithm is arguably one of the most interesting meta-heuristic algorithms to handle complex real-world problems. In this chapter, we will dig deep and explore the mechanisms used in this algorithm. We would get familiar with the shuffled frog-leaping algorithm's terminology and see how one can implement this algorithm in the Python programming language. Finally, we will explore the potential merits and drawbacks of this algorithm.

10.1 Introduction

Nowadays, social media is saturated with internet sensations, that is, *internet memes*. Regardless of our background, sense of humor, or comedic taste, we all have enjoyed and laughed with these memes. Though they have become part of our internet culture over the years, the reality is memes are much more than a few topical-themed pictures with funny jokes that may go over our heads from time to time. In fact, there is a scientific discipline dedicated to understanding memes and how they work.

Although there is undoubtedly no universally accepted definition here (Chvaja, 2020), memetic generally refers to a relatively new scientific branch that tends to explain and understand the evolution of cultural patterns through studying information and culture based on a Darwinian paradigm (Kantorovich, 2014). In fact, the works of evolutionary biologists such as Edward O. Wilson and Richard Dawkins have pioneered the idea that human behavior and culture can be seen as a result of the same evolutionary processes that shaped and formed the behavior of other species in nature, and as such, these phenomena could be studied by applying the same fundamental principles and methods that are commonly associated to animal studies (Chvaja, 2020).

Here *memes* are units of human culture that would be replicated within a society to preserve or promote desirable properties in a given community, the same way a good gene would be passed over generations through the evolutionary process to

DOI: 10.1201/9781003424765-10

enhance the surviving odds of a given species. Simply put, these are nothing but ideas, behaving patterns, styles, or generally a piece of information that gets spread from one individual to others. If a community finds a particular meme attractive, sound, or desirable, the community members start to imitate and implement it in their daily lives. As such, a meme would be preserved as part of the community's cultural DNA so long as it is deemed helpful by the said society, otherwise, the public would toss it aside and eventually move on to the next topical thing.

From a computational standpoint, this mechanism closely resembles the evolutionary mechanism, which has proven to be a game-changing idea for creating practical meta-heuristic optimization algorithms such as the genetic algorithm (Holland, 1975). This is, of course, no mere coincidence that these two mechanisms follow the same ideas, for, as we have seen earlier, the memetic process is an expansion of the evolutionary paradigm to explain the cultural behavior of human communities. But interestingly, there are subtle differences between these two paradigms when they are analyzed from a computational side of things, which could give an extra edge to the memetic paradigm over the evolutionary one.

While they are theorized to explain two seemingly different phenomena, we have seen that genetic and memetic paradigms are subjected to the same evolutionary principles. Nonetheless, some exciting yet subtle differences here are worth noting. For instance, due to its nature, on paper, the memetic paradigm seems to take effect much faster than the evolutionary process (Bozorg-Haddad et al., 2017). Of course, when it comes to the computational imitation of these paradigms to create an optimization engine, time, in its original physical sense, is not necessarily the reflection of this difference per se. Instead, the iteration it takes to capture the evolutionary or memetic paradigm is the measure we should look at here. Note that in conventional evolutionary computers, say the genetic algorithm, passing desirable properties through genes would be restricted to a limited number of offspring chromosomes. But more importantly, in order to emulate this evolution of properties, we have to transition from one generation to another, meaning that creating the evolved population is tied to foregoing the previous generation. In computational science, this could be translated as moving from one iteration to another. However, this is not the case in the memetic paradigm, for here, we could have multiple memes simultaneously influencing the community. At the same time, each portion of society is affected by one of the locally dominating memes. More importantly, these cycles of passing information have a shorter time span than the evolutionary process, which means a number of these memes may influence a community member before passing their properties to the next generation. From a computational standpoint, we ought to embed the main iterative body of computational structure with a series of nested inner loops. The point is that before transiting from one generation to another, which is the equivalent of executing a full iteration, this paradigm would repeatedly alter the properties of the same generation through these inner loops. As can be seen, this means that often enough, these memetic-based algorithms are not faster per se, as they require more callbacks and computation than their evolutionary counterpart due to this very same feature. But these nested inner loops enable such algorithms to use the locally obtained information better to adjust the population before moving from one generation to another. This means that the memetic paradigm could potentially be seen as a more flexible and suitable alternative to carry the computational properties commonly associated with meta-heuristic optimization (Eusuff et al., 2006).

As such, the memetic paradigm strikes as a suitable mechanism to build a search engine for meta-heuristic optimization algorithms, especially considering how its core idea aligns with the general theme of swarm intelligence. The point is that similar to swarm intelligence, this paradigm propagates implementing the locally obtained information to create a self-organizing searching mechanism that gradually improves the properties of its members. With these in mind, Eusuff and Lansey (2003) proposed a novel swarm intelligence-based meta-heuristic optimization algorithm called the shuffled frog-leaping algorithm that utilized the core principles of the memetic paradigm.

The shuffled frog-leaping algorithm is a stochastic, population-based meta-heuristic optimization algorithm that takes inspiration from the foraging behavior of a group of frogs as they search for nourishment in their natural inhabit, say, a swamp or a pond. The idea here is that as the group of frogs attempt to forage for food in the pond, they tend to, in a sense, communicate with one another and coordinate their moves based on other members of their respective communities. To that end, the frogs would form smaller sub-groups called the *memeplexes*. A memeplex is actually a locally formed gathering of frogs that can influence one another. The worst frog in these locally based sub-groups would tend to improve its position by imitating the best meme, or behaving pattern, of either the memeplex or the entire community. This process would be repeated for a specified span of time, through which each of the memeplexes attempts to improve their worst members repeatedly. After this, the frogs would be re-shuffled and create a new set of memeplexes based on their new positions. The whole theme would be repeated until a termination point is reached, at which point the best position of the frog would be reported as the solution to the optimization problem at hand.

In this analogy, the pond denotes the search space, and the frogs represent the search agents that are responsible to enumerate through the search space. Each decision variable here is referred to as a *memotype*, and the amount of food in each position is proportionate to the objective function of the said position. The idea is that the algorithm would simultaneously conduct a series of locally based searches using these memeplexes. These locally formed sub-communities would tend to reorganize themselves as they improve themselves using locally obtained information. The point is that while this may seem a more computationally taxing procedure, it can provide a more thorough search than the conventional approach. As such, the algorithm could have a faster convergence rate, hence less iteration needed to get to the optimum point, than a typical meta-heuristic algorithm.

It should be noted that, like many other meta-heuristic algorithms, the imitation procedure used in the algorithm is an abstract representation of the actual phenomenon, and as such, it may not be considered an accurate mathematical depiction of what actually takes place. And from a computational standpoint, this is beside the point so long as the algorithm is efficient enough. In fact, this is a general principle of computational intelligence-based methods, where it is acceptable to settle for an approximation or an estimation so long as it solves a complex problem with reasonable precision.

Specific nuances in this algorithm would differentiate this algorithm from previously introduced meta-heuristic algorithms. The most notable feature that distinguishes this algorithm is the way it implements a series of simultaneous local searches in each iteration. This brings the idea of parallelized computation to a new level as it equips the algorithm with a much more effective local search paradigm. Under this new paradigm, the algorithm would allocate the searching process to different sub-groups, where each, through a self-organizing process, would attempt to improve their members' properties based on the locally obtained information. Although this is much more computationally taxing as it requires to do series of nested loops in each given iteration, it also creates a much more thorough search engine for the algorithm. As a result, such a paradigm would reduce the odds of premature convergence.

Over the years, many variants of the shuffled frog-leaping algorithm have been proposed in the literature, some of which are chaos-based shuffled frog-leaping algorithm (Li et al., 2008), fast shuffled frog-leaping algorithm (e.g., Wang & Gong, 2013), multi-phase modified shuffled frog-leaping algorithm (Luo & Chen, 2014), differential shuffled frog-leaping algorithm (Naruka et al., 2015), and dynamic shuffled frog-leaping algorithm (Cai et al., 2020), to name a few. That said, the standard shuffled frog-leaping algorithm is still a viable option for real-world optimization problems. In fact, the standard shuffled-frog-leaping algorithm has been successfully used for agricultural engineering (e.g., Fallah-Mehdipour et al., 2013), climatology (e.g., Dai et al., 2018), data science (e.g., Zhao et al., 2016), energy industry (e.g., Hasanien, 2015), environmental engineering (e.g., Mahmoudi et al., 2016), financial management (e.g., Alghazi et al., 2012), geohydrology (e.g., Eusuff et al., 2006), hydraulic design (e.g., Orouji et al., 2016), hydrology (e.g., Mohammadi et al., 2020), image processing (e.g., Pérez-Delgado, 2019), land-use (e.g., Huang & Song, 2019), medical science (e.g., Huang et al., 2019; Liu et al., 2021), project management (e.g., Elbeltagi et al., 2007), robotic science (e.g., Hassanzadeh et al., 2010), and water resources planning and management (e.g., Chung & Lansey, 2009). In the following sections, we will explore the computational structure of the standard shuffled frog-leaping algorithm.

10.2 Algorithmic Structure of the Shuffled Frog-Leaping Algorithm

The shuffled frog-leaping algorithm is a stochastic population-based meta-heuristic algorithm that has a unique take on how to utilize parallel computation to locate what could potentially be the optimum solution. In a sense, the primary searching scheme of this algorithm is based on the common notion of the *divide and conquer* strategy. The idea is to break down a complex problem into smaller and more manageable portions. The algorithm would tackle each portion and solve them individually. The obtained information would be used to see the bigger picture. As such, the algorithm would divide the overwhelming task of enumerating the entire search space into a series of local search parties. Each group would then attempt to improve their properties by sharing this locally obtained information. Note that this improvement strategy is a repetitive process that follows the same computational structure as we have seen in the harmony search algorithm (Geem et al., 2001), that is, in each attempt, the goal is only to improve the worst search agent in the sub-group. Naturally, this is a much more computationally taxing strategy than the traditional practice often cited in a population-based meta-heuristic algorithm that adjusts the entire subset in each attempt. Note that these repetitive procedures

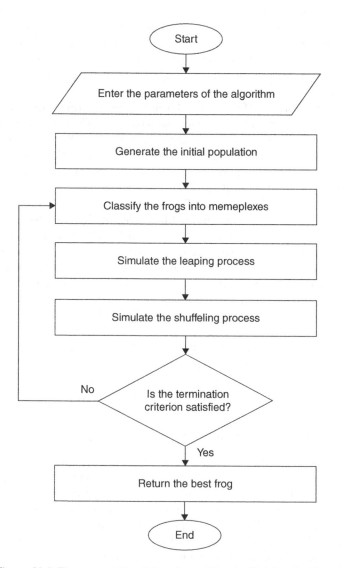

Figure 10.1 The computational flowchart of the shuffled frog-leaping algorithm.

that serve as a self-organizing strategy for the locally formed sub-groups are, in fact, a nested loop within the main structural body of the algorithm. As such, the computational architecture of the shuffled frog-leaping algorithm can be seen as a multilevel iterative searching paradigm. The point is that the algorithm is not obliged to refine the parametric properties of its searching scheme within these inner loops, yet it can conduct a more thorough local search before creating a new population.

The shuffled frog-leaping algorithm would initiate by randomly generating feasible solutions. This initial population of search agents would subsequently be categorized into a series of subsets called memeplexes. Memeplexes would simultaneously be upgraded via locally obtained information in a repetitive process. These upgraded memeplexes would then come together to reform a refined population. This process would be repeated until a termination criterion is met, at which point the best-obtained solution would be returned as the optimum solution. The shuffled frog-leaping algorithm's flowchart is depicted in Figure 10.1. A closer look at the architecture of the shuffled frog-leaping algorithm would reveal that it actually consists of three main stages that are the initiation, foraging, and termination stages. The foraging stage itself consists of three main strategies that are *partitioning*, *leaping process*, and *shuffling*. Using this structure, the algorithm would conduct a thorough search and locate what could be the optimum solution to the problem at hand. The following subsection will discuss each of these stages and their mathematical structures.

10.2.1 Initiation Stage

The shuffled frog-leaping algorithm is a population-based meta-heuristic optimization algorithm, and as such, it works with multiple search agents, here called frogs, that would enumerate through the search space. As we have seen, in an optimization problem with N decision variables, an N-dimension coordination system could be used to represent the search space. In this case, any point within the search space, say X, can be represented mathematically as a $1 \times N$ array as follows:

$$X = \left(x_1, x_2, x_3, \ldots, x_j, \ldots, x_N \right) \tag{10.1}$$

where X represents a frog in the search space of an optimization problem with N decision variables, and x_j represents the value associated with the jth decision variable, or what is technically known as a memotype.

The shuffled frog-leaping algorithm starts with randomly placing a series of frogs within the feasible boundaries of the search space. This bundle of arrays, which in the shuffled frog-leaping algorithm's terminology is referred to as the population, can be mathematically expressed as $M \times N$ matrix, where M denotes the number of frogs or what is technically referred to as the population size. In such a structure, each row represents a single search agent. A population, denoted by *pop*, can be represented as follows:

$$
pop = \begin{bmatrix} X_1 \\ X_2 \\ \vdots \\ X_i \\ \vdots \\ X_M \end{bmatrix} = \begin{bmatrix} x_{1,1} & x_{1,2} & \cdots & x_{1,j} & \cdots & x_{1,N} \\ x_{2,1} & x_{1,2} & \cdots & x_{2,j} & \cdots & x_{2,N} \\ & \vdots & & \vdots & & \\ x_{i,1} & x_{i,2} & \cdots & x_{i,j} & \cdots & x_{i,N} \\ & \vdots & & \vdots & & \\ x_{M,1} & x_{M,2} & \cdots & x_{M,j} & \cdots & x_{M,N} \end{bmatrix} \tag{10.2}
$$

where X_i represents the ith frog in the population, and $x_{i,j}$ denotes the jth memotype of the ith frog.

The initially generated population represents the first generation of frogs. As we progress, the values stored in the *pop* matrix will be altered according to the computational structure of the shuffled frog-leaping algorithm's foraging stage. By the end of this iterative computation process, when the termination criterion is met, one or possibly multiple frogs will converge to the optimum solution.

10.2.2 Foraging Stage

Inspired by the memetic paradigm that helps preserve and replenish a cultural unit in a given community, the foraging stage is the general term used to refer to a bundle of computational processes that transform the older population of frogs into a new refined one. The foraging stage itself consists of three main pillars that are partitioning, leaping process, and shuffling. These mechanisms are used as alteration tools to enhance the population's good qualities and eradicate undesirable properties.

The partitioning mechanism is the first step in the foraging stage of the shuffled frog-leaping algorithm. The idea here is to divide the original population into a series of subcategories here, referred to as memeplexes. Each memeplex would then act as a local community where each member can culturally influence the other members. This is the same idea as forming a meme in a given society, where the community would help preserve and propagate a beneficial culture by encouraging its members to mimic such behavior. Here, these locally formed memeplexes would tend to advocate and promote what they perceive as the meme in that given local community. These locally based enhancement mechanisms would simultaneously improve different portions of the original population without forcing the entire community to follow a single theme, thus preventing the community from being trapped in local optima.

From a mathematical standpoint, partitioning the population starts with sorting it based on the member's performance from the best down to the worst. For instance, in a maximization problem, the frogs with the greater objective functions would assume a higher position than the frogs with lower objective function values. On the contrary, in a minimization problem, the frog with the lowest objective function would assume the highest position in the population, and the frog with the greatest objective function value would be placed at the bottom of the list. The sorted population would then be portioned into sub-groups called memeplexes. Note that the number of memeplexes, denoted by Z, and the number of frogs in each of these memeplexes, represented by Y, are two of the user-defined parameters of the shuffled frog-leaping algorithm. The relation between these parameters and the population size parameter can be expressed as follows:

$$M = Z \times Y \tag{10.3}$$

In order to assign the frogs to their corresponding memeplex, the algorithm would go over the sorted frogs. Here the first frog, or what is the search agent with the best objective function value, would be assigned to the first memeplex. The second frog on the list would be assigned to the second memeplex, which would continue until the Zth frog in the sorted list would be assigned to the Zth memeplex. Next, the algorithm would return back to the first memeplex and assign the $(Z+1)st$ frog to it, and this process would go on until all the frogs are assigned to a memeplex. Figure 10.2 depicts how the partition machine of the shuffled frog-leaping algorithm.

After the sorted frogs of the original population are assigned to their corresponding memeplex using the partitioning mechanism, the algorithm tries to emulate the leaping process of the frogs, which is basically the mechanism to adjust the positioning of the frogs using locally obtained information.

The leaping process starts by identifying the worst and the best frogs in each of these parallel communities. According to the memetic evolution paradigm, the community members would attempt to preserve a desirable cultural pattern by adjusting their behavior to this emerging meme in their community. The shuffled frog-leaping algorithm emulates this notion by adjusting the worst-identified frog in the sub-community to resemble the best-identified frog in the said group, which denotes the meme of that memeplex. This can be seen as a local alteration

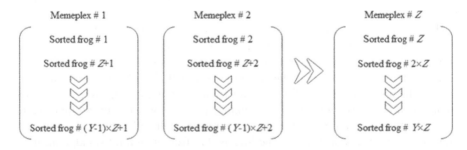

Figure 10.2 The partition machine of the shuffled frog-leaping algorithm.

of a memeplex to push the community's worst members toward the best locally obtained properties. In order to avoid being trapped in local optima, this process needs to have a stochastic nature, creating a mechanism that selects the adjusting member via a stochastic process. Naturally, this process needs to be explicitly designed so that the best members have a better chance of being selected for readjustments than the frogs with less desirable properties. After all, the whole point is to realign the worst member's performance according to the meme of the sub-community.

As such, in each parallel sub-community, another sub-group from here onward, referred to as the sub-memeplex, needs to be selected. Mathematically speaking, this would mean picking a specified number of frogs, denoted by Q, from each memeplexes using a stochastic selection procedure. Note that the size of these sub-memeplexes is another user-defined parameter of the shuffled frog-leaping algorithm. The probability distribution used to select these sub-memeplexes needs to reflect the idea that the worst members are less likely to be tagged in this process than the best members of the sub-community. Given that these frogs are ranked chronically from the best to the worst, the probability of selecting a given member of the community can be mathematically expressed as follows:

$$Pr_i = \frac{2 \times (Y + 1 - i)}{Y \times (Y + 1)} \qquad \forall i \tag{10.4}$$

in which Pr_i denotes the probability of selecting the ith frog in a given memeplex for forming a sub-memeplex. Note that the worst memeplex is ranked lower than the best memeplex, which should be at the top of the list. As such, for instance, the probability of selecting the best member of the sub-community in this process is equal to $2/(Y+1)$, while the odd of selecting the worst member here is $2/(Y^2+Y)$.

After forming this sub-memeplex, the algorithm would again rearrange these stochastically selected frogs from best to worst. Next, the algorithm attempts to improve the properties of the worst member of the sub-memeplex, here denoted by the *Mworst*, using the locally obtained information. To do that, the algorithm would also identify the best member of the sub-memeplex and the best member of the member in the entire population, denoted, respectively, by the *Mbest* and *Pbest*. Note that here the two latter members represent the memes for the said communities. It is worth noting that the algorithm is designed so that the locally obtained information has a higher priority over globally obtained information, that is, the algorithm would first attempt to adjust the worst members of these smaller communities with the locally obtained memes. Only if such attempts were deemed a failure would make the algorithm resort to globally obtained memes to alter the worst frog's properties. The idea is that from a cultural standpoint, it is more likely in a small community to be influenced by the local environment than the outside world.

In order to capture this procedure in a mathematical format, the algorithm tends to compute the leaping size for each memotype of the worst-tagged frog in each sub-memeplexes. This can be mathematically formulated as follows:

$$d_j = Rand \times \left(x_{Mbest,j} - x_{Mworst,j} \right) \qquad \forall j \tag{10.5}$$

$$x'_j = x_{Mworst,j} + d_j \qquad \forall j \tag{10.6}$$

$$X' = \left(x'_1, x'_2, x'_3, \ldots, x'_j, \ldots, x'_N \right) \tag{10.7}$$

where d_j denotes the leaping motion for the jth memotype of the worst identified frog in the sub-memeplex; $x_{Mworst,j}$ represents the jth memotype of the worst identified frog in the sub-memeplex; $x_{Mbest,j}$ denotes the jth memotype of the best-identified frog in the sub-memeplex; $Rand$ is randomly generated value within the range 0–1; x'_j is the jth memotype of the new position for the worst identified frog in the sub-memeplex; and X' denotes the new position for the worst identified frog in the sub-memeplex.

After computing the new position for the worst identified frog in the sub-memeplex, the algorithm would evaluate the performance of the leaping frog against its previous location. If the leap has indeed improved the properties of the said frog, the new position would be assumed by the said frog. However, if the locally obtained information could not encourage the worst identified frog to make an improving leap, the algorithm would have to resort to globally obtained information on its second attempt to improve the position of the said frog. The mathematics of this leaping motion is precisely the same as what we have earlier, except here, rather than being coordinated by the best local meme, the frog would take inspiration from the population meme that is the *Pbest* frog. This can be mathematically modeled as follows:

$$d_j = Rand \times \left(x_{Pbest,j} - x_{Mworst,j} \right) \qquad \forall j \tag{10.8}$$

$$x'_j = x_{Mworst,j} + d_j \qquad \forall j \tag{10.9}$$

$$X' = \left(x'_1, x'_2, x'_3, \ldots, x'_j, \ldots, x'_N \right) \tag{10.10}$$

where $x_{Pbest,j}$ denotes the jth memotype of the best-identified frog in the entire population.

The new position would again be evaluated against the current position of the worst identified frog in the sub-memeplex. If this leaping notion is deemed successful, meaning that it has improved the properties of the frog, the new position would be assumed for the said frog. However, if this were not the case, the algorithm would replace the said frog with a randomly generated feasible solution in one last attempt. This is expressed in mathematical terms as follows:

$$x'_j = x^L_j + Rand \times \left(x^U_j - x^L_j\right) \qquad \forall j \tag{10.11}$$

$$X' = \left(x'_1, x'_2, x'_3, \ldots, x'_j, \ldots, x'_N\right) \tag{10.12}$$

where x^U_j denotes the feasible upper bound for the jth memotype, and x^L_j represents the lower feasible bound for the jth memotype.

Note that you may want to cap the leaping motion of each of these memotypes. After each leaping adjustment, the algorithm could test the motion against two arbitrarily defined boundaries. If the said motion violates any of them, they could be replaced with these boundary values. As such, the new adjusted motions should always be within this feasible range, which can be formulated as follows:

$$d_j = \begin{cases} D_{max} & \text{if} & D_{max} \le d_j \\ D_{min} & \text{if} & d_j \le D_{min} \\ d_j & \text{otherwise} \end{cases} \tag{10.13}$$

in which D_{min} denotes the lower cap on the leaping motion, and D_{max} represents the upper cap on the leaping motion. Note that these two are both user-defined parameters of the shuffled frog-leaping algorithm, should you decide to include them in your calculation.

As stated earlier, the leaping process is basically a repetitive self-adjusting motion that is mainly based on the locally obtained information, which means that it can be seen as an inner nested loop within the basic structure of the shuffled frog-leaping algorithm. As such, after each adjustment of the worst identified frog in the sub-memeplexes, the new frog would eventually replace the said frog in each local community. The algorithm would again rearrange the memeplex members based on their performance from best to worst, resample a new sub-memeplex from the said sub-community, and repeat the whole process. For each of the memeplexes, this procedure would be repeated for a predefined specified number of times, here called the memetic evolution. Note that the number of memetic evolution, denoted by μ is another user-defined parameter of the shuffled frog-leaping algorithm.

After the algorithm is done with emulating the leaping process, it finally moves on to the last step of the frogs' foraging behavior, the shuffling procedure. To this point, the majority of the memetic correction of the leaping frogs was influenced by the locally formed memeplexes. The shuffling process does regroup these memeplexes into a singular community that is the new population. Without this process, any cultural influence from the memes would be, for the most part, limited to these isolated memeplexes, and as such, the algorithm may have suffered from premature convergence. However, thanks to this shuffling procedure, the frogs would come back together to be re-evaluated by the algorithm. Again based on their performance, these frogs would be rearranged from the best performance to the worst. And based on these evaluations, a new set of sub-communities would be formed that, in all likelihood, differs from the previous ones.

10.2.3 Termination Stage

Based on the foraging stage described above, in each iteration, the shuffled frog-leaping algorithm would partition the population into sub-groups, upgrade their positions by emulating the frogs' leaping procedure, and finally regroup the entire memeplex into a new population. As such, the algorithm tends to encourage the community to imitate the locally, or in some cases globally, identified memes. By doing so, constantly improve the entire community toward what could potentially be the optimum solution.

Like other meta-heuristic algorithms, the sequence of operational structures of this algorithm needs to be executed iteratively until a certain termination criterion is met, at which point the execution of the algorithm would be terminated, and the best frog recorded in the memory would be reported as the solution to the optimization problem. Note that without such a termination stage, the algorithm would potentially be executed in an infinite loop. The termination stage would, in effect, determine whether the algorithm has reached what could be the optimum solution.

As the shuffled frog-leaping algorithm is not equipped with an explicitly defined, unique termination mechanism, one could implement the commonly available options, most notably limiting the number of iterations, run time, or perhaps monitoring the improvement made to the best solution in consecutive iterations. Among these options, limiting the number of iterations is arguably the most cited mechanism to create a termination stage for the shuffled frog-leaping algorithm. The idea being the process would be executed only for a specified number of times, a parameter known as the maximum iteration. In any case, it should be noted that the selection of a termination mechanism is also considered one of the algorithm's parameters. Bear in mind that in most cases, these termination mechanisms may require setting up additional parameters.

10.3 Parameter Selection and Fine-Tuning the Shuffled Frog-Leaping Algorithm

From the *no-free-lunch theorem*, one can conclude that fine-tuning an algorithm is essential to get the best performance out of a meta-heuristic algorithm. This would basically ensure that an algorithm is equipped to handle the unique characteristics of a given optimization problem. Of course, it is possible to use our intuition, experience, and default values suggested for an algorithm's parameters as a good starting point, one should bear in mind that fine-tuning these parameters is more than anything a trial-and-error process. Thus, while it is possible to get a good enough result by having an educated guess for setting the parameters of these algorithms, to get the best possible performance, it is necessary to go through this fine-tuning process.

In the case of the shuffled frog-leaping algorithm, these parameters are the number of memeplexes (Z), the number of frogs in each of these memeplexes (Y), the size of sub-memeplexes (Q), the lower (D_{min}) and upper (D_{max}) cap on the leaping motion in case one decides to apply these boundaries, the number of memetic evolution (μ), and of course, opting for the termination criterion, and all the parameters that are associated with these methods. For instance, if limiting the number of iterations has been selected as a termination criterion, the maximum iteration (T) is another parameter that needs to be defined by the user. As can be seen here, the number of parameters of the shuffled frog-leaping algorithm has a relatively reasonable number of parameters. For the most part, however, the role of these parameters on the final outcome is easy to deduce intuitively, even for those with little experience with meta-heuristic optimization, which makes the fine-tuning process a bit more manageable than other population-based optimization algorithms we have seen thus far. That said, to get the absolute best results out of the shuffled frog-leaping algorithm, it is best to dabble with these algorithms first to gain some experience and inside knowledge about such parameters. By doing so, you could better understand how to fine-tune these parameters as your initial guesses and parameter selection strategies become more educated. The pseudocode for the shuffled frog-leaping algorithm is shown in Figure 10.3.

```
Begin
        Set the algorithm's parameter and input the data
        Let Z denote the number of memeplexes and Y represent the number of frogs in each memeplex
        Let μ denote the number of memetic evolution and Q represent the submemeplex size
        Generate the initial population with the population size M
        Evaluate each frog in the population
        While the termination criterion is not met
                Sort population by their performance
                Let Pbest denote the best frog in the population
                Divide the population into Z memeplexes
                For s in range 1 to Z
                        For j in range 1 to μ
                                Select Q frogs randomly from memeplex s to form a submemeplex
                                Let Mwrost denote the worst frog in the submemeplex
                                Let Mbest denote the best frog in the submemeplex
                                Improve Mwrost using Mbest
                                If we cannot improve Mwrost
                                        Improve Mwrost using Pbest
                                End if
                                If still cannot improve Mwrost
                                        Generate a feasible from randomly
                                End if
                        Next j
                Next s
                Merge all memeplexes
        End while
        Report the best frog
End
```

Figure 10.3 Pseudocode for the shuffled frog-leaping algorithm.

10.4 Python Codes

The code to implement the shuffled frog-leaping algorithm can be found below:

```python
import numpy as np

def init_generator(num_variables, pop_size, min_val, max_val):
    return np.random.uniform(min_val,max_val,(pop_size, num_
  variables))

def sorting_pop(pop, obj_func, minimizing):
    ofs = np.apply_along_axis(func1d=obj_func, axis=1, arr=pop)
    if minimizing:
        indeces = np.argsort(ofs)
    else:
        indeces = np.argsort(ofs)[::-1]
    return pop[indeces]

def creating_memeplexes(pop_size, num_memeplex):
    memeplexes=list()
    for i in range(num_memeplex):
        temp=list(range(i,pop_size,num_memeplex))
        memeplexes.append(temp)
    return memeplexes

def creating_submemeplex(memeplex, size_memeplex, size_
  submemeplex):
    Y = np.arange(size_memeplex)
    prob = 2*(1+size_memeplex-Y)/(size_memeplex*(size_memeplex+1))
    const = (1-np.sum(prob))/size_memeplex
    prob += const
    sub_index = np.random.choice(np.arange(size_memeplex),
                                 size_submemeplex, replace=False,
  p=prob)
    return memeplex[sub_index]

def compare(a, b , obj_func, minimizing):
    if minimizing:
        if obj_func(a)<obj_func(b):
            return True
        else:
            return False
    else:
        if obj_func(a)>obj_func(b):
            return True
        else:
            return False

def adjust_movement(x, dmin, dmax):
    x = np.where(x<dmin, dmin, x)
    x = np.where(x>dmax, dmax, x)
    return x
```

```
def shuffled_frog_leaping_algorithm(num_memeplex, size_memeplex,
                                    num_variables, min_val, max_val,
                                    iteration, obj_func,
                                    num_memetic_evelution, dmin=-np.inf,
                                    dmax=np.inf, minimizing=True,
                                    full_result=False):
    NFE_value = 0
    NFE = np.zeros(iteration)
    results = np.zeros(iteration)
    pop_size = num_memeplex*size_memeplex
    pop = init_generator(num_variables, pop_size, min_val, max_val)
    pop = sorting_pop(pop, obj_func, minimizing)
    NFE_value += pop_size
    for i in range(iteration):
        pbest=pop[0]
        memeplexes_indeces = creating_memeplexes(pop_size, num_
        memeplex)
        for memeplexes_index in memeplexes_indeces:
            memeplex = pop[memeplexes_index]
            for j in range(num_memetic_evelution):
                sub_memeplex = creating_submemeplex(memeplex,
                                                    size_memeplex,
                                                    size_
    submemeplex)
                sub_memeplex = sorting_pop(sub_memeplex, obj_func,
                                            minimizing)
                mbest = sub_memeplex[0]
                mworst = sub_memeplex[-1]
                rand = np.random.uniform(0,1, num_variables)
                move_1 = rand*(mbest-mworst)
                move_1 = adjust_movement(move_1, dmin, dmax)
                temp_point_1 = mworst+move_1
                move_2 = rand*(pbest-mworst)
                move_2 = adjust_movement(move_2, dmin, dmax)
                temp_point_2 = mworst+move_2
                if compare(temp_point_1, mworst, obj_func, minimizing):
                    new_x=temp_point_1
                    NFE_value += 1
                elif compare(temp_point_2,mworst,obj_func,minimizing):
                    new_x=temp_point_2
                    NFE_value += 1
                else:
                    new_x=np.random.uniform(min_val, max_val,
                                            num_variables)
                    NFE_value += 1
                pop = np.concatenate(((pop, np.reshape(new_x, (1,-1)))),
                                      axis=0)
        pop = sorting_pop(pop, obj_func, minimizing)[:pop_size]
        NFE[i] = NFE_value
        results[i] = obj_func(pop[0])
    if not full_result:
        return pop[0], obj_func(pop[0])
    else:
        return pop[0], obj_func(pop[0]), results, NFE
```

10.5 Concluding Remarks

Inspired by the foraging behavior of frogs, the shuffled frog-leaping algorithm is still seen as a formidable meta-heuristic algorithm when it comes to tackling complex real-world problems. This algorithm's nuances and features made it an interesting population-based meta-heuristic optimization method from a computational standpoint. For a start, the algorithm is built upon the memetic paradigm, which is the implementation of the evolutionary process in the context of a society's cultural behavior pattern. But, from a computational standpoint, this algorithm's architecture resembles a swarm intelligence-based algorithm, as the bases of the coordinating force in this algorithm are driven mainly by the interaction of the search agents within a local scale. The central premise used to create the structure of the search engine of this algorithm is based upon the principle of divide and conquer. The algorithm would break down the daunting task of enumerating the entire search space into a series of locally based searches conducted by smaller sub-communities. In each sub-group, the members tend to imitate the meme of their local community and improve the properties of their worst members. In the shuffled frog-leaping algorithm, the main focus is on improving and eradicating the worst identified properties in the population set rather than the conventional approach to upgrade the properties of the entire swarm.

The other notable feature of this algorithm is its multilevel searching structure, where the main body of the algorithm is equipped with a series of nested inner loops. The point is that each sub-community tends to self-correct itself using the locally obtained information a specified number of times before all the sub-groups gather together and form the new population set. While this may seem a bit more computationally taxing as the algorithm is doing more computation in each iteration in comparison to the conventional algorithms, the point is these locally based corrections would not only help prevent premature convergence but also would encourage a more thorough search and reduce the need to do additional iteration. While this is certainly not the most straightforward and easy-to-implement computational structure, the lack of an overwhelming number of complicated parameters and an efficient searching strategy makes the shuffled frog-leaping algorithm a viable option when tackling real-world complex optimization problems.

References

Alghazi, A., Selim, S.Z., & Elazouni, A. (2012). Performance of shuffled frog-leaping algorithm in finance-based scheduling. *Journal of Computing in Civil Engineering*, 26(3), 396–408.

Bozorg-Haddad, O., Solgi, M., & Loáiciga, H.A. (2017). *Meta-heuristic and evolutionary algorithms for engineering optimization*. John Wiley & Sons. ISBN: 9781119386995

Cai, J., Zhou, R., & Lei, D. (2020). Dynamic shuffled frog-leaping algorithm for distributed hybrid flow shop scheduling with multiprocessor tasks. *Engineering Applications of Artificial Intelligence*, 90, 103540.

Chung, G. & Lansey, K. (2009). Application of the shuffled frog leaping algorithm for the optimization of a general large-scale water supply system. *Water Resources Management*, 23(4), 797–823.

Chvaja, R. (2020). Why did memetics fail? Comparative case study. *Perspectives on Science*, 28(4), 542–570.

Dai, S., Niu, D., & Han, Y. (2018). Forecasting of energy-related CO_2 emissions in China based on GM(1,1) and least squares support vector machine optimized by modified shuffled frog leaping algorithm for sustainability. *Sustainability*, 10(4), 958.

Elbeltagi, E., Hegazy, T., & Grierson, D. (2007). A modified shuffled frog-leaping optimization algorithm: Applications to project management. *Structure and Infrastructure Engineering*, 3(1), 53–60.

Eusuff, M.M. & Lansey, K.E. (2003). Optimization of water distribution network design using the shuffled frog leaping algorithm. *Journal of Water Resources Planning and Management*, 129(3), 210–225.

Eusuff, M.M., Lansey, K.E., & Pasha, F. (2006). Shuffled frog-leaping algorithm: A memetic meta-heuristic for discrete optimization. *Engineering Optimization*, 38(2), 129–154.

Fallah-Mehdipour, E., Bozorg Haddad, O., & Mariño, M.A. (2013). Extraction of multicrop planning rules in a reservoir system: Application of evolutionary algorithms. *Journal of Irrigation and Drainage Engineering*, 139(6), 490–498.

Geem, Z.W., Kim, J.H., & Loganathan, G.V. (2001). A new heuristic optimization algorithm: Harmony search. *Simulation*, 76(2), 60–68.

Hasanien, H.M. (2015). Shuffled frog leaping algorithm for photovoltaic model identification. *IEEE Transactions on Sustainable Energy*, 6(2), 509–515.

Hassanzadeh, I., Madani, K., & Badamchizadeh, M.A. (2010). Mobile robot path planning based on shuffled frog leaping optimization algorithm. In *Proceedings of the IEEE International Conference on Automation Science and Engineering*, Toronto, ON, Canada.

Holland, J.H. (1975). *Adaptation in natural and artificial systems: An introductory analysis with applications to biology, control, and artificial intelligence.* University of Michigan Press.

Huang, C., Tian, G., Lan, Y., Peng, Y., Ng, E.Y.K., Hao, Y., ... & Che, W. (2019). A new pulse coupled neural network (PCNN) for brain medical image fusion empowered by shuffled frog leaping algorithm. *Frontiers in Neuroscience*, 13, 210.

Huang, Q. & Song, W. (2019). A land-use spatial optimum allocation model coupling a multi-agent system with the shuffled frog leaping algorithm. *Computers, Environment and Urban Systems*, 77, 101360.Kantorovich, A. (2014). An evolutionary view of science: Imitation and memetics. *Social Science Information*, 53(3), 363–373.

Li, Y., Zhou, J., Yang, J., Liu, L., Qin, H., & Yang, L. (2008). The chaos-based shuffled frog leaping algorithm and its application. In *Proceedings of the 4th International Conference on Natural Computation*, Jinan, China.

Liu, L., Chen, X., & Wong, K.C. (2021). Early cancer detection from genome-wide cell-free DNA fragmentation via shuffled frog leaping algorithm and support vector machine. *Bioinformatics*. DOI: 10.1093/bioinformatics/btab236

Luo, J. & Chen, M.R. (2014). Multi-phase modified shuffled frog leaping algorithm with extremal optimization for the MDVRP and the MDVRPTW. *Computers & Industrial Engineering*, 72, 84–97.

Mahmoudi, N., Orouji, H., & Fallah-Mehdipour, E. (2016). Integration of shuffled frog leaping algorithm and support vector regression for prediction of water quality parameters. *Water Resources Management*, 30(7), 2195–2211.

Mohammadi, B., Linh, N.T.T., Pham, Q.B., Ahmed, A.N., Vojteková, J., Guan, Y., ... & El-Shafie, A. (2020). Adaptive neuro-fuzzy inference system coupled with shuffled frog leaping algorithm for predicting river streamflow time series. *Hydrological Sciences Journal*, 65(10), 1738–1751.

Naruka, B., Sharma, T. K., Pant, M., Sharma, S., & Rajpurohit, J. (2015). Differential shuffled frog-leaping algorithm. In *Proceedings of the 4th International Conference on Soft Computing for Problem Solving*, New Delhi, India.

Orouji, H., Mahmoudi, N., Fallah-Mehdipour, E., Pazoki, M., & Biswas, A. (2016). Shuffled frog-leaping algorithm for optimal design of open channels. *Journal of Irrigation and Drainage Engineering*, 142(10), 06016008.

Pérez-Delgado, M.L. (2019). Color image quantization using the shuffled-frog leaping algorithm. *Engineering Applications of Artificial Intelligence*, 79, 142–158.

Wang, L. & Gong, Y. (2013). A fast shuffled frog leaping algorithm. In *Proceedings of the 9th International Conference on Natural Computation*, Shenyang, China.

Zhao, Z., Xu, Q., & Jia, M. (2016). Improved shuffled frog leaping algorithm-based BP neural network and its application in bearing early fault diagnosis. *Neural Computing and Applications*, 27(2), 375–385.

11 Invasive Weed Optimization Algorithm

Summary

Inspired by the life cycle of weeds as they spread through a new environment, the invasive weed optimization algorithm is arguably one of the most interesting meta-heuristic algorithms to handle complex real-world problems. What is interesting about this algorithm from a computation standpoint is that it provides a new take on how swarm intelligence can be used to help navigate the search agents through the search space. In this chapter, we will dig deep and explore the mechanisms used in this algorithm. We would get familiar with the invasive weed optimization algorithm's terminology and see how one can implement this algorithm in the Python programming language. Finally, we will explore the potential merits and drawbacks of this algorithm.

11.1 Introduction

If you are into gardening, nothing is more relaxing than mowing your lawn or taking care of the garden in the backyard. The smell of fresh-cut grass, the excitement of growing your own vegetables, or witnessing a plant grow before your eyes are some of the most joyful experiences you could ever have. This is until you run into a weed problem, then the whole process could turn into a huge nightmare. Weeds are, by nature, aggressive growing plants that spread, well, as they say, like "weeds." Of course, this is not the end of the world per se, but if you are not careful enough or do not take the proper course of action in a timely manner, a small problem could quickly turn into a massive and tedious workload.

From the agronomy standpoint, weed refers to any undesirable plant in a given situation. As such, any plant can be called a weed if the circumstance they grow in does not match what they were intended for. A good example is a set of plant species commonly considered undesirable in human-controlled settings, such as farm fields, gardens, lawns, or parks. Again, it is essential to note that, taxonomically speaking, the term weed can have no biological significance, as the desirability of a plant is purely relative to the context of the environment in which the plant is being grown. That said, due to a lack of valuable crops, medical applicability, or having little to no aesthetic values, certain plant species are widely regarded as

DOI: 10.1201/9781003424765-11

weeds in agricultural cultivation or even home gardening. The term *weed* in this context, which happens to be what we exclusively use for from this point onward, actually identifies those plants that grow or reproduce aggressively and, as such, are considered borderline invasive outside their native habitat.

It is a common perception regarding weed plants that as they are introduced to a new habitat, they can reproduce quickly, grow fast, spread aggressively, and adapt to some of the most hostile environments known for plant vegetation. Though depending on the type of the weeds plants, they can have different reproduction mechanisms, one of the most commonly known ways here is the seed-dispersion mechanism. This process starts when the weed reaches a certain phase in its life cycle where it can produce new seeds that would then be released to the outside world by the plant. Through external environmental factors, such as the wind, water flow, and certain species of birds or insects, these seeds may even travel a great distance until they are placed in a suitable environment with enough nourishment to start growing. Naturally, the more suitable the place, the better the odds are for such seeds to survive or perhaps even thrive in their new habitat. These seeds would then start growing and reach a certain phase where they can produce new seeds. Again, it is essential to note that the environmental conditions say the availability of resources, such as water, the quality of the soil, or receiving enough sunshine, would directly affect the weed's biomass production, that is, the number and quality of their seeds.

There are some theories out there that attempt to explain the generic behavior of weeds as they got introduced into a new environment. One of the notable examples here professes that the weeds' behavior in any environment would always fall within two extreme conditions that are *r-selection* and *k-selection* (Mehrabian & Lucas, 2006). The general theme of the r-selection phase is to "*live short, reproduce quickly, and die young.*" A weed plant enters the r-selection phase to preserve its species through rapid reproduction, as the environment may be too hostile or unstable for the said plant to survive on its own for long periods of time. In such circumstances, the weed produces as many seeds as possible and disperses them as far as possible because the harsh environment prevents the plant from living for a long time. Failing to follow this procedure would certainly doom the species to extinction. As such, the r-selection phase is, in fact, nothing but a survival mechanism for the weeds in case they get introduced to a harsh, unforgiving environment. High fertility rate, short-living life spans, and opting for long-distance dispersal are among the adaptive qualities that help the weeds should they enter the r-selection phase. On the other far end of the spectrum, we have the k-selection phase, where the general theme is basically "*live long, reproduce slowly, and die old.*" A weed plant enters the k-selection phase only when introduced into a hospitable and stable environment with enough resources to support the said species' flourishing. Note that in such an environment, one could usually expect heavy competition as the nourishing environment would have definitely helped host a variety of species, and as such, there is standard competition over the limited available resources. Naturally, in these environments where the capacity to host new members has nearly reached its limits, those fittest to survive would have the upper hand in the said competitions. Having a limited number of offspring, longer living life spans, and opting for short-distance dispersal are some of the most notable adaptive behaviors that help the weeds as they enter the k-selection phase. Again, it is essential to remember that both r-selection and k-selection are far ends of the spectrum that describes the adaptive behavior of weeds as they are introduced to a new environment. As such, depending on the environmental circumstance, the weed plant may behave somewhere between these two extreme conditions.

As demonstrated, through their course of the evolutionary process, the so-called weed plants have been equipped with such efficient mechanisms that enable them to survive and often even thrive in situations that are considered to be hostile for most other plants. Inspired by this adaptive quality, Mehrabian and Lucas (2006) theorized a novel meta-heuristic algorithm called the invasive weed optimization algorithm that mimics the behavior of a given weed species as they get introduced to a new environment. From a computational standpoint, the invasive weed optimization algorithm can be categorized as a stochastic population-based algorithm that is based on the idea of a local direct search. Being based on the principles of the direct search would mean that, like other meta-heuristic algorithms, the invasive weed optimization algorithm is also not bound by problems that are often associated with high dimensionality, multimodality, epistasis, non-differentiability, and discontinuous search space imposed by constraints (Du & Swamy, 2016; Bozorg-Haddad et al., 2017). But more importantly, in a sense, the structure of this algorithm also resembles, to some extent, the idea of swarm intelligence. The point is that while in opposition to conventional swarm intelligence-based algorithms, the search agents do not necessarily coordinate their moves based on direct communication with one another, their relative performance against their contemporary counterpart search agents would dictate how their reproduction mechanism of each member would work, as such, having this relative dependency between the search agents' performances would give this algorithm a swarm intelligence-based feeling.

The invasive weed optimization algorithm is based on an abstract representation of the mechanism we explored earlier. The idea is that the algorithm would initiate its search by randomly generating a series of weeds within the feasible area of the search space. Each weed would then continue to create a set of seeds that would be dispersed within the vicinity of the parent weed. The seed would go on to grow and create a set of seeds of its own and disperse them in the environment. At any given point, when the number of weeds exceeds the environment's capacity to host a new plant, the exceeding weed with inferior genes would be terminated as they cannot survive in the competing environment. This whole process would be repeated until a termination criterion is met, at which point the best weed encountered thus far would be returned as the solution to the optimization problem at hand. In this analogy, the weeds represent the search agents that enumerate through the search space, which here is denoted as the environment where these weeds are inhabiting. The desirability of a place within this space is measured with the quality of the objective function or the fitness function in constrained optimization problems.

The invasive weed optimization algorithm has a fairly straightforward computational structure. The lack of complex mathematical formulas or having over-the-top intricate architecture makes it easy to understand and implement when it comes to handling complex real-world optimization problems. More importantly, this algorithm has few parameters to begin with, and for the most part, understanding how these parameters would affect the final result should be fairly intuitive for those with a bit of background in meta-heuristic computation. That said, being a meta-heuristic algorithm, the invasive weed optimization algorithm still suffers from the same generic issues that are commonly associated with this branch of optimization, such as having no guarantee to reach the optimum solution or being trapped in a local optimum solution.

Over the years, many variants of the invasive weed optimization algorithm have been proposed in the literature, some of which are chaotic invasive weed optimization algorithm (Ahmadi & Mojallali, 2012), adaptive invasive weed optimization algorithm (Peng et al., 2015), and chaos-based invasive weed optimization algorithm (Misaghi & Yaghoobi, 2019), to name a few. That said, the standard invasive weed optimization algorithm is still a viable option for real-world optimization problems. In fact, the standard invasive weed optimization algorithm has been successfully used for aerospace science (e.g., Jafari & Montazeri-Gh, 2013), computer science (e.g., Su et al., 2014), data science (e.g., Liu & Nie, 2021), electrical engineering (e.g., Karimkashi & Kishk, 2010), energy industry (e.g., Barisal & Prusty, 2015), financial management (e.g., Ahmadi et al., 2017), hydrology (e.g., Hamedi et al., 2016), mechanical engineering (e.g., Yazdani & Ghodsi, 2017), medical science (e.g., Soulami et al., 2019), mining engineering (e.g., Huang et al., 2019), nuclear engineering (e.g., Rahmani et al., 2021), robotic science (e.g., Mohanty & Parhi, 2014; Panda et al., 2018), and water resources planning and management (e.g., Asgari et al., 2016; Azizipour et al., 2016). In the following sections, we will explore the computational structure of the standard invasive weed optimization algorithm.

11.2 Algorithmic Structure of the Invasive Weed Optimization Algorithm

As a population-based meta-heuristic algorithm enumerates through the search space through a set of search agents. But what distinguishes this algorithm from the other meta-heuristic algorithm that we have seen thus far, for the most part, is how these search agents are coordinated and updated throughout the search process. First, it is essential to note that while, on paper, each search agent is actually participating in the creation of the next generation of search agents, as you can expect, not all agents should have the same role in the process. The point is that, similar to the abstract analogy we had for the weed life cycle, the stronger weeds inhibiting a more desirable habitat would have a better chance of creating offspring. So the invasive weed optimization algorithm built upon this idea by creating a mechanism that uses the relative performance of search agents against their contemporary counterparts to compute how involved each agent should be in creating the next generation of weeds. The other notable thing here is that these

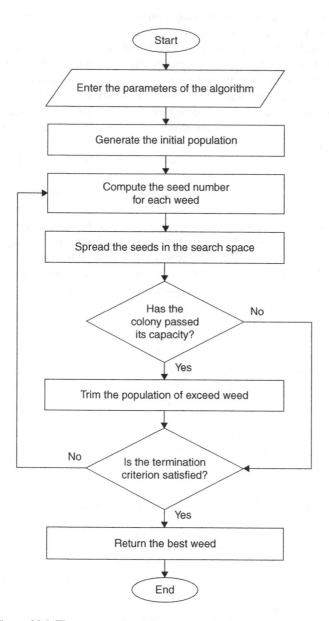

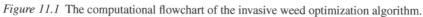

Figure 11.1 The computational flowchart of the invasive weed optimization algorithm.

mechanisms should account for the desperation mechanisms of the seeds in the search agent. If the parenting weed happens to have a good performance, the dispersion should not create seeds that are placed far from their origin point. In this situation, the algorithm needs the dispersion mechanism to act more like a refined local search and search the areas that could hold the optimum solution. On the contrary, if the weed is not doing as well, the algorithm would expect the dispersion mechanism to place the seed as far away from the original point as possible. This would give the algorithm a more exploratory feel as it searches a broader space. In addition to these features, the algorithms generally need to account for a smooth transition between the exploration and exploitation phases. As the algorithm goes through each iteration, the radius of this dispersion should gradually be reduced gradually, transitioning the search from a more expansive space into a more local and focused search. The algorithm uses a heavily statistical-based architecture that uses probabilistic computation to place the seeds within the search space.

The invasive weed optimization algorithm's flowchart is depicted in Figure 11.1. A closer look at the architecture of the invasive weed optimization algorithm would reveal that it actually consists of three main stages that are the initiation, invasion, and termination stages. The invasion stage itself consists of three main strategies that are *reproduction*, *spatial dispersion*, and *competitive exclusion*. Using this structure, the algorithm would conduct a thorough search and locate what could be the optimum solution to the problem at hand. The following subsection will discuss each of these stages and their mathematical structures.

11.2.1 Initiation Stage

The invasive weed optimization algorithm is a population-based meta-heuristic optimization algorithm, and as such, it works with multiple search agents, here called weeds, that would enumerate through the search space. As we have seen, in an optimization problem with N decision variables, an N-dimension coordination system could be used to represent the search space. In this case, any point within the search space, say X, can be represented mathematically as a $1 \times N$ array as follows:

$$X = \left(x_1, x_2, x_3, \ldots, x_j, \ldots, x_N \right) \tag{11.1}$$

where X represents a weed plant in the search space of an optimization problem with N decision variables, and x_j represents the value associated with the jth decision variable.

The invasive weed optimization algorithm starts with randomly placing a series of weed plants within the feasible boundaries of the search space. This bundle of arrays, the invasive weed optimization algorithm's terminology referred to as the population, can be mathematically expressed as $M \times N$ matrix, where M denotes the number of weed plants or what is technically referred to as the population size. In such a structure, each row represents a single search agent. A population, denoted by *pop*, can be represented as follows:

$$
pop = \begin{bmatrix} X_1 \\ X_2 \\ \vdots \\ X_i \\ \vdots \\ X_M \end{bmatrix} = \begin{bmatrix} x_{1,1} & x_{1,2} & \cdots & x_{1,j} & \cdots & x_{1,N} \\ x_{2,1} & x_{1,2} & \cdots & x_{2,j} & \cdots & x_{2,N} \\ & & \vdots & & & \\ x_{i,1} & x_{i,2} & \cdots & x_{i,j} & \cdots & x_{i,N} \\ & & \vdots & & & \\ x_{M,1} & x_{M,2} & \cdots & x_{M,j} & \cdots & x_{M,N} \end{bmatrix} \tag{11.2}
$$

where X_i represents the ith weed plant in the population, and $x_{i,j}$ denotes the jth decision variable of the ith weed plant.

The initially generated population represents the first generation of weed plants. As we progress, the values stored in the *pop* matrix will be altered according to the computational structure of the invasive weed optimization algorithm's invasion stage. By the end of this iterative computation process, when the termination criterion is met, one or possibly multiple weed plants will converge to the optimum solution.

11.2.2 Invasion Stage

Inspired by the invasive behavior of weed plants as they get introduced into a new environment, the algorithm creates a new set of search agents by mimicking the weed's life cycle. The invasive stage itself consists of three main pillars that are reproduction, spatial dispersion of seeds, and competitive exclusion. These mechanisms are used as alteration tools to enhance the population's good qualities and eradicate undesirable properties.

The first step in the invasion stage is to emulate the reproduction of weed plants. The idea is that each virtual weed can produce a number of seeds proportional to its environmental condition, which in the mathematical context is expressed as the objective function associated with the said weed plant. Naturally, the more desirable the environmental conditions are, the more productive a weed plant could get.

The invasive weed optimization algorithm is designed so that the number of seeds for these weed plants always varies between two constant values, λ_{max} and λ_{min}, which denote the maximum and the minimum number of produced seeds, respectively. Note that these are user-defined parameters of the algorithm that show how many seeds would be produced by the best and worst weed plants in a given population set. Here, the best weed in the population would produce the maximum number of seeds, while the worst weed yields the minimum biomass product. As for the rest of the weeds in the population, their seed production is linearly proportional to their objective function values. This can be mathematically expressed as follows:

$$\mu_i = \frac{\lambda_{max} - \lambda_{min}}{Xbest - Xworst} \times f(X_i) \qquad \forall i \qquad (11.3)$$

in which μ_i denotes the number of seeds produced by the ith weed plant; $Xbest$ and $Xworst$ represent the best and worst weed plant in the current population, respectively; X_i denotes the ith weed plant; and lastly, $f()$ represents the objective function. Of course, it should go without saying that desirable conditions (i.e., identifying the best and worst weed plants in the population set) have a different interpretation for maximization and minimization problems. For instance, higher objective functions are considered more desirable in a maximization problem, while we are looking for lower values for the objective function in a minimization problem. The reproduction mechanism of the weed plants for both maximization and minimization problems is depicted in Figure 11.2.

The reproduction mechanism described above gives the invasive weed optimization algorithm a slight edge over a more traditional stochastic-based meta-heuristic algorithm. The significance of this mechanism is that while on paper, all agents are participating in composing the next generation of weeds, in practice, the more elite members of the said community have a more active role than the inferior weed plants. Thus, we are creating a better chance for the population to move toward what could potentially be the right direction. This was not always the case in a more conventional stochastic-based meta-heuristic algorithm, as in most cases, the architecture of these algorithms is based on selecting the participating member in a more random manner. While the algorithm was explicitly designed so that the better members would have a better chance of being selected in such a process, the

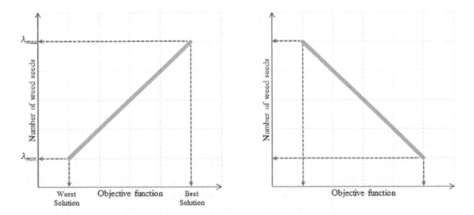

Figure 11.2 The relationship between the number of produced seeds for each weed plant and their objective functions in a (a) maximization, and (b) minimization problem.

bottom line was that there was always a chance, as slim as it was, that the worst members would be selected and divert the search from the more promising areas. Of course, abandoning this selection procedure could lead to premature convergence or cause the algorithm to be trapped in local optima. All in all, at least in theory, the reproduction mechanism embedded in this algorithm seems like an elegant solution to address these issues.

After reproduction, the produced seeds must be dispersed throughout the search space. The invasive weed optimization algorithm has a stochastic procedure to carry out this crucial task. The main feature of this mechanism is that it should be designed in a way that the dispersion radius would gradually decrease as the algorithm progresses in each given iteration. At the beginning of the search process, we want the search engine to resemble the exploration phase to cover a broader range. As we progress, we need to transit into the exploration phase as the search engine must focus more on the local search in the vicinity of the selected points in the space. Thus, to emulate this smooth transition between the exploration and exploitations phases, we need the dispersion radiuses to be more pronounced when the algorithm is just initiated compared to the final iterations of the algorithm.

To capture this from a computational standpoint, the algorithm uses a stochastic dispersion feature that places the seeds within the search space that follows a normal distribution with a varying standard division. Having a normal distribution to place the seeds randomly away from their origin points ensures that the seeds would be placed within the reasonable vicinity of the parent weed plant. And while it is still probable to explore further distances, it is far more likely to place the seed within a reasonable distance from the origin point. But even more critical is that we can control the traveled distance for these seeds by scaling the normal distribution function. The point is that by changing the value of the standard division in the normal distribution, one can easily scale the distribution of the data. Thus, the algorithm could gradually decrease the distribution scale to create a smooth transition from the exploration phase to the exploitation one. With that in mind, the algorithm ensures that the standard division gradually decreases between two certain constant values in a linear fashion. In case the termination criterion is based on limiting the number of iterations, this can be mathematically expressed as follows:

$$\sigma_t = \sigma_{final} + \left[\left(\frac{T-t}{T} \right)^\beta \times \left(\sigma_{initial} - \sigma_{final} \right) \right] \qquad \forall t \qquad (11.4)$$

where σ_t denotes the standard deviation value at the iteration t; T is the maximum number of iteration, a user-defined parameter for this algorithm; t represents the counter for the current iteration; β denotes the non-linear modulation index, which is another user-defined parameter that controls the pace in which the standard deviation decreases; $\sigma_{initial}$ and σ_{final} are the standard deviation values at the moment of imitation and the end of iterations, respectively. These two also happen to be some of the user-defined parameters of the invasive weed algorithm that should be selected based on the span of the search space. It should be noted that the above formula can be altered in a way that would also apply to other termination criteria. For instance, if the idea is to run the algorithm within a specific time span, the maximum time limit would replace the maximum number of iterations, while the current iteration count would be swapped with the current run time value.

Using the computed standard deviation, the algorithm would use the following probabilistic formulation to disperse the new seeds within the search space:

$$x_{i,j}^{new} = Norm\left(0, \sigma_t\right) + x_{i,j} \qquad \forall i, j \qquad\qquad (11.5)$$

where $Norm(0, \sigma_t)$ denotes a function that generates a random value that is drawn from a normal distribution function with the mean 0 and the standard deviation of σ_t; $x_{i,j}$ represents the current value for the jth decision variable of the ith weed plant; while $x_{i,j}^{new}$ denotes the new value for the jth decision variable of the ith weed plant.

Note that the same procedure would be used to place every seed for all weed plants in each given iteration.

As can be seen here, the algorithm would, in effect, create new tentative weeds through the reproduction and spatial dispersion of the seeds. While these weeds are placed randomly, they are formulated in a way that is more likely to be dispersed within the vicinity of their parent weed plants. As such, they are conducting a series of simultaneous local searches in each given iteration. The significant problem here is that as these new weeds get cumulatively staged on top of the original population, the number of generated weeds could easily get out of hand, given that the weed population is actually increasing exponentially. As such, the algorithm is equipped with a competitive exclusion mechanism to thin the population by removing the undesirable weed plants from the set. This process would emulate selective competition, where the fittest members of the species would survive the battle over limited natural resources. To implant this, the algorithm would check the population size at the end of each iteration. If the population happens to surpass a predefined constant value, here denoted by M_{max}, the algorithm will eliminate the exceeding weeds simply by removing the least fit values. The maximum number of weed plants is another user-defined parameter of the invasive weed optimization algorithm. Given that all weed plants are actively participating in the composition of the new generation, it is crucial to have this mechanism in place to eliminate the undesirable tentative solutions and keep things in check. More importantly, this would ensure that the computation task would not get out of hand, as without a competitive exclusion procedure, the number of computations in each iteration could quickly get overwhelmingly taxing within a few iterations.

11.2.3 Termination Stage

Based on the invasion stage described above, in each iteration, the invasive weed optimization algorithm would create a new set of seeds that would be dispersed across the search space and even continue to reproduce their own set of seeds. At any given time, if the population of the weed plants surpasses the defined limits, the algorithm would eliminate the exceeding undesirable plants. Through this stochastic process, the algorithm conducts a series of direct local searches, which gradually tend to be more focused as the algorithm progresses with each iteration until eventually locating what could potentially be the optimum solution.

Like other meta-heuristic algorithms, the sequence of operational structures of this algorithm needs to be executed iteratively until a certain termination criterion is met, at which point the execution of the algorithm would be terminated, and the best weed plant would be reported as the solution to the optimization problem. Note that without such a termination stage, the algorithm would potentially be executed in an infinite loop. The termination stage would, in effect, determine whether the algorithm has reached what could be the optimum solution.

As the invasive weed optimization algorithm is not equipped with an explicitly defined, unique termination mechanism, one could implement the commonly available options, most notably limiting the number of iterations, run time, or perhaps monitoring the improvement made to the best solution in consecutive iterations. Among these options, limiting the number of iterations is arguably the most cited mechanism to create a termination stage for the invasive weed optimization algorithm. The idea is that the process would be executed only for a specified number of times, a parameter known as the maximum iteration. In any case, it should be noted that the selection of a termination mechanism is also considered one of the algorithm's parameters. Bear in mind that in most cases, these termination mechanisms may require setting up additional parameters.

11.3 Parameter Selection and Fine-Tuning the Invasive Weed Optimization Algorithm

From the *no-free-lunch theorem*, one can conclude that fine-tuning an algorithm is essential to get the best performance out of a meta-heuristic algorithm. This would basically ensure that an algorithm is equipped to handle the unique characteristics of a given optimization problem. Of course, it is possible to use our intuition, experience, and default values suggested for an algorithm's parameters as a good starting point, one should bear in mind that fine-tuning these parameters is, more than anything, a trial-and-error process. Thus, while it is possible to get a good enough result by having an educated guess for setting the parameters of these algorithms, to get the best possible performance, it is necessary to go through this fine-tuning process.

In the case of the invasive weed optimization algorithm, these parameters are population size (M), which denote the maximum (λ_{max}) and minimum (λ_{min}) number of produced seeds, non-linear modulation index (β), standard deviation values at the moment of imitation ($\sigma_{initial}$) and the end of iterations (σ_{final}), the maximum number of weed plants (M_{max}), and of course, opting for the termination criterion, and all the parameters that are associated with these methods. For instance, if limiting the number of iterations has been selected as a termination criterion, the maximum iteration (T) is another parameter that needs to be defined by the user. While the invasive weed optimization algorithm is not the best option when it comes to the number of parameters per se, for the most part, the role of these parameters on

```
Begin
        Set the algorithm's parameter and input the data
        Generate the initial population with the population size M
        Let M_max denote the maximum permitted population size
        While the termination criterion is not met
                Evaluate the solutions
                For i in range 1 to M
                        Compute the number of seeds for the ith weed, denoted by μ_i
                        For j in range 1 to μ_i
                                Generate a new solution within the vicinity of the ith weed
                                Add the tentative solution to the tentative offspring set
                        Next j
                Next i
                Merge the current population with the tentative offspring set
                If the number of new solutions exceeds M_max
                        Trim the exceeding undesirable weeds from the merged population
                End if
        End while
        Report the best weed
End
```

Figure 11.3 Pseudocode for the invasive weed optimization algorithm.

the final outcome is easy to deduce intuitively, even for those with little experience with meta-heuristic optimization, which makes the fine-tuning process a bit more manageable than other population-based optimization algorithms we have seen thus far. That said, to get the absolute best results out of the invasive weed optimization algorithm, it is best to dabble with these algorithms first to gain some experience and inside knowledge about such parameters. By doing so, you could better understand how to fine-tune these parameters as your initial guesses and parameter selection strategies become more educated. The pseudocode for the invasive weed optimization algorithm is shown in Figure 11.3.

11.4 Python Codes

The code to implement the invasive weed optimization algorithm can be found below:

```python
import numpy as np

def init_generator(num_variables, pop_size, min_val, max_val):
    return np.random.uniform(min_val, max_val, (pop_size, num_
  variables))

def sorting_pop(pop, obj_func, minimizing):
    results = np.apply_along_axis(func1d=obj_func, axis=1, arr=pop)
    if minimizing:
        indeces = np.argsort(results)
    else:
        indeces = np.argsort(results)[::-1]
    return pop[indeces]
```

```
def seed_counts(sorted_pop, max_seed, min_seed, obj_func,
  minimizing):
    results = np.apply_along_axis(func1d=obj_func, axis=1,
                                  arr=sorted_pop)
    best, worst = results[0], results[-1]
    values=((max_seed-min_seed)/np.abs(best-worst))*np.
  abs(results-worst)
    if minimizing:
        values = values[::-1]
    values = np.round(values, 0)
    values[0], values[-1] = max_seed, min_seed
    values = values.astype(int)
    return values

def std_generator(std_init, std_final, beta, iteration):
    values = np.arange(iteration)
    a = ((iteration-values)**beta/iteration**beta)
    return a *(std_init-std_final) + std_final

def pop_control(sorted_pop, max_pop):
    if len(sorted_pop)>max_pop:
        return sorted_pop[:max_pop]
    else:
        return sorted_pop

def new_weed_generator(x, num_variables, seed_value, std):
    return np.random.normal(0,std,(seed_value, num_variables))+x

def invasive_weed_optimization(min_val, max_val, num_variables,
                               iteration, pop_size, obj_func,
                               max_seed, min_seed, std_init, std_final,
                               beta, max_pop=None,
                               minimizing = True, full_result=False):
    NFE_value = 0
    NFE = np.zeros(iteration)
    results = np.zeros(iteration)
    if max_pop==None:
        max_pop=pop_size
    pop = init_generator(num_variables, pop_size, min_val, max_val)
    NFE_value += pop_size
    pop = sorting_pop(pop, obj_func, minimizing)
    std_values = std_generator(std_init, std_final, beta, iteration)
    for i in range(iteration):
        seed_values = seed_counts(pop, max_seed, min_seed,
                                  obj_func, minimizing)
        NFE_value += np.sum(seed_values)
        std = std_values[i]
        for j in range(len(seed_values)):
            new_weeds = new_weed_generator(pop[j], num_variables,
                                           seed_values[j], std)
            pop = np.concatenate((pop, new_weeds), axis = 0)
```

```
        pop = sorting_pop(pop, obj_func, minimizing)
        pop = pop_control(pop, max_pop)
        results[i] = obj_func(pop[0])
        NFE[i] = NFE_value
if not full_result:
    return pop[0], obj_func(pop[0])
else:
    return pop[0], obj_func(pop[0]), results, NFE
```

11.5 Concluding Remarks

Inspired by the generic mechanism by which weed plants would aggressively inhabit a new environment, the invasive weed optimization algorithm still holds itself as a formidable meta-heuristic algorithm when it comes to tackling complex real-world problems. The nuances and the features used in this algorithm made it an interesting stochastic population-based meta-heuristic optimization method from a computational standpoint. The most notable feature in the structure of this algorithm is that it, in effect, advocated for a new way to interpret the application of swarm intelligence for optimization purposes. The idea is that within this structure, the search agents are not directly communicating with one another during the search to coordinate their movements, but rather their relative performance against the population set would dictate how involved they should be when composing the new generation. The other notable feature is that the architecture of this algorithm is, for the most part, simply a series of simultaneous stochastic-based local searches. The algorithm finds an elegant way to smoothly transit through these structures from the exploration phase to the exploitation phase. The last thing we could take from this algorithm is how it handles the population. Contrary to what we have seen in most population-based meta-heuristic algorithms, here we have a dynamic way to keep track of the search agents. As such, the number of search agents may increase as the algorithm progresses until the algorithm puts a cap on the number of newly generated search agents. The point is that in the exploration phase, where we are doing a wider search, we could settle for fewer search agents than in the exploitation phase, where we need to conduct a more refined local search. As such, opting for this dynamic population control would allow the algorithm to use the computational capacity of the system more efficiently. That said, the algorithm still suffers from the generic flaws commonly associated with meta-heuristic algorithms, such as being trapped in local optima or lacking any guarantee to reach the optimum solution. Overall, a relatively straightforward and easy-to-implement computational structure, lack of an overwhelming number of complicated parameters, and an efficient searching strategy make invasive weed optimization algorithms a viable option when tackling real-world complex optimization problems.

References

Ahmadi, M. & Mojallali, H. (2012). Chaotic invasive weed optimization algorithm with application to parameter estimation of chaotic systems. *Chaos, Solitons & Fractals*, 45(9–10), 1108–1120.

Ahmadi, P., Nazari, M.H., & Hosseinian, S.H. (2017). Optimal resources planning of residential complex energy system in a day-ahead market based on invasive weed optimization algorithm. *Engineering, Technology & Applied Science Research*, 7(5), 1934–1939.

Asgari, H. R., Bozorg Haddad, O., Pazoki, M., & Loáiciga, H.A. (2016). Weed optimization algorithm for optimal reservoir operation. *Journal of Irrigation and Drainage Engineering*, 142(2), 04015055.

Azizipour, M., Ghalenoei, V., Afshar, M.H., & Solis, S.S. (2016). Optimal operation of hydropower reservoir systems using weed optimization algorithm. *Water Resources Management*, 30(11), 3995–4009.

Barisal, A.K. & Prusty, R.C. (2015). Large scale economic dispatch of power systems using oppositional invasive weed optimization. *Applied Soft Computing*, 29, 122–137.

Bozorg-Haddad, O., Solgi, M., & Loáiciga, H.A. (2017). *Meta-heuristic and evolutionary algorithms for engineering optimization*. John Wiley & Sons. ISBN: 9781119386995

Du, K.L. & Swamy, M.N.S. (2016). *Search and optimization by metaheuristics: Techniques and algorithms by nature*. Springer International Publishing Switzerland. ISBN: 9783319411910

Hamedi, F., Bozorg-Haddad, O., Pazoki, M., Asgari, H.R., Parsa, M., & Loáiciga, H.A. (2016). Parameter estimation of extended non-linear Muskingum models with the weed optimization algorithm. *Journal of Irrigation and Drainage Engineering*, 142(12), 04016059.

Huang, L., Asteris, P.G., Koopialipoor, M., Armaghani, D.J., & Tahir, M.M. (2019). Invasive weed optimization technique-based ANN to the prediction of rock tensile strength. *Applied Sciences*, 9(24), 5372.

Jafari, S. & Montazeri-Gh, M. (2013). Invasive weed optimization for turbojet engine fuel controller gain tuning. *International Journal of Aerospace Sciences*, 2(3), 138–147.

Karimkashi, S. & Kishk, A.A. (2010). Invasive weed optimization and its features in electromagnetics. *IEEE Transactions on Antennas and Propagation*, 58(4), 1269–1278.

Liu, B. & Nie, L. (2021). Gradient based invasive weed optimization algorithm for the training of deep neural network. *Multimedia Tools and Applications*, 80, 22795–22819.

Mehrabian, A.R. & Lucas, C. (2006). A novel numerical optimization algorithm inspired from weed colonization. *Ecological Informatics*, 1(4), 355–366.

Misaghi, M. & Yaghoobi, M. (2019). Improved invasive weed optimization algorithm (IWO) based on chaos theory for optimal design of PID controller. *Journal of Computational Design and Engineering*, 6(3), 284–295.

Mohanty, P.K. & Parhi, D.R. (2014). A new efficient optimal path planner for mobile robot based on Invasive Weed Optimization algorithm. *Frontiers of Mechanical Engineering*, 9(4), 317–330.

Panda, M.R., Das, P.K., Dutta, S., & Pradhan, S.K. (2018). Optimal path planning for mobile robots using oppositional invasive weed optimization. *Computational Intelligence*, 34(4), 1072–1100.

Peng, S., Ouyang, A.J., & Zhang, J.J. (2015). An adaptive invasive weed optimization algorithm. *International Journal of Pattern Recognition and Artificial Intelligence*, 29(02), 1559004.

Rahmani, Y., Shahvari, Y., & Kia, F. (2021). Application of invasive weed optimization algorithm for optimizing the reloading pattern of a VVER-1000 reactor (in transient cycles). *Nuclear Engineering and Design*, 376, 111105.

Soulami, K.B., Ghribi, E., Saidi, M.N., Tamtaoui, A., & Kaabouch, N. (2019). Breast cancer: Segmentation of mammograms using invasive weed optimization and SUSAN algorithms. In *Proceeding of 2019 IEEE International Conference on Electro Information Technology*, Brookings, SD, USA.

Su, K., Ma, L., Guo, X., & Sun, Y. (2014). An efficient discrete invasive weed optimization algorithm for web services selection. *Journal of Software*, 9(3), 709–715.

Yazdani, M. & Ghodsi, R. (2017). Invasive weed optimization algorithm for minimizing total weighted earliness and tardiness penalties on a single machine under aging effect. *International Robotics and Automation Journal*, 2(1), 1–5.

12 Biogeography-Based Optimization Algorithm

Summary

Inspired by the biographic principle governing species' distribution patterns in an ecosystem, the biogeography-based optimization algorithm arguably presents one of the most interesting computational structures of meta-heuristic algorithms. One of the most intriguing features of this algorithm is that it finds an elegant way to bridge the gap between two different worlds of evolutionary computation and swarm intelligence. In this chapter, we will dig deep and explore the mechanisms used in this algorithm. We would get familiar with the biogeography-based optimization algorithm's terminology and see how one can implement this algorithm in the Python programming language. Finally, we will explore the potential merits and drawbacks of this algorithm.

12.1 Introduction

Biogeography is the discipline of biology that aims to explain the present and past distribution patterns of biological diversity and species distribution within an ecosystem. The earliest documented attempt to tackle these issues could be traced back to the 19th century when scientists such as the British naturalists Alfred Russel Wallace and even Charles Darwin put forth exciting theories about the subject (Simon, 2008). But the earliest implementation of biogeography was mainly limited to descriptive and historical examinations that lacked certain solid mathematical foundations. This was until the mid-1960s when McArthur and Wilson (1967) started to theorize a mathematical-based model for biogeographic studies. However, it should be noted that the primary focus of their study was to understand the extinction and immigration patterns of certain species through understanding their spatial distribution within neighboring islands. Their work actually pioneered a series of studies on this subject matter and turned biogeography into one of the mainstream disciplines of biology (Fernandez-Palacios et al., 2015).

Although the field of biogeography has undoubtedly come a long way from where it started, one cannot and should not forget the brilliancy of McArthur and Wilson's school of thought in describing the dynamic immigration patterns of species to understand the evolutionary process of a given ecosystem. Arguably, the

DOI: 10.1201/9781003424765-12

driving force of their model was to establish an equilibrium of some sort between the *emigration* and *immigration* patterns of certain species and use this to explain how certain species may thrive or get extinct within an ecosystem. To that end, a given ecosystem would get divided into a series of *islands*. Note that within this context, the term island is not being used in its literal form, but rather, it refers to any geography-isolated distinct area of the ecosystem. As such, from this point onward, it seems more logical to use the more genetic term *habitat* to refer to the same concept. The technical metric *habitat suitability index* is used to quantify the capacity of a given habitat to host diverse biological species (Wesche et al., 1987). Geographical areas with a higher habitat suitability index are more suitable for handling a more significant number of species than a habitat with a lower habitat suitability index. The bio-geographical factors contributing to the quality of the habitat suitability index are collectively referred to as *suitability index variables*.

Studies show that while species would be lured toward habitats with higher habitat suitability index at first, as the population of the inhabiting community rises, the habitat's appeal starts to plummet. This means that, in a way, population density can be seen as one of the suitability index variables. The point is that while a place with high habitat suitability index is certainly enriched enough to host a diverse and compact community, as the number of residences increases, the competition over limited available resources would eventually get intense enough to the point that certain members of the community could not fully benefit from this lucrative features. As such, after an enriched habitat surpasses its capacity to host new members, simply because of the emerging competition to access available resources, some of the inhabiting residences would tend to emigrate to nearby habitats as the current habitat does not seem enticing enough for such members. Note that this does not mean that certain species would fully migrate from one point to another, but rather it simply indicates that some of the members would depart the main group to inhabit a new place within the ecosystem. It is important to note that based on this paradigm, habitats with higher habitat suitability index tend to be more inflexible toward change. This is mainly because they have already reached a certain level of biodiversity and are impervious toward accepting new members. At the same time, it can be seen that such communities have more influence on the biodiversity of the nearby communities, as the members of the said habitats tend to migrate from their original habitat to nearby places.

By the same token, the immigrating members of the saturated communities may even settle in a habitat with lower habitat suitability index. The idea is that while the new community may not be as preferable as their original habitat from the suitability index variables standpoint, less competition for gaining access to natural resources in these new places may be able to compensate for that. The very same act of settlement of new species within a new habitat could raise the value of the habitat suitability index, as hosting new species would increase the biodiversity of a new habitat. However, if the new habitat does not have the necessary capacity to host these species, they will eventually go extinct. The most extreme case here would be a deserted habitat where the said place is not hosting any species. While hosting no species would undoubtedly mean that the emigration rate is literally zero, the lack of competition in such deserted places would certainly lure new species to inhabit such places. Again, it is crucial to notice that habitats with lower habitat suitability index tend to be more flexible toward accepting change. This is mainly because they are deserted habitats that can accept any new species. At the same time, it can be deduced that such communities do not influence the biodiversity of the nearby communities, as they have no members to emigrate to nearby habitats.

Of course, these extreme examples are the far side of the spectrum that describes the status of a given habitat. The idea is that based on this paradigm, a habitat could fall somewhere between these two cases, where there is an equilibrium between the immigration (i.e., settling in a place) and emigration (i.e., moving from a place) of species. These can be measured via the immigration rate (λ) and emigration rate (μ). Note that the habitat suitability index is proportional to the said habitat's biological diversity in this paradigm. This means that as the population increases from a deserted habitat into a fully saturated community, the dominating movements of species transit from immigration to emigration. The schematic theme of these movements is depicted in Figure 12.1. In this diagram, I denotes the maximum possible immigration rate in a given ecosystem, while E represents the maximum rate of emigration. S_{max} denotes the maximum number of species a habitat can support, while $S_{equilibrium}$ is the point at which the immigration and emigration rates are equal. Reaching the equilibrium point simply means that the population of the given habitat would remain the same so long as the equilibrium is held. It should be noted that while linear curves are used in this depiction, in reality, the shape of these immigration and emigration rate curve lines could be way more intricate. However, the same described principle would be applicable there as well.

Inspired by the basic principles of the mathematical representation of biogeographic models, Simon (2008) theorized a novel meta-heuristic optimization method called the biogeography-based optimization algorithm. This stochastic population-based optimization algorithm mimics the evolutionary biogeographic process governing species' movement within a geographic domain. What is interesting about this algorithm is that it utilizes the principles of evolutionary computing, as it is equipped with some of the more traditional operators of this optimization branch. The structure of this algorithm also resembles, to some degree, the algorithmic architecture of swarm intelligence-based algorithms. What distinguishes this algorithm from the algorithms we have explored thus far is that the biogeography-based optimization algorithm has found a unique and elegant way to bridge the gap between these two mainstream schools of thought in meta-heuristic optimization. While this is certainly not the most straightforward or efficient algorithm, it is a viable option to handle real-world optimization problems. More importantly, the nuances and novelties used in the structure of this algorithm make it quite interesting to study from a computational standpoint.

As stated biogeography-based optimization algorithm is a population-based meta-heuristic algorithm, and as such, it works with multiple search agents that enumerate through the search space. Based on the analogy of the biogeography-based optimization algorithm, each search agent is referred to as habitat. The

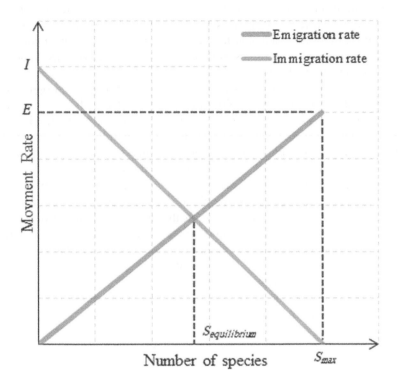

Figure 12.1 Species relocating rate based on the habitats' population.

decision variables for each of these habitats are called suitability index variables. Similar to what we have seen earlier, these variables would determine how desirable a habit could be. The desirability of these habitats is measured with the habitat suitability index. From an optimization standpoint, the said value is proportional to the computed objective functions of these habitats.

The algorithm initiates by randomly generating a set of habitats. Based on the relative desirability of these habitats, the immigration and emigration rates would be calculated for each of these habitats. Similar to what we have seen earlier, these two values would basically determine how these habitats are prone to change. The habitats with higher habitat suitability index have higher rates of emigration and lower rates of immigration. As such, they are less flexible in accepting any change, while they have the tendency to influence other habitats. By the same token, the habitats that are associated with lower habitat suitability index would be assigned with lower rates of emigration and higher rates of immigration. As such, they are more flexible when it comes to accepting any change, while they have little to no influence over the other habitats. Using these rates, the algorithm would stochastically alter the habitats' suitability index variables. The key feature in this process is that habitats with higher habitat suitability index are less likely to be affected, while the habitats with lower habitat suitability index would be more prone to adapt to any changes from the better habitats. This upgrading procedure would be basically repeated until a certain termination criterion is met, at which point the algorithm would be terminated, and the best-observed habitat would be returned as the solution to the optimization problem at hand.

As a meta-heuristic optimization approach, the biogeography-based optimization algorithm inherits the generic qualities that are associated with this branch of optimization. This means that, on the one hand, the algorithm is also not bound by problems often associated with high dimensionality, multimodality, epistasis, non-differentiability, and discontinuous search space imposed by constraints (Du & Swamy, 2016; Bozorg-Haddad et al., 2017). On the other hand, it is crucial to remember that, like other meta-heuristic algorithms, there can be no guarantee of reaching the optimum solution, or the algorithm could get trapped in a local optimum solution. On a more specific note, while this algorithm has few parameters, which makes the fine-tuning process much more manageable, like many other evolutionary computational-based algorithms, the role of these parameters is not intuitively clear, which could make things a bit challenging for those with less experienced with meta-heuristic optimization.

Over the years, many variants of the biogeography-based optimization algorithm have been proposed in the literature, some of which are oppositional biogeography-based optimization (Ergezer et al., 2009), perturb biogeography-based optimization (Li et al., 2011), linearized biogeography-based optimization (Simon et al., 2014), localized biogeography-based optimization (Zheng et al., 2014), and metropolis biogeography-based optimization (Al-Roomi & El-Hawary, 2016), to name a few. That said, the standard biogeography-based optimization algorithm is still considered a viable option to handle real-world optimization problems. In fact, the standard biogeography-based optimization algorithm has been successfully used for civil engineering (e.g., Aydogdu, 2017), computer science (e.g., Sangaiah et al., 2020), earthquake engineering (e.g., Zheng et al., 2014), electrochemical engineering (e.g., Mukherjee & Chakraborty, 2013), energy industry (e.g., Bhattacharya & Chattopadhyay, 2009), food industry (e.g., Zhang et al., 2016), hydrology (e.g., Roy et al., 2021), manufacturing management (e.g., Tamjidy et al., 2015), medical science (e.g., Yang et al., 2016; Zhang et al., 2019), mining engineering (e.g., Giri et al., 2021), railway engineering (e.g., Zheng et al., 2014), remote sensing (e.g., Gupta et al., 2011), and water resources planning and management (e.g., Bozorg-Haddad et al., 2016). In the following sections, we will explore the computational structure of the standard biogeography-based optimization algorithm.

12.2 Algorithmic Structure of the Biogeography-Based Optimization Algorithm

The biogeography-based optimizing algorithm is a stochastic population meta-heuristic algorithm, and as such, it uses multiple search agents that simultaneously enumerate throughout the search space to locate what could possibly be the optimum solution. What is interesting about the architecture of this algorithm is that it found a unique way to bridge the gap between evolutionary computation and swarm intelligence. As we see in this section, the main exploratory engine of this algorithm, which in this case is a pure random alteration in the components of the search agents, is mainly inspired by conventional evolutionary computation-based algorithms such as the genetic algorithm. While arguably, this is not the most comprehensive way to explore the search space, from the computational standpoint, it is undoubtedly one of the most straightforward ways to assemble and execute an exploratory operator. As for converging the solutions, the biogeography-based optimizing algorithm resorts to a stochastic mechanism that, for the most part, resembles the idea of swarm intelligence. The idea behind this is that the algorithm would use the relative performance of the search agates to determine how likely it is for a search agent to be altered or used as a reference point to alter the components of other search agents. As one can expect, the general principle here is that the agents that are performing better are less likely to be altered but rather be used as an aspiration to alter the components of the more inferior search agents stochastically. As can be seen, the algorithm does not necessarily use the obtained information to coordinate the agents' movement like what we have come to expect in more conventional swarm intelligence-based algorithms, but the idea of utilizing the performance of the entire group in a way that has been used here would certainly qualify this algorithm to be considered a swarm intelligence-based algorithms.

The biogeography-based optimizing algorithm's flowchart is depicted in Figure 12.2. A closer look at the architecture of the biogeography-based optimizing algorithm would reveal that it actually consists of three main stages that are the initiation, migration, and termination stages. The migration stage itself consists of two main strategies, which are *relocation* and *mutation*. Using this structure, the algorithm would conduct a thorough search and locate what could be the optimum solution to the problem at hand. The following subsection will discuss each of these stages and their mathematical structures.

12.2.1 Initiation Stage

The biogeography-based optimization algorithm is a population-based meta-heuristic optimization algorithm, and as such, it works with multiple search agents, here called habitats, that would enumerate through the search space. As we have seen, an N-dimension coordination system could represent the search space in an optimization problem with N decision variables, here called suitability index variables. In this case, any point within the search space, say X, can be represented mathematically as a $1{\times}N$ array as follows:

$$X = \left(x_1, x_2, x_3, \ldots, x_j, \ldots, x_N\right) \tag{12.1}$$

where X represents a habitat in the search space of an optimization problem with N decision variables; and x_j represents the value associated with the jth suitability index variable.

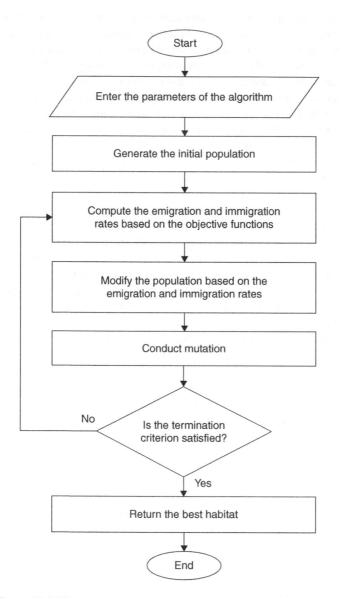

Figure 12.2 The computational flowchart of the biogeography-based optimization algorithm.

The biogeography-based optimization algorithm starts with randomly placing a series of habitats within the feasible boundaries of the search space. This bundle of arrays, which biogeography-based optimization algorithm's terminology is referred to as the population, can be mathematically expressed as $M \times N$ matrix, where M

denotes the number of habitats or what is technically referred to as the population size. In such a structure, each row represents a single search agent. A population, denoted by *pop*, can be represented as follows:

$$
pop = \begin{bmatrix} X_1 \\ X_2 \\ \vdots \\ X_i \\ \vdots \\ X_M \end{bmatrix} = \begin{bmatrix} x_{1,1} & x_{1,2} & \cdots & x_{1,j} & \cdots & x_{1,N} \\ x_{2,1} & x_{1,2} & \cdots & x_{2,j} & \cdots & x_{2,N} \\ & & \vdots & & & \\ x_{i,1} & x_{i,2} & \cdots & x_{i,j} & \cdots & x_{i,N} \\ & & \vdots & & & \\ x_{M,1} & x_{M,2} & \cdots & x_{M,j} & \cdots & x_{M,N} \end{bmatrix}
\tag{12.2}
$$

where X_i represents the ith habitat in the population, and $x_{i,j}$ denotes the jth suitability index variable of the ith habitat.

The initially generated population represents the first generation of habitats. As we progress, the values stored in the *pop* matrix will be altered according to the computational structure of the biogeography-based optimization algorithm's migration stage. By the end of this iterative computation process, when the termination criterion is met, one or possibly multiple habitats could converge to the optimum solution.

12.2.2 Migration Stage

By emulating the fundamental principle governing species' biographical migration within an ecosystem, the biogeography-based optimization algorithm attempts to alter the components of the search agents within each iteration. The migration stage consists of two main pillars: Relocation and mutation. These stochastic mechanisms are basically emplaced as alteration tools to enhance the population's good qualities and eradicate undesirable properties.

The first step in the migration stage is to emulate the relocation process of species between habitats. The basic idea is to create a stochastic mechanism that alters the suitability index variables based on the habitat's suitability index. The general principle of biogeography was that more desirable habitats have already reached saturation, given that they cannot host any new migrating species. This would, in turn, mean that such habitats are impervious to accepting any change from other habitats. However, these habitats would continue to influence the other habitats on the ecosystem, as the species inhabiting these communities would continually emigrate to nearby habitats. By the same token, habitats with lower habitats suitability index would be more flexible when it comes to accepting any outside change, given that their deserted status gives them the opportunity to host any immigrating community. In the meantime, these communities are less likely to influence the nearby communities, as their population is so dispersed that it is less likely for the inhabiting species to emigrate to nearby habitats.

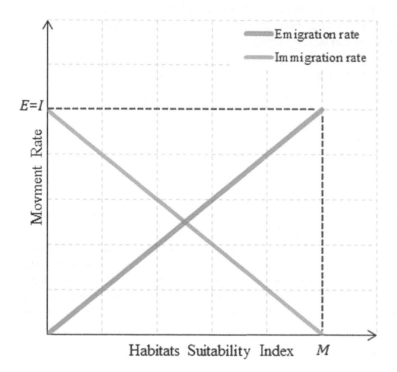

Figure 12.3 The relationship between movement ratings and habitats suitability index.

As we have seen, there is always a balance between the immigration and emigration rates for each habitat, which is proportional to the habitats suitability index. These rates would determine the probability by which a habitat can be influenced by other habitats or their alteration, respectively. To capture this notion mathematically, the biological-based optimization algorithm uses a specific version of the immigration and emigration curves we have seen earlier, where the maximum possible immigration rate (I) is equal to the maximum possible emigration rate (E). This variation of the relocation rate for species in an ecosystem is depicted in Figure 12.3. As seen in this variation, the maximum number of species a habitat can support (S_{max}) has been replaced by the population size (M). Note that, in this case, the following condition would always be held by all the habitats:

$$\lambda_i + \mu_i = E = I \qquad \forall i \tag{12.3}$$

where λ_i denotes the immigration rate of the ith habitat and μ_i represents the emigration rate of the ith habitat.

From a mathematical standpoint, repositioning the population starts with sorting the population based on their performance from the worst down to the best.

For instance, in a maximization problem, the habitats with the greater objective functions would assume a lower position than the habitats with lower objective function values. On the contrary, in a minimization problem, the habitats with the lowest objective function would assume the bottom position on the population, and the habitat with the greatest objective function value would be placed at the top of the list. Using this ascending ranking paradigm, the bottom habitat always has the best habitats suitability index, while the top habitat in the sorted list is the worst identified habitat in the ecosystem. As such, the algorithm could now safely implement the rankings of the sorted population to compute the immigration and emigration rates for each of these habitats. This could be mathematically expressed as follows:

$$\lambda_i = I \times \left(1 - \frac{i}{M}\right) \qquad \forall i \tag{12.4}$$

$$\mu_i = E \times \frac{i}{M} \qquad \forall i \tag{12.5}$$

in which i denotes the rank of the habitat in the sorted list. I and E are user-defined algorithm parameters that should be selected from 0 to 1.

Note that both immigration (λ_i) and emigration (μ_i) rates of the ith habitat also have a probabilistic interpretation that is used in the computational architecture of this algorithm. The idea is that, in practice, the computed values show how likely it is for the said habitat to experience immigration or emigration, respectively.

In order to implement this idea, for each of the habitats, a value denoted by *Rand1* would be selected randomly from the range [0, 1]. If the drawn value happens to be less than the computed immigration rate of the said habitat, that particular habitat will experience immigration. Note that it is more likely for the habitats with lower habitat suitability index values to be selected in this process. Being selected for immigration means that one of the variables of the selected habitat would be replaced with the corresponding variable of another habitat in the set. To select this value, the algorithm would follow a similar procedure. Again for each of the habitats in the set, another random value, here denoted by *Rand2*, would be drawn from the range [0, 1]. Note that here, habitats with higher habitats suitability index values are more likely to be selected in this process. Being selected for emigration indicates that one of the variables of the selected habitat would be used to replace the corresponding variable of the immigrating habitat. After both immigrating and emigrating habitats are selected using the above stochastic mechanism, one of the variables would be randomly selected to be replaced in the immigration habitat. This procedure of altering an immigrating habituate, denoted by $X_{immigration}$, using an emigrating habitat, denoted by $X_{emigration}$ can be mathematically expressed as follows:

$$X_{immigration} = \left(x_{k,1}, x_{k,2}, x_{k,3}, \ldots, x_{k,j}, \ldots, x_{k,N} \right) \tag{12.6}$$

$$X_{emigration} = \left(x_{e,1}, x_{e,2}, x_{e,3}, \ldots, x_{e,j}, \ldots, x_{e,N} \right) \tag{12.7}$$

$$r = IntRand\left(1, N\right) \tag{12.8}$$

$$X_{relocated} = \left(x_{k,1}, x_{k,2}, x_{k,3}, \ldots, x_{e,r}, \ldots, x_{k,N} \right) \tag{12.9}$$

in which k and e denote the indices of immigrating and emigrating habitats, respectively; $IntRand(1, N)$ denotes a function that would randomly select an integer from the range 1 to N; and $X_{relocated}$ represents the new relocated. Note that the relocated habitat would be replacing the immigrating one.

As can be seen that both immigrating and emigrating habitats would be selected stochastically. In this case, the emigrating habitat would be used to alter the properties of the immigrating habitat, as the former, in all likelihood, have more desirable properties than the latter habitat. Of course, being a stochastic process, there is no guarantee that the immigrating habitat is always inferior to the emigrating one. And even if the emigrating habitat is a more desirable solution than the immigrating one, there can be no guarantee that the above process is always an improving movement. But, what is interesting about this process is that it is structured in a way so that it can always preserve the best habitat as is. In computational science lingo, this feature is referred to as *elitism*, and the biogeography-based optimization algorithm is an elitist algorithm by nature. This means that the best solution in the set would always remain untampered. The reason is that the probability of a habitat being selected for alteration is computed via the immigration rate. As seen in Equation (12.4), the elite solution in each population that is the best habitat in that given set could never be selected as the probability of being tapped for this process is always equal to zero.

After the population has been altered through the repositioning of the habitats, the algorithm would continue the immigration stage by emulating a mutation procedure. The idea is that the habitats may experience some rapid alteration due to external stressors such as facing a pandemic in the inhabiting community or experiencing a natural hazard such as volcanic eruptions, earthquakes, hurricanes, tornados, tsunamis, wildfires, landslides, or floods, to name a few. These events can be seen as pure random alterations in the properties of a habitat. The idea behind this last alteration in the population set is to encourage exploration and prevent the algorithm from being trapped in a local optimum. To emulate this, the biogeography-based optimization algorithm uses a typical mutation algorithm, similar to what we usually associate with a conventional evolutionary-based algorithm such as the genetic algorithm. In this context, the mutation is simply a purely probabilistic-based operator that randomly alters a tentative solution's properties to increase diversity among the population set (Du & Swamy, 2016).

The main idea to emulate mutation is to randomly locate a set of variables that needs to be altered by replacing their assigned values. To do that, we first need to assign a proper value to the mutation probability parameter, denoted by P_m. This is another user-defined parameter of the biogeography-based optimization algorithm that should be selected from the range [0, 1]. Similar to what we have seen earlier, a randomly generated number within the range of 0–1 would be assigned to any decision variables in the population set. If the said value is below the threshold that is defined by the mutation probability parameter, the said value will be replaced by the mutation operator. The values selected for mutation would be replaced by randomly feasible values. In other words, the replacing value for the selected decision variables would be randomly generated within their feasible range's lower and upper limits. This gives the missing values an equal opportunity to be reintroduced into the mix. This can be mathematically expressed as follows:

$$x_{i,j}^{new} = L_j + Rand \times \left(U_j - L_j\right) \tag{12.10}$$

in which $x_{i,j}^{new}$ represents the jth decision variable of the ith habitat that has been selected randomly for mutation; $Rand$ is a random number that ranges from 0 to 1; U_j and L_j represent the upper and lower feasible boundaries of the jth decision variable, respectively. Note that in order to preserve its elitist characteristic, the biogeography-based optimization algorithm ensures that the best habitat would not be tampered with in this process as well. As such, the mutation probability for the elite habitat is assumed to be zero. This means that the best habitat would never be selected to go through mutation.

12.2.3 Termination Stage

Based on the migration stage described above, in each iteration, the biogeography-based optimization algorithm would stochastically alter some of the components of the habitats, thus updating the position of these search agents within the search space. Given that the less desirable habitats are less impervious to being altered in this process than those with greater habitats' suitability index, the process eventually creates a smooth change between the explorations to the exploitation phase, when the habitats start to have more or less similar properties.

Like other meta-heuristic algorithms, the sequence of operational structures of this algorithm needs to be executed iteratively until a certain termination criterion is met, at which point the execution of the algorithm would be terminated, and the best habitat recorded in the memory would be reported as the solution to the optimization problem. Note that without such a termination stage, the algorithm would potentially be executed in an infinite loop. The termination stage would, in effect, determine whether the algorithm has reached what could be the optimum solution.

As the biogeography-based optimization algorithm is not equipped with an explicitly defined, unique termination mechanism, one could implement the commonly available options, most notably limiting the number of iterations, run time, or perhaps monitoring the improvement made to the best solution in consecutive iterations. Among these options, limiting the number of iterations is arguably the most cited mechanism to create a termination stage for the biogeography-based optimization algorithm. The idea being the process would be executed only for a specified number of times, a parameter known as the maximum iteration. In any case, it should be noted that the selection of the termination mechanism is also considered one of the algorithm's parameters. Bear in mind that in most cases, these termination mechanisms may require setting up additional parameters.

12.3 Parameter Selection and Fine-Tuning the Biogeography-Based Optimization Algorithm

From the *no-free-lunch theorem*, one can conclude that fine-tuning an algorithm is essential to get the best performance out of a meta-heuristic algorithm. This would basically ensure that an algorithm is equipped to handle the unique characteristics

```
Begin
        Set the algorithm's parameter and input the data
        Let M denote the population size
        Let N denote the number of decision variables
        Generate the initial population
        While the termination criterion is not met
                Evaluate solutions
                Rank the habitats in ascending order so that the best habitat is placed last
                Compute the emigration (μ) and immigration (λ) probability for each habitat
                For i in range 1 to M-1
                        Let Rand1 denote a random number between 0 and 1
                        If Rand1 < λi
                                For r in range 1 to M
                                        Let Rand2 denote a random number between 0 and 1
                                        If Rand2 < μr
                                                Randomly select a decision variable index j
                                                xi,j would be replaced by xr,j
                                        End if
                                Next r
                        End if
                Next i
                Let Pm denote the mutation probability
                For i in range 1 to M-1
                        For j in range 1 to N
                                Let Rand3 denote a random number between 0 and 1
                                If Rand3 < Pm
                                        Replace xi,j with a randomly generated feasible value
                                End if
                        Next j
                Next i
        End while
        Report the best solution
End
```

Figure 12.4 Pseudocode for the biogeography-based optimization algorithm.

of a given optimization problem. Of course, it is possible to use our intuition, experience, and default values suggested for an algorithm's parameters as a good starting point, one should bear in mind that fine-tuning these parameters is more than anything a trial-and-error process. Thus, while it is possible to get a good enough result by having an educated guess for setting the parameters of these algorithms, to get the best possible performance, it is necessary to go through this fine-tuning process.

In the case of the biogeography-based optimization algorithm, these parameters are population size (M), maximum possible immigration rate (I), maximum possible emigration rate (E), mutation probability (P_m), and of course, opting for the termination criterion, and all the parameters that are associated with these methods. For instance, if limiting the number of iterations has been selected as a termination criterion, the maximum iteration (T) is another parameter that needs to be defined by the user. As can be seen here, while the biogeography-based optimization algorithm has very few parameters, like most evolutionary-based algorithms, for the most part, the role of these parameters on the final outcome is not easy to deduce intuitively, especially for those with little experience with meta-heuristic optimization, which makes the fine-tuning process a bit more challenging. That said, to get the best results out of the biogeography-based optimization algorithm, it is best to dabble with these algorithms first to gain some experience and knowledge about such parameters. By doing so, you could better understand how to fine-tune these parameters as your initial guesses and parameter selection strategies become more educated. The pseudocode for the biogeography-based optimization algorithm is shown in Figure 12.4.

12.4 Python Codes

The code to implement the biogeography-based optimization algorithm can be found below:

```python
import numpy as np

def init_genrator(pop_size, num_variables, min_val, max_val):
    return np.random.uniform(min_val, max_val, (pop_size, num_
  variables))

def sorting_pop(pop, obj_func, minimizing):
    results = np.apply_along_axis(func1d=obj_func, axis=1, arr=pop)
    indeces = np.argsort(results)
    if minimizing:
        indeces = indeces[::-1]
    return pop[indeces]
```

```
def emigeration_probabilty(pop_size, max_emig_prob):
    return max_emig_prob*(np.arange(pop_size)+1)/pop_size

def imigeration_probabilty(pop_size, max_imig_prob):
    return max_imig_prob*(1-((np.arange(pop_size)+1)/pop_size))

def imigeration(pop, pop_size, num_variables,
                imigration_probs, emigration_probs):
    rand_values = np.random.uniform(0,1, pop_size)
    indeces = np.argwhere(rand_values<imigration_probs)
    for i in indeces.flat:
        rands = np.random.uniform(0,1, pop_size)
        index = np.argwhere(rands<emigration_probs)
        for j in index.flat:
            position = np.random.randint(0, num_variables)
            pop[i][position] = pop[j][position]
    return pop

def mutate(pop, pop_size, num_variables, min_val, max_val, mut_prob):
    mutation_prob_values = np.random.uniform(0,1,pop_size)
    mutation_prob_values[-1] = 1
    mut_indeces = np.argwhere(mutation_prob_values<mut_prob)
    for mut_index in mut_indeces:
        position = np.random.randint(0,num_variables)
     pop[mut_index[0]][position]=np.random.uniform(min_val,max_val)
    return pop

def biogeography_based_optimization(pop_size, num_variables, min_val,
                                    max_val, max_imig_prob,
                                    max_emig_prob,
                                    mut_prob, iteration, obj_func,
                                    minimizing = True,
                                    full_result=False):
    NFE_value = 0
    NFE = np.zeros(iteration)
    results = np.zeros(iteration)
    pop = init_genrator(pop_size, num_variables, min_val, max_val)
    for i in range(iteration):
        pop = sorting_pop(pop, obj_func, minimizing)
        imigration_probs = imigeration_probabilty(pop_size,
                                                  max_imig_prob)
        emigration_probs = emigeration_probabilty(pop_size,
                                                  max_emig_prob)
        pop = imigeration(pop, pop_size, num_variables,
                          imigration_probs, emigration_probs)
        pop = mutate(pop, pop_size, num_variables, min_val,
                     max_val, mut_prob)
        NFE_value += pop_size
        NFE[i] = NFE_value
        results[i] = obj_func(pop[-1])
    if not full_result:
        return pop[-1], obj_func(pop[-1])
    else:
        return pop[-1], obj_func(pop[-1]), results, NFE
```

12.5 Concluding Remarks

Inspired by the biogeographic process's basic principles governing the migration patterns of different species as they move between different habitats in a given ecosystem, the biogeography-based optimization algorithm has theorized a novel approach to meta-heuristic optimization that walks the thin line between the realms of evolutionary computation and swarm intelligence. The nuances and the features used in this algorithm have made it an interesting stochastic population-based meta-heuristic optimization method from a computational standpoint.

As stated, one of the most prominent features that made this algorithm unique is that it bridges two worlds of evolutionary computation and swarm intelligence. This certainly makes the algorithm interesting from a computational standpoint. And when it comes to efficiency, the fine-tuned algorithm should be able to perform similarly to most meta-heuristic algorithms. That said, one could argue that the biogeography-based optimization algorithms did not necessarily inherit the best qualities of both worlds of evolutionary computation and swarm intelligence. For one, when it comes to swarm intelligence, the structure of the biogeography-based optimization algorithm does not permit the algorithm to fully utilize the idea of parallelized computation. In fact, embedding this idea may require tweaking some of the main computational architectures of the algorithm. And while this does not mean that the algorithm is inherently slow per se, it simply states that it is not utilizing its full potential. As for the evolutionary operator used in this algorithm to encourage the exploration phase during the search process, it is essential to note that while this is an efficient and straightforward enough approach, the main problem is that, like most evolutionary computations, the parameter selection for the operators may sometimes feel a bit counter-intuitive given that the impacts of some of the parameters used in such operators may not seem easy to grasp for those less experienced.

That said, this algorithm is not without its redeeming features. The most notable one would be that the algorithm is inherently equipped with an elitism mechanism that always preserves the best solution obtained thus far. This would ensure that we would not miss the best-observed solution throughout the search, which means that we are not required to append additional features to have this effect. Overall, a relatively straightforward and easy-to-implement computational structure, a lack of an overwhelming number of complicated parameters, and an efficient searching strategy make the biogeography-based optimization algorithm a viable option when tackling real-world complex optimization problems.

References

Al-Roomi, A.R. & El-Hawary, M.E. (2016). Metropolis biogeography-based optimization. *Information Sciences*, 360, 73–95.

Aydogdu, I. (2017). Cost optimization of reinforced concrete cantilever retaining walls under seismic loading using a biogeography-based optimization algorithm with Levy flights. *Engineering Optimization*, 49(3), 381–400.

Bhattacharya, A. & Chattopadhyay, P. K. (2009). Biogeography-based optimization for different economic load dispatch problems. *IEEE Transactions on Power Systems*, 25(2), 1064–1077.

Bozorg-Haddad, O.B., Hosseini-Moghari, S.M., & Loáiciga, H.A. (2016). Biogeography-based optimization algorithm for optimal operation of reservoir systems. *Journal of Water Resources Planning and Management*, 142(1), 04015034.

Bozorg-Haddad, O.B., Solgi, M., & Loáiciga, H.A. (2017). *Meta-heuristic and evolutionary algorithms for engineering optimization.* John Wiley & Sons. ISBN: 9781119386995

Du, K.L. & Swamy, M.N.S. (2016). *Search and optimization by metaheuristics: Techniques and algorithms inspired by nature.* Springer International Publishing Switzerland. ISBN: 9783319411910

Ergezer, M., Simon, D., & Du, D. (2009). Oppositional biogeography-based optimization. In *Proceeding of IEEE International Conference on Systems, Man and Cybernetics*, San Antonio, TX, USA.

Fernandez-Palacios, J.M., Kueffer, C., & Drake, D. (2015). A new golden era in island biogeography. *Frontiers of Biogeography*, 7(1), 14–20.

Giri, P.K., De, S.S., Dehuri, S., & Cho, S.B. (2021). Biogeography based optimization for mining rules to assess credit risk. *Intelligent Systems in Accounting, Finance and Management*, 28(1), 35–51.

Gupta, S., Arora, A., Panchal, V.K., & Goel, S. (2011). Extended biogeography based optimization for natural terrain feature classification from satellite remote sensing images. In *Proceeding of International Conference on Contemporary Computing*, Berlin, Heidelberg.

Li, X., Wang, J., Zhou, J., & Yin, M. (2011). A perturb biogeography based optimization with mutation for global numerical optimization. *Applied Mathematics and Computation*, 218(2), 598–609.

MacArthur, R.H. & Wilson, E.O. (1967). *The theory of island biogeography.* Princeton University Press, Princeton, NJ.

Mukherjee, R. & Chakraborty, S. (2013). Selection of the optimal electrochemical machining process parameters using biogeography-based optimization algorithm. *The International Journal of Advanced Manufacturing Technology*, 64(5), 781–791.

Roy, B., Singh, M.P., & Singh, A. (2021). A novel approach for rainfall-runoff modelling using a biogeography-based optimization technique. *International Journal of River Basin Management*, 19(1), 67–80.

Sangaiah, A.K., Bian, G.B., Bozorgi, S.M., Suraki, M.Y., Hosseinabadi, A.A.R., & Shareh, M.B. (2020). A novel quality-of-service-aware web services composition using biogeography-based optimization algorithm. *Soft Computing*, 24(11), 8125–8137.

Simon, D. (2008). Biogeography-based optimization. *IEEE Transactions on Evolutionary Computation*, 12(6), 702–713.

Simon, D., Omran, M.G., & Clerc, M. (2014). Linearized biogeography-based optimization with re-initialization and local search. *Information Sciences*, 267, 140–157.

Tamjidy, M., Paslar, S., Baharudin, B.H.T., Hong, T.S., & Ariffin, M.K.A. (2015). Biogeography based optimization (BBO) algorithm to minimise non-productive time during hole-making process. *International Journal of Production Research*, 53(6), 1880–1894.

Wesche, T.A., Goertler, C.M., & Hubert, W.A. (1987). Modified habitat suitability index model for brown trout in southeastern Wyoming. *North American Journal of Fisheries Management*, 7(2), 232–237.

Yang, G., Zhang, Y., Yang, J., Ji, G., Dong, Z., Wang, S., ... & Wang, Q. (2016). Automated classification of brain images using wavelet-energy and biogeography-based optimization. *Multimedia Tools and Applications*, 75(23), 15601–15617.

Zhang, X., Wang, D., & Chen, H. (2019). Improved biogeography-based optimization algorithm and its application to clustering optimization and medical image segmentation. *IEEE Access*, 7, 28810–28825.

Zhang, Y., Phillips, P., Wang, S., Ji, G., Yang, J., & Wu, J. (2016). Fruit classification by biogeography-based optimization and feedforward neural network. *Expert Systems*, 33(3), 239–253.

Zheng, Y.J., Ling, H.F., Shi, H.H., Chen, H.S., & Chen, S.Y. (2014). Emergency railway wagon scheduling by hybrid biogeography-based optimization. *Computers & Operations Research*, 43, 1–8.

13 Cuckoo Search Algorithm

Summary

Inspired by the parasitism reproduction strategy of some cuckoo bird species, the cuckoo search algorithm holds itself as a practical yet straightforward swarm intelligence-based meta-heuristic optimization algorithm that can be quite efficient when it comes to handling real-world complex optimization problems. The cuckoo search algorithm is known to have a simple computational architecture that has few parameters in its structure, which is not a common sight in meta-heuristic algorithms. In this chapter, we will dig deep and explore the mechanisms used in this algorithm. We would get familiar with the cuckoo search algorithm's terminology and see how one can implement this algorithm in the Python programming language. Finally, we will explore the potential merits and drawbacks of this algorithm.

13.1 Introduction

For those who live in Britain, hearing the male cuckoo bird in late March and early April is a sign that spring has arrived. The sound made by these species is as charming as the bird themselves to spectators. However, the hard truth is that most cuckoo birds, these seemingly innocent creatures, are known to have one of the most aggressive, gruesome reproduction strategies.

Cuckoo birds are arguably one of the most notorious brood parasites (Krüger & Davies, 2002). Most cuckoo birds do not bother themselves with creating nests. So, when it is time for the female cuckoo to lay an egg, it must locate a suitable communal nest to hide its own egg among the eggs other species lay (Payne, 1977). The main idea is that the parent bird is basically outsourcing its parental duties, including handling the incubation and feeding of the chicklings to other birds by tricking them into thinking that they are looking after one of their own. However, finding a foster parent and, more importantly, deceiving them is not an easy task by any means. Different species of cuckoo birds have evolved so that they can mimic the recognizable patterns of their host's eggs to a remarkable degree (Lahti, 2006). This adaptive feature is crucial for this species' survival because the foster

DOI: 10.1201/9781003424765-13

parent would immediately abandon any unfamiliar egg. As such, the female cuckoo patrols the habitat for a suitable nest where the foster parents have already laid their own eggs. As the foster parent is out looking to forage for food, the cuckoo bird would sneak in and lay their own egg. Timing is everything in this process because the eggs should hatch roughly simultaneously. If the foster parent does not get suspicious, it will continue to guard the nest, where the vicious parasitism cycle of the cuckoo's life truly begins.

If the cuckoo's egg goes undetected by the foster, one of two things would usually occur. The most common thing is that the cuckoo's egg would hatch first. The blind, featherless cuckoo chickling instinctively knows that it needs to remove any competition to survive (Mikulica et al., 2017). As such, it starts to thrust the foster parent's eggs out of the nest, sometimes right in front of their eyes. Alternatively, there are times when other eggs hatch earlier than the imposter bird's egg. Again, as soon as the cuckoo egg hatches, the chickling would trust any other offspring out of the nest. It is crucial that this process happens the way it does, given that the foster parent would usually abandon a nest with one egg, but instinctively it would not be able to abandon a newborn baby. From this point onward, the foster parent's life is totally dominated by this imposter. The foster parent would continue feeding the cuckoo bird until it is mature enough to live independently.

Inspired by the brooding parasitism behavior of cuckoo birds, Yang and Deb (2009) theorized a novel meta-heuristic algorithm called the cuckoo search algorithm. Built upon the principles of swarm intelligence, the cuckoo search algorithm could be described as a stochastic population-based algorithm that emulates the brood parasitism behavior of cuckoo birds. In this analogy, each search agent acts as an individual cuckoo bird as they impose on other species to survive. Being based on the principles of the direct search would simply mean that, like other meta-heuristic algorithms, the cuckoo search algorithm is also not bound by problems that are often associated with high dimensionality, multimodality, epistasis, non-differentiability, and discontinuous search space imposed by constraints (Yang, 2010; Du & Swamy, 2016). Perhaps the most distinctive feature of this algorithm is that it has a fairly straightforward computational structure with very few parameters. These features make this algorithm a formidable choice to tackle optimization problems, even for those with little experience with meta-heuristic optimization algorithms.

The cuckoo search algorithm is based on an abstract interpretation of the cuckoo birds' brood parasitism behavior. The idea is that the algorithm would initiate its search by randomly generating a series of search agents, here referred to as cuckoos, within the feasible area of the search space. These agents would patrol the search space using the *Lévy flight*. A Lévy flight is a special form of the *random walk* where the step lengths are drawn from a Lévy distribution (Yaghoubzadeh-Bavandpour et al., 2022). It was argued that from a mathematical point of view, such step sizes are much more efficient for exploration moves than purely random walks (Yang, 2010). The cuckoos would find a nest within this search space to lay their eggs. There is a possibility here that the foster parent would detect these eggs as not one of its own, in which case the eggs would be removed from the process. Otherwise, the eggs would hatch into new cuckoo birds, which continue the searching process. This whole process would be repeated until a termination criterion is met, at which point the best point encountered thus far would be returned as the solution to the optimization problem at hand.

Over the years, many variants of the cuckoo search algorithm have been proposed in the literature, some of which are self-adaptive parameter cuckoo search algorithms (Li & Yin, 2015), chaotic cuckoo search (e.g., Wang et al., 2016), island-based cuckoo search (Abed-alguni, 2019), and reinforced cuckoo search algorithm (Thirugnanasambandam et al., 2019), to name a few. That said, the standard cuckoo search algorithm is still considered a viable option to handle real-world optimization problems. In fact, the standard cuckoo search algorithm has been successfully used for agricultural engineering (e.g., Chrouta et al., 2018), civil engineering (e.g., Tran-Ngoc et al., 2019), climatology (e.g., Garg, 2015; Wang et al., 2015), computer science (e.g., Cai et al., 2020), energy industry (e.g., Basu & Chowdhury, 2013), image processing (e.g., Chakraborty et al., 2017), manufacturing management (e.g., Yildiz, 2013), medical science (e.g., Mohapatra et al., 2015), mining engineering (e.g., Fouladgar et al., 2017), plant pathology (e.g., Cristin et al., 2020), robotic science (e.g., Mohanty & Parhi, 2016), structural engineering (e.g., Kaveh & Bakhshpoori, 2013), and water resources planning and management (e.g., Ming et al., 2015). In the following sections, we will explore the computational structure of the cuckoo search algorithm.

13.2 Algorithmic Structure of the Cuckoo Search Algorithm

In order to create a functional algorithm, we first need to extract and idealize some of governing principles of the cuckoo birds' parasitism brooding. The main idea here is to establish a simplified and abstract mathematical representation of the said procedure that would be utilized later on to assemble the search engine of the cuckoo search algorithm. These simplifying assumptions are as follows:

I   Each cuckoo bird would lay one egg at a time in a randomly selected nest within the feasible part of the search space.
II In each iteration, only the best eggs would be able to hatch. This notion depicts the competitive nature of the cuckoo bird's life cycle.
III To account for the process in which the foster parent would detect the cuckoo bird's egg, we assume the process is basically a random procedure in which there is a fixed possibility for the host to detect the parasite bird.
IV For simplicity's sake, we assume there is no distinction between the egg, nest, and the mature cuckoo bird.

Based on these principles, the cuckoo search algorithm creates a search engine that coordinates the movement of the search agents and navigates them to the point that could potentially be the optimum solution. It should be noted that the self-organizing nature of the search agents' movement in this algorithm qualifies it as a swarm intelligence-based optimization algorithm.

The cuckoo search algorithm's flowchart is depicted in Figure 13.1. A closer look at the architecture of the cuckoo search algorithm would reveal that it actually consists of three main stages that are the initiation, brood parasitism, and termination stages. Using this structure, the algorithm would conduct a thorough search and locate what could be the optimum solution to the problem at hand.

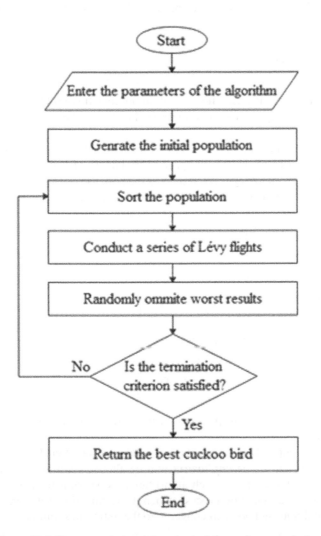

Figure 13.1 The computational flowchart of the cuckoo search algorithm.

The following subsection will discuss each of these stages and their mathematical structures.

13.2.1 Initiation Stage

The cuckoo search algorithm is a population-based meta-heuristic optimization algorithm, and as such, it works with multiple search agents, here called cuckoos, that would enumerate through the search space. As we have seen, in an optimization problem with N decision variables, an N-dimension coordination system could be used to represent the search space. In this case, any point within the search space, say X, can be represented mathematically as a $1 \times N$ array as follows:

$$X = \left(x_1, x_2, x_3, \ldots, x_j, \ldots, x_N \right) \tag{13.1}$$

where X represents a cuckoo in the search space of an optimization problem with N decision variables, and x_j represents the value associated with the jth decision variable.

The cuckoo search algorithm starts by randomly placing a series of organisms within the feasible boundaries of the search space. This bundle of arrays, which cuckoo search algorithm's terminology is referred to as the population, can be mathematically expressed as $M \times N$ matrix, where M denotes the number of particles or what is technically referred to as the population size. In such a structure, each row represents a single search agent. A population, denoted by *pop*, can be represented as follows:

$$pop = \begin{bmatrix} X_1 \\ X_2 \\ \vdots \\ X_i \\ \vdots \\ X_M \end{bmatrix} = \begin{bmatrix} x_{1,1} & x_{1,2} & \cdots & x_{1,j} & \cdots & x_{1,N} \\ x_{2,1} & x_{1,2} & \cdots & x_{2,j} & \cdots & x_{2,N} \\ & & \vdots & & & \\ x_{i,1} & x_{i,2} & \cdots & x_{i,j} & \cdots & x_{i,N} \\ & & \vdots & & & \\ x_{M,1} & x_{M,2} & \cdots & x_{M,j} & \cdots & x_{M,N} \end{bmatrix} \tag{13.2}$$

where X_i represents the ith cuckoo in the population, and $x_{i,j}$ denotes the jth decision variable of the ith cuckoo.

The initially generated population represents the initial position of cuckoos. As we progress, the values stored in the *pop* matrix will be altered according to the computational structure of the cuckoo search algorithm's brood parasitism stage. By the end of this iterative computation process, when the termination criterion is met, one or possibly multiple cuckoos will converge to the optimum solution.

13.2.2 Brood Parasitism Stage

By mimicking the life cycle of cuckoo birds, in this stage, the algorithm would tend to update the population set. The brood parasitism stage consists of two pillars that simulate the birds flying within the search space and the procedure of egg-laying. These mechanisms are used as alteration tools to create a new set of tentative solutions. Using the greedy strategy, the algorithm would compare these solutions with their original counterparts and replace them if they perform better.

The first step here in this stage is for the algorithm to readjust the position of each cuckoo in the set based on the Lévy flight. This idea is introduced to emulate the flying patterns of the cuckoo birds as they scavenge the area for a suitable nest to lay their eggs. In fact, numerous studies have pointed out that the flying patterns of many bird species and insects, say the fruit flies, resemble the characteristics of a Lévy flight (Yang, 2010). As such, this procedure has been used as the core idea for repositioning the search agent within the search space.

A Lévy flight is a special form of the *random walk* where the step lengths are drawn from a Lévy distribution. The proponents of implementing the Lévy flight instead of the usual random walk argue that this is an efficient way to explore the search space as the step lengths drawn via this procedure are more heavy-tailed (Yang, 2010). As a result, it is more likely to generate longer steps and cover more ground with this mechanism. A special case of the Lévy distribution can be defined as follows:

$$L(s, \gamma, \mu) \approx \sqrt{\frac{\gamma}{2\pi}} \frac{1}{s^{3/2}} \tag{13.3}$$

in which s, γ, and μ are all parameters of the Lévy distribution.

From an implementation standpoint, however, generating a random step from the Lévy distribution consists of two main steps that are opting for the direction and selecting the step size. The *Mantegna algorithm* is one of the most straightforward ways to generate a step size that bears the stochastic properties of a Lévy distribution (Mantegna, 1994). Based on this algorithm, the step size, here denoted by l, can be computed as follows:

$$l = \frac{u}{|v|^{1/\beta}} \tag{13.4}$$

in which β is a parameter of the Lévy distribution, and both u and v are randomly generated values that are drawn from normal distributions that are defined as follows:

$$u \sim Norm(0, \sigma_u^2) \tag{13.5}$$

$$v \sim Norm\left(0, \sigma_v^2\right) \tag{13.6}$$

$$\sigma_u = \left[\frac{\Gamma\left(1+\beta\right) \times \sin\left(\frac{\pi\beta}{2}\right)}{\Gamma\left(\frac{1+\beta}{2}\right) \times \beta \times 2^{\frac{\beta-1}{2}}}\right]^{1/\beta} \tag{13.7}$$

$$\sigma_v = 1 \tag{13.8}$$

where *Norm* denotes the normal distribution.

Using the principles of the Lévy flight, the cuckoo search algorithm would reposition the cuckoo algorithms within the search space as follows:

$$x'_{i,j} = x_{i,j} + \alpha \times l \tag{13.9}$$

in which α denotes the step size scale, which is one of the user-defined parameters of the algorithm.

After the tentative solutions are constructed using the above-described procedure, the algorithm randomly selects a cuckoo in the population set and compares it against the tentative solutions. For instance, imagine that for the ith cuckoo, the algorithm has randomly picked the rth cuckoo, denoted by X_r, for comparison purposes. After the ith cuckoo is altered using the Lévy flight operation, the generated tentative solution would be compared against the rth selected cuckoo. If the tentative solution is deemed more suitable, the algorithm will swap the rth solution with the tentative solution. It goes without saying that making such a comparison has a different interpretation for maximization and minimization problems. For instance, higher objective functions are considered more desirable in a maximization problem, while we are looking for lower values for the objective function in a minimization problem.

Aside from making such comparisons, the algorithm would conduct a stochastic procedure for each population set member, allowing the algorithm to drop the worst member of the population set and replace it with a new search agent. The idea here is to emulate that in nature, cuckoos that are not skilled enough to deceive the foster parent may have their eggs abandoned by the host birds. To implement this idea, after each comparison, the algorithm would generate a purely random number, denoted by *Rand*, from the range 0 to 1. The algorithm then compares the said value against a predefined constant value called the detection probability, denoted by P_a. The detection probability is another user-defined parameter of the cuckoo search algorithm. Suppose the generated value is below the said threshold. In that case, the algorithm will continue by replacing the worst member of the population set with a tentative solution generated by applying the Lévy flight operation on the worst cuckoo. It should be noted that while this parameter, like any parameter in a meta-heuristic optimization algorithm, needs to be fine-tuned for a given problem, the value 0.25 suggested for this parameter is a good starting point (Yang, 2010).

13.2.3 Termination Stage

Based on the above procedure, the algorithm would go over the agents in the populations set one by one to imitate how each cuckoo would lay its egg within the search space. What is important here is that the new tentative solution generated in this step would only be accepted if it can outperform a randomly selected member of the population set. In a way, the algorithm implements the basic idea of a greedy strategy to ensure that only improving moves can take effect. There is also the possibility of rejecting the worst solution in the set and replacing it with a newly generated solution. Like other meta-heuristic algorithms, the sequence of operational structures of this algorithm needs to be executed iteratively until a certain termination criterion is met, at which point the execution of the algorithm would be terminated, and the best cuckoo recorded in the memory would be reported as the solution to the optimization problem. Note that without such a termination stage, the algorithm would potentially be executed in an infinite loop. The termination stage would, in effect, determine whether the algorithm has reached what could be the optimum solution.

As the cuckoo search algorithm is not equipped with an explicitly defined, unique termination mechanism, one could implement the commonly available options, most notably limiting the number of iterations, run time, or perhaps monitoring the improvement made to the best solution in consecutive iterations. Among these options, limiting the number of iterations is arguably the most cited mechanism to create a termination stage for the cuckoo search algorithm. The idea being the process would be executed only for a specified number of times, a parameter known as the maximum iteration. In any case, it should be noted that the selection of the termination mechanism is also considered one of the algorithm's parameters. Bear in mind that in most cases, these termination mechanisms may require setting up additional parameters.

13.3 Parameter Selection and Fine-Tuning the Cuckoo Search Algorithm

From the *no-free-lunch theorem*, one can conclude that fine-tuning an algorithm is essential to get the best performance out of a meta-heuristic algorithm. This would basically ensure that an algorithm is equipped to handle the unique characteristics of a given optimization problem. Of course, it is possible to use our intuition, experience, and default values suggested for an algorithm's parameters as a good starting point, one should bear in mind that fine-tuning these parameters is more than anything a trial-and-error process. Thus, while it is possible to get a good enough result by having an educated guess for setting the parameters of these algorithms, to get the best possible performance, it is necessary to go through this fine-tuning process.

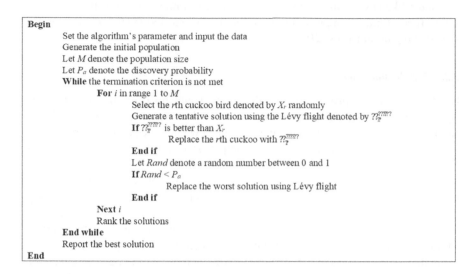

Figure 13.2 Pseudocode for the cuckoo search algorithm.

In the case of the cuckoo search algorithm has specifically designed to have nearly the minimum number of parameters, which are the population size (*M*), step size scale (*α*), the detection probability (*P_a*), and of course, opting for the termination criterion, and all the parameters that are associated with these methods. For instance, if limiting the number of iterations has been selected as a termination criterion, the maximum iteration (*T*) is another parameter that needs to be defined by the user. As can be seen here, not only does the cuckoo search algorithm have very few parameters to begin with, but for the most part, the role of these parameters on the final outcome is easy to deduce intuitively. This makes the fine-tuning process less challenging for those less experienced. That said, to get the absolute best results out of the cuckoo search algorithm, it is best to dabble with these algorithms first to gain some experience and inside knowledge about such parameters. By doing so, you could better understand how to fine-tune these parameters as your initial guesses and parameter selection strategies become more educated. The pseudocode for the cuckoo search algorithm is shown in Figure 13.2.

13.4 Python Codes

The code to implement the cuckoo search algorithm can be found below:

```python
import numpy as np

from scipy import special

def init_generator(pop_size, num_variables, min_val, max_val):
    return np.random.uniform(min_val, max_val,(pop_size, num_variables))
```

```python
def levy_flight(x, beta, step_size):
    sigma_v = 1
    num_u = (special.gamma(beta+1)*np.sin(np.pi*beta*.5)
    denom_u = special.gamma((1+beta)*.5)*beta*2**((beta-1)*.5))
    sigma_u = (num_u)/( denom_u)**(1/beta)
    v = np.random.normal(0,sigma_v,size=x.shape)
    u = np.random.normal(0,sigma_u,size=x.shape)
    s = u/(np.abs(v)**(1/beta))
    return x + step_size*s

def sorting_pop(pop, obj_func, minimizing):
    results = np.apply_along_axis(obj_func, 1, pop)
    indeces = np.argsort(results)
    if not minimizing:
        indeces = indeces[::-1]
    return pop[indeces]

def evaluate(a, b, obj_func, minimizing):
    if minimizing:
        if obj_func(b)<obj_func(a):
            return True
        else:
            return False
    else:
        if obj_func(b)>obj_func(a):
            return True
        else:
            return False

def cuckoo_search_algorithm(pop_size, num_variables, min_val, max_val,
                            step_size, beta,
                            prob_a, obj_func, iteration, minimizing,
                            full_result=False):
    NFE = np.zeros(iteration)
    NFE_value = 0
    results = np.zeros(iteration)
    pop = init_generator(pop_size, num_variables, min_val, max_val)
    pop = sorting_pop(pop, obj_func, minimizing)
    for i in range(iteration):
        for j in range(pop_size):
            m, n = np.random.randint(1, pop_size, 2)
            new_cuckoo = levy_flight(pop[m], beta, step_size)
            if evaluate(pop[n], new_cuckoo, obj_func, minimizing):
                pop[n] = new_cuckoo
                pop = sorting_pop(pop, obj_func, minimizing)
            rand = np.random.uniform()
            if rand<prob_a:
                pop[-1] = levy_flight(pop[-1], beta, step_size)
                pop = sorting_pop(pop, obj_func, minimizing)
            NFE_value += 2
        NFE[i] = NFE_value
        results[i] = obj_func(pop[0])
    if not full_result:
        return pop[0], obj_func(pop[0])
    else:
        return pop[0], obj_func(pop[0]), results, NFE
```

13.5 Concluding Remarks

Inspired by the life cycle of cuckoo birds, the search algorithm has established itself as a formidable meta-heuristic algorithm for tackling complex real-world problems. A straightforward computational structure is perhaps one of the notable features of this algorithm. The unique structure of this algorithm enables it to take the idea of elitism to the next level. Not only any move here is based on the greedy stagey, which basically forbids any non-improving move, but the algorithm could also randomly drop the worst solutions in each iteration. By doing so, the algorithm slowly pushed the entire population forward to have better properties. The other notable feature is embedding the Lévy flight into the algorithm's search agent repositioning mechanism, arguably more efficient than a basic random walk procedure, which is the main option for most meta-heuristic optimization algorithms. But perhaps the most notable feature of this algorithm is the limited number of parameters used in this algorithm's computational structure. While this certainly facilitates the calibration and implementation of the algorithm in general, there is a slight drawback that comes with the territory. The main idea of having parameters in an algorithm is to control its performance and fine-tune the procedure's features to make it more suitable for a specific problem. Lacking these parameters means that the user may be unable to adjust the algorithm's search mechanism. For instance, in this case, the user cannot control the transition between the exploration and exploitation phase of the search process. This does not indicate any exploration or exploitation phase during the search procedure. In fact, as the algorithm initiates, the movement is more exploratory, but as the search progresses and the solutions get closer together, the search is more in line with the idea of the exploitation phase. But the important thing to note here is that the user has no control over the pace of this transition, given that there are practically no parameters to control the searching properties of the algorithm. As such, the only way to avoid being trapped in local optima or premature convergence is to increase the population size or the number of iterations. Even with these changes, we cannot guarantee we will avoid these pitfalls altogether. With that said, all in all, from a practical standpoint, the cuckoo search algorithm can certainly be considered a viable option when it comes to tackling real-world complex optimization problems.

References

Abed-alguni, B.H. (2019). Island-based cuckoo search with highly disruptive polynomial mutation. *International Journal of Artificial Intelligence*, 17(1), 57–82.

Basu, M. & Chowdhury, A. (2013). Cuckoo search algorithm for economic dispatch. *Energy*, 60, 99–108.

Cai, X., Niu, Y., Geng, S., Zhang, J., Cui, Z., Li, J., & Chen, J. (2020). An under-sampled software defect prediction method based on hybrid multi-objective cuckoo search. *Concurrency and Computation: Practice and Experience*, 32(5), e5478.

Chakraborty, S., Chatterjee, S., Dey, N., Ashour, A. S., Ashour, A. S., Shi, F., & Mali, K. (2017). Modified cuckoo search algorithm in microscopic image segmentation of hippocampus. *Microscopy Research and Technique*, 80(10), 1051–1072.

Chrouta, J., Chakchouk, W., Zaafouri, A., & Jemli, M. (2018). Modeling and control of an irrigation station process using heterogeneous cuckoo search algorithm and fuzzy logic controller. *IEEE Transactions on Industry Applications*, 55(1), 976–990.

Cristin, R., Kumar, B. S., Priya, C., & Karthick, K. (2020). Deep neural network based Rider-Cuckoo Search Algorithm for plant disease detection. *Artificial Intelligence Review*, 53(7), 4993–5018.

Du, K.L. & Swamy, M.N.S. (2016). *Search and optimization by metaheuristics: Techniques and algorithms inspired by nature.* Springer International Publishing Switzerland. ISBN: 9783319411910

Fouladgar, N., Hasanipanah, M., & Amnieh, H.B. (2017). Application of cuckoo search algorithm to estimate peak particle velocity in mine blasting. Engineering with *Computers*, 33(2), 181–189.

Garg, H. (2015). An approach for solving constrained reliability-redundancy allocation problems using cuckoo search algorithm. *Beni-Suef University Journal of Basic and Applied Sciences*, 4(1), 14–25.

Kaveh, A. & Bakhshpoori, T. (2013). Optimum design of steel frames using Cuckoo Search algorithm with Lévy flights. *The Structural Design of Tall and Special Buildings*, 22(13), 1023–1036.

Krüger, O. & Davies, N.B. (2002). The evolution of cuckoo parasitism: A comparative analysis. *Proceedings of the Royal Society of London. Series B: Biological Sciences*, 269(1489), 375–381.

Lahti, D.C. (2006). Persistence of egg recognition in the absence of cuckoo brood parasitism: Pattern and mechanism. *Evolution*, 60(1), 157–168.

Li, X., & Yin, M. (2015). Modified cuckoo search algorithm with self-adaptive parameter method. *Information Sciences*, 298, 80–97.

Mantegna, R.N. (1994). Fast, accurate algorithm for numerical simulation of Levy stable stochastic processes. *Physical Review E*, 49(5), 4677.

Mikulica, O., Grim, T., Schulze-Hagen, K., Stokke, B.G., & Davies, N. (2017). *The cuckoo: The uninvited guest.* Wild Nature Press.

Ming, B., Chang, J.X., Huang, Q., Wang, Y.M., & Huang, S.Z. (2015). Optimal operation of multi-reservoir system based-on cuckoo search algorithm. *Water Resources Management*, 29(15), 5671–5687.

Mohanty, P.K. & Parhi, D.R. (2016). Optimal path planning for a mobile robot using cuckoo search algorithm. *Journal of Experimental & Theoretical Artificial Intelligence*, 28(1–2), 35–52.

Mohapatra, P., Chakravarty, S., & Dash, P.K. (2015). An improved cuckoo search based extreme learning machine for medical data classification. *Swarm and Evolutionary Computation*, 24, 25–49.

Payne, R.B. (1977). The ecology of brood parasitism in birds. *Annual Review of Ecology and Systematics*, 8(1), 1–28.

Thirugnanasambandam, K., Prakash, S., Subramanian, V., Pothula, S., & Thirumal, V. (2019). Reinforced cuckoo search algorithm-based multimodal optimization. *Applied Intelligence*, 49(6), 2059–2083.

Tran-Ngoc, H., Khatir, S., De Roeck, G., Bui-Tien, T., & Wahab, M.A. (2019). An efficient artificial neural network for damage detection in bridges and beam-like structures by improving training parameters using cuckoo search algorithm. *Engineering Structures*, 199, 109637.

Wang, G.G., Deb, S., Gandomi, A.H., Zhang, Z., & Alavi, A.H. (2016). Chaotic cuckoo search. *Soft Computing*, 20(9), 3349–3362.

Wang, J., Jiang, H., Wu, Y., & Dong, Y. (2015). Forecasting solar radiation using an optimized hybrid model by Cuckoo Search algorithm. *Energy*, 81, 627–644.

Yaghoubzadeh-Bavandpour, A., Bozorg-Haddad, O., Zolghadr-Asli, B., & Gandomi, A.H. (2022). Improving approaches for meta-heuristic algorithms: A brief overview. In Bozorg-Haddad, O., Zolghadr-Asli, B. eds. *Computational intelligence for water and environmental sciences*, Springer Singapore, 35–61.

Yang, X.S. (2010). *Nature-inspired metaheuristic algorithms*. Luniver Press. ISBN: 9781905986286

Yang, X.S. & Deb, S. (2009). Cuckoo search via Lévy flights. In *Proceeding of the World Congress on Nature & Biologically Inspired Computing*, Coimbatore, India.

Yildiz, A.R. (2013). Cuckoo search algorithm for the selection of optimal machining parameters in milling operations. *The International Journal of Advanced Manufacturing Technology*, 64(1), 55–61.

14 Firefly Algorithm

Summary

Inspired by the mating ritual of fireflies emitting bright flashing lights to attract their partners, the firefly algorithm holds itself as a formidable swarm intelligence-based meta-heuristic optimization algorithm that can be quite efficient when it comes to handling real-world complex optimization problems. In this chapter, we will dig deep and explore the mechanisms used in this algorithm. We would get familiar with the firefly algorithm's terminology and see how one can implement this algorithm in the Python programming language. Finally, we will explore the potential merits and drawbacks of this algorithm.

14.1 Introduction

If you are lucky enough to encounter the fireflies' synchronized glowing as they illuminate the dark summer nights, you would know that it is, bar far, one of the most mesmerizing and magical sights in nature. It is such a surreal phenomenon for those who observe it firsthand to the extent that it feels like something that came straight out of a fairytale. However, the fireflies' harmonized flashing of lights in the dark surrounding of a hot summer night is much more than an aesthetically pleasing phenomenon, for it is an evolutionary feature to help this species adapt to their environment.

The *Lampyridae*, which are commonly known as fireflies, glowworms, or lightning bugs, are a family of insects in the beetle order. It is estimated that there are over 2,000 discovered species of fireflies worldwide (Ohba, 2004). Although these species are known for their conspicuous use of *bioluminescence* during twilight, the functionality of these often short and rhythmic flashings is still unclear (Buck & Buck, 1966). In fact, it is believed that different firefly species may use this signaling mechanism for different purposes. For instance, female *Photuris* fireflies would use this as a prey mechanism by adjusting their flashing signals to mimic the glowing patterns of other firefly species to lure their males as potential prays (Lloyd, 1975). That said, most agree with a reasonable degree of certainty that there are at least three recognized functionalities for these communicating patterns,

DOI: 10.1201/9781003424765-14

namely, attracting mating partners, luring potential prey, and discouraging possible predators (Yang, 2010; Du & Swamy, 2016).

As stated, one of the well-recognized applications of illuminating these rhythmic flashes for communication is enticing potential mating partners. Generically speaking, the mating ritual of fireflies hinges on the partners identifying and, in turn, imitating their partners' flashing patterns. Different fireflies have unique and distinctive illuminating patterns. The mating partners use these flashing signals to ensure finding a suitable match. Usually, the female tends to identify and, in turn, duplicate the male distinct flashing patterns. Naturally, for this interaction to happen, the emitted lighting patterns of fireflies should first be able to reach the other fireflies. Thus, the more pronounced these lighting patterns are, the better the odds of finding a mating partner. However, it should be noted that the distance between the source of these flashing lights and the spectator also plays a crucial role here as well. The reason is the environment would absorb that emitted light, and as such, the intensity of received light would decrease proportionally to the distance it travels. Actually, it is known that the light intensity decreases proportionate to the inversed squared distance between the light sources and the observer, given that the light follows the *inverse square law*. This means that fireflies would only be able to identify and attract other fireflies that are located within their effective line of sight, that is, in most cases, somewhere around several hundred meters at night (Yang, 2010).

One could see that the core governing principle of the above-described phenomenon, for all intent and purposes, resembles the idea of swarm intelligence. The point is that fireflies are using this simple flashing mechanism to communicate with one another and form a decentralized and self-organizing community. As such, Yang (2009) theorized a novel meta-heuristic optimization algorithm called the firefly algorithm that uses the idealized interpretation of the mating ritual of fireflies as a source of inspiration to create the search engine of the said algorithm. The firefly algorithm is a stochastic population-based algorithm that is based on the fundamental principles of swarm intelligence-based computation. A closer inspection of the algorithmic architecture of the firefly algorithm would reveal that this algorithm is, in fact, a mixture of a generalized variation of the particle swarm optimization algorithm and pure random search strategy. That said, not only are there distinctive features in the searching mechanism of this algorithm that make it quite interesting, but from a computational standpoint, this is a formidable option for handling real-world optimization problems.

As stated, the firefly algorithm is a population-based meta-heuristic algorithm, and as such, it works with multiple search agents that simultaneously enumerate through the search space. Based on the analogy of the firefly algorithm, each of these search agents is referred to as fireflies. In order to imitate the mating ritual of the fireflies, the algorithm needs to emulate the flashing patterns of these virtual fireflies. To that end, the brightness of the emitted flashing patterns is assumed to be proportional to the objective function of each of these fireflies. The algorithm would keep track of the fireflies' movements as they coordinate their fly based on their mating partners.

To implement this idea in practice, the algorithm initiates its search by creating a random set of fireflies. Each of these fireflies would look at the other members of the community as a potential mating partner and adjust their position within the search space based on the brightness of their partners and the distance between them. Basically, fireflies that emit brighter flashing signals would appear more appealing to others, and as such, the other fireflies would coordinate their movement to be placed near such fireflies. Of course, as the distance between two given fireflies increases, such movements seem less enticing for the fireflies as the light that reaches the spectator would decrease proportioned to the inverse secured distance between these mating partners. Using this simple mechanism, the algorithm would adjust the position of all the fireflies within the search space. This whole process would then be repeated until a certain termination criterion is met, at which point the best-observed firefly would be returned as the solution to the optimization problem at hand.

Given that the firefly algorithm is a moderated interoperation of the generalized particle swarm optimization and pure random search, it was argued that it is often possible to create a more efficient searching strategy out of the firefly algorithm than both these approaches by fine-tuning its parameter (Yang, 2010). More importantly, the unique structure of the firefly algorithm allows us to fully take advantage of the parallelized computation as each of these search agents seemingly works almost independently as they enumerate through the search space (Yang, 2010). But, one of the notable drawbacks of this algorithm also roots in this unique architecture. The main problem here is the repositioning of each firefly needs to be coordinated against all the fireflies in the community. This means the algorithm would need to review the entire population for every firefly in each iteration. It is worth noting that the readjustment formulation used in this algorithm is fairly simple to the point that such a computer is not particularly challenging for modern computers when the number of fireflies in the community is reasonable. The other cited drawback of this algorithm is based on the notion that, unlike the particle swarm optimization algorithm, the firefly movement in this algorithm is solely based on locally identified optima and not the global point (Du & Swamy, 2016). While this feature is indeed quite helpful when it comes to multimodal optimization, where identifying local optimum solutions is also one of the search engine's priorities (Yang, 2010), it certainly limits the exploration rate of the algorithm when it comes to identifying the global optimum solution.

Over the years, many variants of the firefly algorithm were proposed in the literature, some of which are the chaos-based firefly algorithm (Gandomi et al., 2013), adaptive control parameters firefly algorithm (Wang et al., 2017), mutated firefly algorithm (Arora et al., 2014), and wise step strategy firefly algorithm (Yu et al., 2014), to name a few. That said, the standard firefly algorithm is still considered a viable option to handle real-world optimization problems. In fact, the standard firefly algorithm has been successfully used for climatology (e.g., Santos et al., 2013), data science (e.g., Alweshah, 2014; Wang et al., 2018), energy industry (e.g., dos Santos Coelho & Mariani, 2013), environmental engineering (e.g., Riahi-Madvar et al., 2020), financial management (e.g., Kazem et al., 2013), hydro-climatology (e.g., Tao et al., 2018), image processing (e.g., Horng, 2012), manufacturing management (e.g., Ahmed et al., 2019), hydrology (e.g., Hamadneh, 2020), materials engineering (e.g., Avendaño-Franco & Romero, 2016), mechanical engineering (e.g., Baykasoğlu & Ozsoydan, 2015), medical science (e.g., Peng et al., 2021), project management (e.g., Wang et al., 2020), structural engineering (e.g., Talatahari et al., 2014), and water resources planning and management (e.g., Garousi-Nejad et al., 2016; Wang et al., 2017; Arefinia et al., 2022). In the following sections, we will explore the computational structure of the standard firefly algorithm.

14.2 Algorithmic Structure of the Firefly Algorithm

In order to create a functional algorithm, we first need to extract and idealize some of governing principles of the fireflies mating ritual. The main idea here is to establish a simplified and abstract mathematical representation of the said procedure that would be utilized later on to assemble the search engine of the firefly algorithm. These simplifying assumptions are as follows:

I   The intensity of the emitted flashes, or fireflies' brightness, is proportionate to the desirability of the corresponding search agent within the search space. The better the objective function of the said agent, the brighter the corresponding firefly would be.

II All represented virtual fireflies in this algorithm are assumed to be *unisex*. As such, it is possible for all the fireflies to engage in the mating ritual with one another.

III The intensity of the emitted brightness is the factor that determines the movement of the fireflies. The general rule of thumb here is that the fireflies would be attracted to the brighter firefly. As such, when paired together, the less glowing firefly moves toward the one that emits brighter flashes.

Based on these principles, the firefly algorithm creates a search engine that coordinates the movement of the search agents and navigates them to the point that could potentially be the optimum solution. It should be noted that the self-organizing nature of the search agents' movement in this algorithm qualifies it to be categorized as a swarm intelligence-based optimization algorithm.

The firefly algorithm's flowchart is depicted in Figure 14.1. A closer look at the architecture of the firefly algorithm would reveal that it actually consists of three main stages that are the initiation, mating, and termination stages. Using this structure, the algorithm would conduct a thorough search and locate what could be the optimum solution to the problem at hand. The following subsection will discuss each of these stages and their mathematical structures.

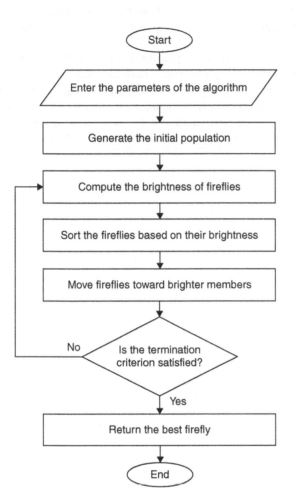

Figure 14.1 The computational flowchart of the firefly algorithm

14.2.1 Initiation Stage

The firefly algorithm is a population-based meta-heuristic optimization algorithm, and as such, it works with multiple search agents, here called fireflies, that are simultaneously enumerated through the search space. As we have seen, in an optimization problem with N decision variables, an N-dimension coordination system could be used to represent the search space. In this case, any point within the search space, say X, can be represented mathematically as a $1 \times N$ array as follows:

$$X = \left(x_1, x_2, x_3, \ldots, x_j, \ldots, x_N\right) \tag{14.1}$$

where X represents a firefly in the search space of an optimization problem with N decision variables, and x_j represents the value associated with the jth decision variable.

The firefly algorithm starts by randomly placing a series of fireflies within the feasible boundaries of the search space. This bundle of arrays, which in the firefly algorithm's terminology is referred to as the population, can be mathematically expressed as $M \times N$ matrix, where M denotes the number of fireflies or what is technically referred to as the population size. In such a structure, each row represents a single search agent. A population, denoted by *pop*, can be represented as follows:

$$pop = \begin{bmatrix} X_1 \\ X_2 \\ \vdots \\ X_i \\ \vdots \\ X_M \end{bmatrix} = \begin{bmatrix} x_{1,1} & x_{1,2} & \cdots & x_{1,j} & \cdots & x_{1,N} \\ x_{2,1} & x_{1,2} & \cdots & x_{2,j} & \cdots & x_{2,N} \\ & & \vdots & & & \\ x_{i,1} & x_{i,2} & \cdots & x_{i,j} & \cdots & x_{i,N} \\ & & \vdots & & & \\ x_{M,1} & x_{M,2} & \cdots & x_{M,j} & \cdots & x_{M,N} \end{bmatrix} \tag{14.2}$$

where X_i represents the ith firefly in the population, and $x_{i,j}$ denotes the jth decision variable of the ith firefly.

The initially generated population represents the first generation of fireflies. As we progress, the values stored in the *pop* matrix will be altered according to the computational structure of the firefly algorithm's mating stage. By the end of this iterative computation process, when the termination criterion is met, one or possibly multiple fireflies could converge to the optimum solution.

14.2.2 Mating Stage

By emulating the fundamental principle that governs the mating process of fireflies, the firefly algorithm attempts to alter the components of the search agents within each iteration. Here, like many other population-based metaheuristic algorithms, each individual search agent is, in fact, conducting a local parallelized search. But what distinguishes this algorithm from other population-based algorithms we have got to know thus far is that here each community member that is relatively more desirable than its selected mating partner would actively participate in the repositioning procedure of the said agent. While this is indeed a more computationally taxing strategy than the conventional approach we have seen thus far, it can potentially create a more thorough local search than such algorithms.

The algorithm needs to be structured to have an inner nested loop to implement this idea. The idea is that for each firefly, the algorithm would need to go over the entire population set as a potential mating partner for the selected firefly. Only those fireflies considered more desirable than the selected agent could participate in the realignment of their position. To determine whether a firefly is more desirable or what is referred to in firefly algorithm terminology as a brighter firefly, the algorithm would simply compare the objective function values of these search agents. The better the objective function of these search agents, the more desirable the corresponding firefly would be. Needless to say, the interpretation of being better depends on the type of optimization problem we are dealing with. For the maximization problem, for instance, the higher values for the objective function are considered better. Conversely, in a minimization problem, the lower values for objective function are deemed more desirable.

Suppose a firefly is deemed to be appealing for the mating process. In that case, the other partner would coordinate its move to reposition itself closer to the selected mating partner. The effort that each firefly is willing to put into such a realignment is proportionate to the distance between these two potential partners. Basically, the closer these two fireflies are, the more willing the partner would be to calibrate its current position. The firefly algorithm uses the idea of *mutual attractiveness* to quantify this willingness to coordinate one's current position.

Mutual attractiveness, or attractiveness for short, is simply a quantitative measure to show how willing a firefly is to realign itself with another firefly that was deemed suitable earlier on by the algorithm as a potential mating partner. Based on the idealized mating principles we have established previously, this value should be proportionate to the distance between these two paired fireflies. The general rule of thumb was that the less distance between these two partners, the more willing the said firefly would be to readjust its position to get closer to its partner. A potentially viable approach to calculate this attractiveness is as follows:

$$\beta(r) = \beta_0 e^{-\gamma r^m} \qquad m \geq 1 \tag{14.3}$$

in which r denotes the distance between the paired fireflies; $\beta(r)$ represents the attractiveness of a potential mating partner that is placed at an r distance of the given firefly; γ denotes the light absorption coefficient, which happens to one of the user-defined parameters of the firefly algorithm; β_0 is the attractiveness of two mating partners that are near it each other at a distance equal to zero, which is another parameter of the algorithm; and finally m is the exponent coefficient, which is again a user-defined parameter for the firefly algorithm.

Alternatively, the following formulation could be used instead to compute the attractiveness values:

$$\beta(r) = \frac{\beta_0}{1 + \gamma r^m} \qquad m \geq 1 \tag{14.4}$$

For simplicity's sake, it is recommended that the m coefficient be set to 2, while β_0 could be assumed to be 1 (Yang, 2009). As for the light absorption coefficient, while the feasible boundary is [0, ∞], in most practical cases, the fine-tuned value for this parameter typically varies within the range of 0.1–10 (Yang, 2010).

But as can be seen in both Equations (14.3) and (14.4), in order to compute the attractiveness measure, one must first measure the distance between two potential mating partners. Often, measuring the *Euclidean distance* between two search agents is the most conventional approach to computing this distance. For two given fireflies X_l and X_k, this distance can be mathematically measured as follows:

$$r_{l,k} = \|X_l - X_k\| = \sqrt{\sum_{j=1}^{N} \left(x_{l,j} - x_{r,j}\right)^2} \qquad \forall l,k \tag{14.5}$$

where $r_{l,k}$ denotes the Euclidean distance between the *l*th and *k*th fireflies.

After the algorithm quantifies the tension between two given fireflies, it needs to realign the position of the firefly with the lower brightness with respect to the better-performing firefly. The position of the moving firefly can be computed as follows:

$$X_k^{new} = X_k + [r_{l,k} \times (X_l - X_k)] + [\alpha \times (Rand - 0.5)] \tag{14.6}$$

in which X_k denotes the firefly with lower brightness; X_l denotes the better-performing firefly; *Rand* is a randomly generated value within the range 0–1; X_k^{new} denotes the newly aligned position of the *k*th firefly; and finally, α represents the randomization parameter, another parameter of this algorithm that ranges between 0 and 1. Note that the above formula only should be applied to those fireflies that are matched with a firefly that can outperform them. If the firefly is matched with an inferior firefly, the algorithm will not realign its position, given that it is already in a more suitable position as is. If both matched fireflies emit brightness at the same intensity, you could skip any realignment or make random alterations to the firefly's position (Yang, 2010).

14.2.3 Termination Stage

Based on the migration stage described above, in each iteration, the firefly algorithm would stochastically update the position of each individual firefly in the population set. Given the elitist nature of the standard variation of the firefly algorithm, the most desirable solution obtained thus far would remain intact throughout each. Using these series of parallelized local searches, the algorithm would eventually converge to what could be the optimum solution.

Like other meta-heuristic algorithms, the sequence of operational structures of this algorithm needs to be executed iteratively until a certain termination criterion is met, at which point the execution of the algorithm would be terminated, and the best firefly recorded in the memory would be reported as the solution to the optimization problem. Note that without such a termination stage, the algorithm would

potentially be executed in an infinite loop. The termination stage would, in effect, determine whether the algorithm has reached what could be the optimum solution.

As the firefly algorithm is not equipped with an explicitly defined, unique termination mechanism, one could implement the commonly available options, most notably limiting the number of iterations, run time, or perhaps monitoring the improvement made to the best solution in consecutive iterations. Among these options, limiting the number of iterations is arguably the most cited mechanism to create a termination stage for the firefly algorithm. The idea being the process would be executed only for a specified number of times, a parameter known as the maximum iteration. In any case, it should be noted that the selection of the termination mechanism is also considered one of the algorithm's parameters. Bear in mind that in most cases, these termination mechanisms may require setting up additional parameters.

14.3 Parameter Selection and Fine-Tuning the Firefly Algorithm

From the *no-free-lunch theorem*, one can conclude that fine-tuning an algorithm is essential to get the best performance out of a meta-heuristic algorithm. This would basically ensure that an algorithm is equipped to handle the unique characteristics of a given optimization problem. Of course, it is possible to use our intuition, experience, and default values suggested for an algorithm's parameters as a good starting point, one should bear in mind that fine-tuning these parameters is more than anything a trial-and-error process. Thus, while it is possible to get a good enough result by having an educated guess for setting the parameters of these algorithms, to get the best possible performance, it is necessary to go through this fine-tuning process.

In the case of the firefly algorithm, these parameters are population size (M), light absorption coefficient (γ), the attractiveness of two mating partners that are near it each other at a distance equal to zero (β_0), exponent coefficient (m), randomization parameter (α), and of courses, opting for the termination criterion, and all the parameters that are associated with these methods. For instance, if limiting the number of iterations has been selected as a termination criterion, the maximum iteration (T) is another parameter that needs to be defined by the user. As can be seen here, not only the firefly algorithm is riddled with a relatively high number of parameters, but for the most part, the role of these parameters on the final outcome is not easy to deduce intuitively, especially for those with little experience with meta-heuristic optimization, which makes the fine-tuning process a bit more challenging. That said, to get the absolute best results out of the firefly algorithm, it is best to dabble with these algorithms first to gain some experience and inside knowledge about such parameters. By doing so, you could better understand how to fine-tune these parameters as your initial guesses and parameter selection strategies become more educated. The pseudocode for the firefly algorithm is shown in Figure 14.2.

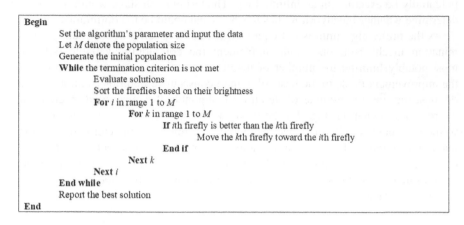

```
Begin
        Set the algorithm's parameter and input the data
        Let M denote the population size
        Generate the initial population
        While the termination criterion is not met
                Evaluate solutions
                Sort the fireflies based on their brightness
                For i in range 1 to M
                        For k in range 1 to M
                                If ith firefly is better than the kth firefly
                                        Move the kth firefly toward the ith firefly
                                End if
                        Next k
                Next i
        End while
        Report the best solution
End
```

Figure 14.2 Pseudocode for the firefly algorithm.

14.4 Python Codes

The code to implement the firefly algorithm can be found below:

```python
import numpy as np

def init_generator(pop_size, num_variables, min_val, max_val):
    return np.random.uniform(min_val, max_val, (pop_size, num_variables))

def distance(x, y):
    return np.sqrt(np.sum((x-y)**2))

def attractivness(dist_val, beta_0, gamma, m):
    return beta_0*np.exp(-gamma*dist_val**m)

def evaluate(a, b, obj_func, minimizing):
    if minimizing:
        if obj_func(a)<obj_func(b):
            return True
        else:
            return False
    else:
        if obj_func(a)>obj_func(b):
            return True
        else:
            return False
def improve_position(a, b, obj_func, beta_0, gamma, m, minimizing):
    if evaluate(b, a, obj_func, minimizing):
        dist_val = distance(a, b)
        rand_value = np.random.normal()*(np.random.uniform()-.5)
        return a+attractivness(dist_val,beta_0,gamma,m)*(b-a)+rand_value
    else:
        return a
```

```
def sorting_pop(pop, obj_func, minimizing):
    results = np.apply_along_axis(obj_func, 1, pop)
    indeces = np.argsort(results)
    if not minimizing:
        indeces = indeces[::-1]
    return pop[indeces]
def firefly_algorithm(pop_size, num_variables, min_val, max_val,
                      beta_0, gamma, m, iteration, obj_func,
                      minimizing=True, full_result=False):
    NFE = np.zeros(iteration)
    results = np.zeros(iteration)
    NFE_value = 0
    pop = init_generator(pop_size, num_variables, min_val, max_val)
    pop = sorting_pop(pop, obj_func, minimizing)
    NFE_value += pop_size
    for k in range(iteration):
        for i in range(pop_size):
            for j in range(pop_size):
                pop [i] = improve_position(pop[i], pop[j], obj_func,
                                           beta_0, gamma, m,
minimizing)
            NFE_value += pop_size
        pop = sorting_pop(pop, obj_func, minimizing)
        NFE[k] = NFE_value
        results[k] = obj_func(pop[0])
    if not full_result:
        return pop[0], obj_func(pop[0])
    else:
        return pop[0], obj_func(pop[0]), results, NFE
```

14.5 Concluding Remarks

The search engine of the firefly algorithm is inspired by an idealized interpretation of the processes by which fireflies use their bioluminescent abilities to attract their mating partners. This algorithm can be categorized as a stochastic population-based meta-heuristic algorithm built on the swarm intelligence principle. Upon closer investigation of its algorithmic architecture, it can be deduced that the firefly algorithm is, in fact, a combination of a generalized variation of the particle swarm optimization algorithm and a pure random search, and in fact, it can be switched between these two extreme interpretations by adjusting the parameters of the algorithm.

One notable feature of this model is how it utilizes the idea of parallelized computation. While most population-based algorithms benefit from parallelized computations, the firefly algorithm elevates this idea simply by creating a structure that allows each search agnate to working simultaneously but almost independently from other search agents in the population set. To build this structure, the algorithm would need to go over the entire population for each search agent in each iteration. While this creates a much more exhaustive search, it could also be more computationally taxing. The other notable feature contributing to this algorithm's unique structure is that the entire search engine of this algorithm is basically a series of localized searches that are working in parallel. This would make the algorithm an ideal option to handle multimodal optimization problems when the main objective is to identify all optimum solutions, whether local optima or global ones. On the other hand, the same feature may cause the algorithm to be trapped in local optima, as the exploration rate of this algorithm could be somewhat limited if the algorithm is not tuned correctly to handle the certain landscape of a given search space. As a last note on this algorithm, one should bear in mind that while the algorithm has a fairly straightforward computational structure that follows simple mathematics and easy-to-understand principles, it is riddled with a lot of parameters that often are not intuitively clear how to fine-tune them for those with less experience with meta-heuristic optimization algorithms. All in all, however, one can safely state that the firefly algorithm can be considered a viable option when it comes to tackling real-world complex optimization problems.

References

Ahmed, H.A., Zolkipli, M.F., & Ahmad, M. (2019). A novel efficient substitution-box design based on firefly algorithm and discrete chaotic map. *Neural Computing and Applications*, 31(11), 7201–7210.

Alweshah, M. (2014). Firefly algorithm with artificial neural network for time series problems. *Research Journal of Applied Sciences, Engineering and Technology*, 7(19), 3978–3982.

Arefinia, A., Bozorg-Haddad, O., Oliazadeh, A., Zolghadr-Asli, B., & Keller, A.A. (2022). Firefly Algorithms (FAs): Application in water resource systems. In *Computational intelligence for water and environmental sciences*. Springer Nature Singapore, 103–118.

Arora, S., Singh, S., Singh, S., & Sharma, B. (2014). Mutated firefly algorithm. In *Proceeding of 2014 International Conference on Parallel*, Distributed and Grid Computing, Solan, India.

Avendaño-Franco, G. & Romero, A.H. (2016). Firefly algorithm for structural search. *Journal of Chemical Theory and Computation*, 12(7), 3416–3428.

Baykasoğlu, A. & Ozsoydan, F.B. (2015). Adaptive firefly algorithm with chaos for mechanical design optimization problems. *Applied Soft Computing*, 36, 152–164.

Buck, J. & Buck, E. (1966). Biology of synchronous flashing of fireflies. *Nature*, 211(5049), 562–564.

dos Santos Coelho, L. & Mariani, V.C. (2013). Improved firefly algorithm approach applied to chiller loading for energy conservation. *Energy and Buildings*, 59, 273–278.

Du, K.L. & Swamy, M.N.S. (2016). *Search and optimization by metaheuristics: Techniques and algorithms inspired by nature*. Springer International Publishing Switzerland. ISBN: 9783319411910

Gandomi, A.H., Yang, X.S., Talatahari, S., & Alavi, A.H. (2013). Firefly algorithm with chaos. *Communications in Nonlinear Science and Numerical Simulation*, 18(1), 89–98.

Garousi-Nejad, I., Bozorg-Haddad, O., & Loáiciga, H.A. (2016). Modified firefly algorithm for solving multireservoir operation in continuous and discrete domains. *Journal of Water Resources Planning and Management*, 142(9), 04016029.

Hamadneh, N.N. (2020). Dead sea water levels analysis using artificial neural networks and firefly algorithm. *International Journal of Swarm Intelligence Research*, 11(3), 19–29.

Horng, M.H. (2012). Vector quantization using the firefly algorithm for image compression. *Expert Systems with Applications*, 39(1), 1078–1091.

Kazem, A., Sharifi, E., Hussain, F.K., Saberi, M., & Hussain, O.K. (2013). Support vector regression with chaos-based firefly algorithm for stock market price forecasting. *Applied Soft Computing*, 13(2), 947–958.

Lloyd, J.E. (1975). Aggressive mimicry in Photuris fireflies: Signal repertoires by femmes fatales. *Science*, 187(4175), 452–453.

Ohba, N. (2004). Flash communication systems of Japanese fireflies. *Integrative and Comparative Biology*, 44(3), 225–233.

Peng, H., Zhu, W., Deng, C., Yu, K., & Wu, Z. (2021). Composite firefly algorithm for breast cancer recognition. *Concurrency and Computation*: *Practice and Experience*, 33(5), e6032.

Riahi-Madvar, H., Dehghani, M., Parmar, K.S., Nabipour, N., & Shamshirband, S. (2020). Improvements in the explicit estimation of pollutant dispersion coefficient in rivers by subset selection of maximum dissimilarity hybridized with ANFIS-firefly algorithm (FFA). *IEEE Access*, 8, 60314–60337.

Santos, A.F.D., Campos Velho, H.F.D., Luz, E.F., Freitas, S.R., Grell, G., & Gan, M.A. (2013). Firefly optimization to determine the precipitation field on South America. *Inverse Problems in Science and Engineering*, 21(3), 451–466.

Talatahari, S., Gandomi, A.H., & Yun, G.J. (2014). Optimum design of tower structures using firefly algorithm. *The Structural Design of Tall and Special Buildings*, 23(5), 350–361.

Tao, H., Diop, L., Bodian, A., Djaman, K., Ndiaye, P.M., & Yaseen, Z.M. (2018). Reference evapotranspiration prediction using hybridized fuzzy model with firefly algorithm: Regional case study in Burkina Faso. *Agricultural Water Management*, 208, 140–151.

Wang, H., Wang, W., Cui, L., Sun, H., Zhao, J., Wang, Y., & Xue, Y. (2018). A hybrid multi-objective firefly algorithm for big data optimization. *Applied Soft Computing*, 69, 806–815.

Wang, H., Zhou, X., Sun, H., Yu, X., Zhao, J., Zhang, H., & Cui, L. (2017). Firefly algorithm with adaptive control parameters. *Soft Computing*, 21(17), 5091–5102.

Wang, Z., Liu, D., & Jolfaei, A. (2020). Resource allocation solution for sensor networks using improved chaotic firefly algorithm in IoT environment. *Computer Communications*, 156, 91–100.

Yang, X.S. (2009). Firefly algorithms for multimodal optimization. In *P*roceeding of the 5th International Symposium on Stochastic Algorithms, Sapporo, Japan.

Yang, X.S. (2010). *Nature-inspired metaheuristic algorithms*. Luniver Press. ISBN: 9781905986286

Yu, S., Su, S., Lu, Q., & Huang, L. (2014). A novel wise step strategy for firefly algorithm. *International Journal of Computer Mathematics*, 91(12), 2507–2513.

15 Gravitational Search Algorithm

Summary

Inspired by an idealized interpretation of Newton's law of universal gravitation, the gravitational search algorithm is a formidable swarm intelligence-based meta-heuristic optimization algorithm that can be quite efficient when handling real-world complex optimization problems. In this chapter, we will dig deep and explore the mechanisms used in this algorithm. We would get familiar with the gravitational search algorithm's terminology and see how one can implement this algorithm in the Python programming language. Finally, we will explore the potential merits and drawbacks of this algorithm.

15.1 Introduction

Have you ever looked at the night sky and pondered upon the wonders of the universe? If you have, know that you are not alone. In fact, for centuries, the cosmos and its mysteries have fascinated some of humankind's brightest and most curious minds. The likes of *Galileo Galilei, Francesco Maria Grimaldi, Giovanni Battista Riccioli*, and *Robert Hooke* have made a tremendous contribution to the way we perceive the cosmos and how celestial bodies interact with one another in deep space. But when it comes to monumental milestones that caused a tectonic shift in the way, we see the governing rules of the universe, one name pops to mind, *Isaac Newton*, and of course, the story of the apple that changed everything.

Though many believe the story has been mythicized over the years, the story goes along these lines; while Sir Isaac was passing through his garden, he had an epiphany when he saw an apple fall from a tree. Evidently, this simple event inspired him to formulate what became known as the *law of universal gravitation*. The gist of this fundamental idea was that every particle attracts every other particle in the universe with a *force* that is directly proportional to the product of their masses and inversely proportional to the square of the *distance* between their centers. In plain language, Newton's law of universal gravitation simply states that if an objective is assumed to be represented with a point in space, these points would collectively tend to move toward one another along the line intersecting each of the paired points. The gravitational force here represents the magnitude of the

DOI: 10.1201/9781003424765-15

attraction between these paint and is proportional to the product of the two points' masses and inversely proportional to the square of the distance between them. This can be mathematically expressed as follows:

$$F = G\frac{m_1 \times m_2}{r^2} \tag{15.1}$$

where F represents the gravitation force acting between two objects [N], m_1 and m_2 are the masses of two given objects [kg], r denotes the distance between these two point objects [m], and G is Newton's gravitational constant [Nm²/kg²].

Here any mass object can be mathematically represented with an array that specifies the position of the center of this mass within the space. What is important to note here is that a vector can denote any imposed force on such an object. However, if there are multiple objects that are interacting with one another in this space, the resultant force would be computed by summing up these vectors (Figure 15.1). This means that all the mass particles in space affect one another, and these effects are directly proportional to how massive these objects are and inversely proportional to how far they are from one another.

Coupling this idea with *Newton's laws of motion* would give us the bigger picture of how these forces could cause motion between these objects. The *first law* simply states that an object at rest will stay at rest, and an object in motion will stay in motion unless acted on by a net external force. In the context of gravitational force, this could be interpreted as an object that would stay dormant or continue on its trajectory unless another object's gravitational force interrupts this motion.

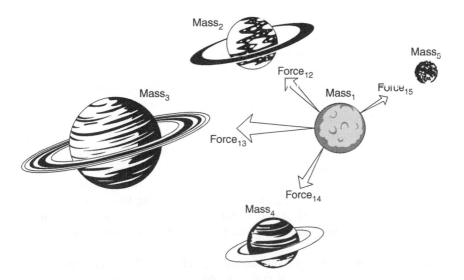

Figure 15.1 Resultant gravitational force for a given particle in space.

The *second law* states that the rate of change of an object's momentum over time is directly proportional to the force applied and occurs in the same direction as the applied force. To put this in the context of the gravitational force, the second law simply states the particle that is being exposed to the gravitational force of another particle would be directed toward the said object with an acceleration that is calculated as follows:

$$a = \frac{F}{m} \tag{15.2}$$

where a denotes the particles' acceleration [m/s^2]. What is essential about calculating this acceleration rate is that you could simply compute its position in the next time step by having the object's velocity, acceleration, and current position at a given time. In other words, using these concepts, one could easily simulate an object's motion caused by another particle's gravitational force.

Last, the third law of *Newtonian motion* states that all forces between two objects exist in equal magnitudes and opposite directions. In the context of the gravitational force, the force imposed on two particles in the vicinity of one another is equal but in the opposite direction. All in all, it is worth noting that while Newton's gravitational law has since been superseded by Einstein's ingenious *general relativity theory*, to this day, it is an excellent way to get an approximation of gravitational force in many practical cases.

Inspired by Newton's law of universal gravitation and laws of motion, Rashedi et al. (2009) theorized a stochastic population-based meta-heuristic optimization algorithm, namely, the gravitational search algorithm, that is rooted in the fundamental principles of swarm intelligence-based computation. At the core, the gravitational search algorithm uses a computational architecture that, to some extent, resembles the algorithmic structure of the particle swarm optimization, a well-known and efficient population-based meta-heuristic algorithm. But the nuances and novelties of the gravitational search algorithm certainly distinguish it from the algorithms we have seen thus far. For instance, rather than using each particle's memory and the best-encountered position to coordinate the repositioning process of the search agents, this algorithm emulates the gravitational force to coordinate the movements of the search agents within the search space. One of the most immediate effects of such recalibration of the searching strategy is that here all particles actively influence the repositioning of other search agents. This simply means that each search agent is conducting a simultaneous parallel local search of its own. While this is undoubtedly more computationally taxing, it can arguably create a much more exhaustive search engine, potentially improving the algorithm's performance. The other subtle effect of this strategy is that an adequately fine-tuned algorithm could, on paper, make a smooth transition from the exploration phase into the exploitation one.

The gravitational search algorithm starts its search by randomly placing a set of search agents, called particles, within the feasible span of the search space. These particles are also assigned random velocities and accelerations. The algorithm would then continue to compute the mass of each given particle, which is proportional to its objective function. Using the principles of Newton's law of universal gravitation, the algorithm would then compute the gravitational force of these particles and how they impose external forces on other particles in the set. The general rule of thumb here is the particles would tend to move toward better positions. After computing the accelerations and velocities of these particles, the algorithm would use the generic principles of Newtonian motion to reposition the particles in space. This process would be repeated until a certain termination criterion is met, at which point the best particle observed thus far would be reported as the optimum to the problem at hand. It is worth noting that while it has been argued that the gravitational search algorithm is, in fact, not strictly following the principles of the law of universal gravitation (Gauci et al., 2012), the bottom line is that from a computational standpoint, this is indeed a formidable option when it comes to handling real-world optimization problems.

Over the years, many variants of the gravitational search algorithm have been proposed in the literature, some of which are grouping gravitational search algorithm (Dowlatshahi & Nezamabadi-Pour, 2014), chaotic *kbest* gravitational search algorithm (Mittal et al., 2016), locally informed gravitational search algorithm (Sun et al., 2016), self-adaptive gravitational algorithm (Lei et al., 2020), and the sine chaotic gravitational search algorithm (Jiang et al., 2020), to name a few. That said, the standard gravitational search algorithm is still considered a viable option to handle real-world optimization problems. In fact, the standard gravitational search algorithm has been successfully used for civil engineering (e.g., Momeni et al., 2021), computer science (e.g., Zibanezhad et al., 2009), data mining (e.g., Zahiri, 2012), data science (e.g., Gonzalez et al., 2015), economy (e.g., Behrang et al., 2011), energy industry (e.g., Li & Zhou, 2011), hydrology (e.g., Niu et al., 2020), image processing (e.g., Mittal & Saraswat, 2018), mechanical engineering (e.g., Yildiz et al., 2016), structural engineering (e.g., Khajehzadeh et al., 2013), text mining (e.g., Mosa, 2019), and water resources and management (e.g., Moeini et al., 2017; Niu et al., 2019). In the following sections, we will explore the computational structure of the standard gravitational search algorithm.

15.2 Algorithmic Structure of the Gravitational Search Algorithm

As a population-based algorithm, the gravitational search algorithm would keep track of several search agents as they enumerate through the search space. Given that this algorithm is based on the principle of swarm intelligence, the search agents would tend to coordinate their moves based on the information that is collectively gathered by the entire set. Resultantly, search agents' decentralized, self-organizing behavior would eventually lead to a gradual improvement of the entire set. What is interesting about the basic structure of this algorithm is that it creates a series of simultaneous and somewhat independent local searches. This is mainly because each search agent's repositioning is coordinated against the entire

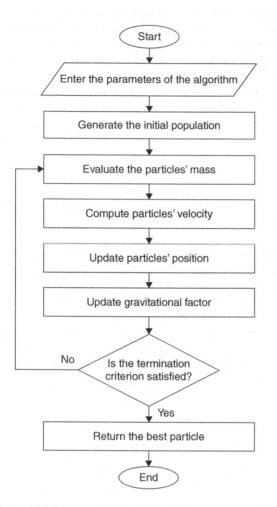

Figure 15.2 The computational flowchart of the gravitational search algorithm.

set. From a computing standpoint, this is more demanding than most conventional swarm intelligence-based optimization algorithms, such as the particle swarm optimization algorithm. The other notable effect of such a structure is that it can help the algorithm transition smoothly from exploration to exploitation.

The gravitational search algorithm's flowchart is depicted in Figure 15.2. A closer look at the architecture of the gravitational search algorithm would reveal that it actually consists of three main stages that are the initiation, repositioning, and termination stages. Using this structure, the algorithm would conduct a thorough search and locate what could be the optimum solution to the problem at hand.

These stages and their mathematical structures will be discussed in the following subsections.

15.2.1 Initiation Stage

The gravitational search algorithm is a population-based meta-heuristic optimization algorithm, and as such, it works with multiple search agents, here called particles, that are simultaneously, and to some extent independently, enumerating through the search space. As we have seen, in an optimization problem with N decision variables, an N-dimension coordination system could be used to represent the search space. In this case, any point within the search space, say X, can be represented mathematically as a $1{\times}N$ array as follows:

$$X = \left(x_1, x_2, x_3, \ldots, x_j, \ldots, x_N \right)$$

(15.3)

where X represents a particle in the search space of an optimization problem with N decision variables, and x_j represents the value associated with the jth decision variable.

The gravitational search algorithm starts with randomly placing a series of particles within the feasible boundaries of the search space. This bundle of arrays, which in the gravitational search algorithm's terminology is referred to as the population, can be mathematically expressed as $M \times N$ matrix, where M denotes the number of particles or what is technically referred to as the population size. In such a structure, each row represents a single search agent. A population, denoted by *pop*, can be represented as follows:

$$pop = \begin{bmatrix} X_1 \\ X_2 \\ \vdots \\ X_i \\ \vdots \\ X_M \end{bmatrix} = \begin{bmatrix} x_{1,1} & x_{1,2} & \cdots & x_{1,j} & \cdots & x_{1,N} \\ x_{2,1} & x_{1,2} & \cdots & x_{2,j} & \cdots & x_{2,N} \\ & & & \vdots & & \\ x_{i,1} & x_{i,2} & \cdots & x_{i,j} & \cdots & x_{i,N} \\ & & & \vdots & & \\ x_{M,1} & x_{M,2} & \cdots & x_{M,j} & \cdots & x_{M,N} \end{bmatrix}$$

(15.4)

where X_i represents the ith particle in the population, and $x_{i,j}$ denotes the jth decision variable of the ith particle.

The initially generated population represents the initial position of particles. As we progress, the values stored in the *pop* matrix will be altered according to the computational structure of the gravitational search algorithm's repositioning stage. By the end of this iterative computation process, when the termination criterion is met, one or possibly multiple particles could converge to the optimum solution.

15.2.2 Repositioning Stage

The repositioning stage is where the gravitational search algorithm implements an idealized interpretation of Newton's law of universal gravitation to coordinate the movement of the particles within the search space. The general theme here is to ensure that particles create a gravitational field that is in proportion to their mass, which itself is linked to the search agent objective function value. These gravitational fields would attract an object toward the other particles. Naturally, the greater these forces are, the faster the motion would become. Conversely, the further the distance between these particles, the less enforcing these gravitational forces would become. One of the most brilliant subtleties in how this algorithm is structured is that it ensures that the particles with greater mass are harder to gain *inertia*. As such, they are far more impervious to repositioning than particles with less mass. The other notable effect of this structure on the search strategy is that at the beginning of the search, the particle movements of particles are much more prominent, as the difference between masses is, by nature, more pronounced. As the algorithm progresses and the particles get closer to one another, the motions become more refined. All in all, this algorithm's unique structure ensures a smooth transition from the exploration to the exploitation phase.

To implement the idea of gravitational forces for repositioning the particles, the algorithm first needs to evaluate the performance of all the particles and identify the best and worst particles in the set. Of course, it should go without saying that the desirability of a particle (i.e., identifying the best and worst particles in the population set) has a different interpretation for maximization and minimization problems. For instance, higher objective functions are considered more desirable in a maximization problem, while we are looking for lower values for the objective function in a minimization problem. The best and worst particles identified in the population set are denoted by X_{Best} and X_{Worst}, respectively. The algorithm would then use these values to normalize the objective function values of the particles in the set. This could be mathematically achieved by implanting the following formula:

$$\vartheta_i = \frac{f(X_i) - X_{Worst}}{X_{Wbest} - X_{Worst}} \qquad \forall i \tag{15.5}$$

in which ϑ_i denotes the normalized fitness value of the ith particle, and $f()$ represents the objective function. Note that, from a mathematical standpoint, the values would always be within the range of 0–1, where the worst particle would have a normalized fitness of 0, and the best particle's normalized fitness would be equal to 1.

The algorithm would then call these normalized values to compute the mass of each particle using the following equation:

$$m_i = \frac{\vartheta_i}{\sum_{i=1}^{M} \vartheta_i} \qquad \forall i \tag{15.6}$$

where m_i denotes the mass of the *i*th particle. Note that the mass of each particle is a non-negative value that ranges from 0 to 1. This value is directly proportional to how desirable the said particle is.

As we have discussed, any two given paired particles in this space would attract one another with a force that is directly proportional to the product of their masses and inversely proportional to the square of the distance between their centers. In the context of the gravitational search algorithm, the algorithm next needs to compute the posed force between any two given paired particles. This can be mathematically expressed as follows:

$$F_{l,k} = \Omega \frac{m_l \times m_k}{r_{l,k} + \varepsilon} \left(X_k - X_l \right) \qquad \forall l,k \tag{15.7}$$

in which $F_{l,k}$ denotes the force acting on the *l*th particle imposed by the *k*th particle; Ω represents the gravitational factor parameter, which is one of the parameters of the algorithm; ε is a small positive constant; and $r_{l,k}$ denotes the distance between the two said particles. Often, measuring the *Euclidean distance* between two search agents is considered a straightforward approach to computing this distance. For two given particles, X_l and X_k, this distance can be mathematically resented as follows:

$$r_{l,k} = \|X_l - X_k\| = \sqrt{\sum_{j=1}^{N} \left(x_{l,j} - x_{r,j} \right)^2} \qquad \forall l,k \tag{15.8}$$

Adding ε to the denominator in Equation (15.7) was merely a mathematical trick to avoid *division by zero*.

The next step would be to compute the acceleration of each individual particle within the search space. As we have seen, acceleration is the resultant force divided over the mass of an object. By the same token, the acceleration vector of any given particle could be calculated as follows:

$$a_i = \frac{\sum_{k=1}^{M} Rand \times F_{i,k}}{m_i} \qquad \forall i \tag{15.9}$$

where a_i denotes the acceleration of the *i*th particle, and *Rand* is a randomly generated number from the range 0 to 1. The idea of introducing this random component here is to expand the exploration capabilities of the algorithm.

Based on the first law of Newtonian motion, an object tends to stay dormant or continue with its motion unless an external force is introduced to the picture. In other words, the main idea here is that according to Newton's law of motion, an object tends to keep its inertia. Thus, when an external force is applied to a moving object, the new trajectory is, in fact, the resultant vector in which the former trajectory line of the object is added to the motion imposed by the external vector.

By the same token, in the gravitational search algorithm, any object tends to keep track of its current motion through a variable called velocity. Thus, the motions imposed by the gravitational force of other particles would need to be added to this trajectory line. For the velocity of the particles, the following formula could be used:

$$v_i^{new} = Rand \times v_i + a_i \qquad \forall i \qquad\qquad (15.10)$$

where v_i is the velocity of the ith particle, and v_i^{new} denotes the adjusted velocity vector of the ith particle.

Using the principles of Newtonian motion, the new position of the particles could be calculated as follows:

$$X_i^{new} = X_i + v_i^{new} \qquad \forall i \qquad\qquad (15.10)$$

in which X_i^{new} denotes the new position of the ith particle. It should be noted that for the first iteration, given that you would need the velocity vectors of the particles in the previous iteration (i.e., iteration zero), you could simply use a randomly generated vector as the old velocity vector.

As a final note on the gravitational factor parameter, it should be noted that this is the parameter by which the algorithm controls the expansion of the searching premise of the particles. The higher the value of this parameter, the broader the searching domain would become for the particles. Like any efficient meta-heuristic algorithm, the gravitation search algorithm must smoothly transition from the exploration phase into the exploitation phase. By doing so, the algorithm converges to what could be the optimum solution. In order to achieve this effect, the algorithm needs to reduce the searching domain of the particles gradually. While, in and of itself, the algorithm's unique structure is creating such an effect to some extent, to amplify this notion, the gravitational search algorithm would dynamically change the values of the gravitational factor parameter in each iteration. In case the termination criterion is based on limiting the number of iterations, this can be mathematically expressed as follows:

$$\Omega_t = \Omega_0 \times e^{-\frac{C \times t}{T}} \qquad \forall t \qquad\qquad (15.11)$$

in which Ω_t denotes the gravitational factor in the tth iteration; Ω_0 is the initiation value for the gravitational factor, another parameter of the algorithm; C denotes the controlling coefficient, which happens to be a user-defined parameter of the algorithm; T is the maximum number of iteration, a user-defined parameter for this algorithm; and t represents the counter for the current iteration. It should be noted that the above formula can be altered in a way that would also apply to other termination criteria. For instance, if the idea is to run the algorithm within a specific time span, the maximum time limit would replace the maximum number of iterations, while the current iteration count would be swapped with the current run time value.

15.2.3 Termination Stage

Based on the repositioning stage described above, in each iteration, the gravitational search algorithm would stochastically update the position of each individual particle in the population set. The governing rule of motion in this algorithm ensures that the more desirable particles are less prone to be repositioned, while particles placed in less desirable places in the search space would experience more pronounced motions. This unique feature would ensure that the algorithm makes a smooth and gradual transition between the exploration phase and exploitation phase, given that in the first iterations, particles would experience more drastic repositioning, while eventually, as the population would gradually become more homogenous, the repositioning of the particles transpose more into a local search within the area with the most potential to have the optimum solution.

Like other meta-heuristic algorithms, the sequence of operational structures of this algorithm needs to be executed iteratively until a certain termination criterion is met, at which point the execution of the algorithm would be terminated, and the best particle recorded in the memory would be reported as the solution to the optimization problem. Note that without such a termination stage, the algorithm would potentially be executed in an infinite loop. The termination stage would, in effect, determine whether the algorithm has reached what could be the optimum solution.

As the gravitational search algorithm is not equipped with an explicitly defined, unique termination mechanism, one could implement the commonly available options, most notably limiting the number of iterations, run time, or perhaps monitoring the improvement made to the best solution in consecutive iterations. Among these options, limiting the number of iterations is arguably the most cited mechanism to create a termination stage for the gravitational search algorithm. The idea being the process would be executed only for a specified number of times, a parameter known as the maximum iteration. In any case, it should be noted that the selection of a termination mechanism is also considered one of the algorithm's parameters. Bear in mind that in most cases, these termination mechanisms may require setting up additional parameters.

15.3 Parameter Selection and Fine-Tuning the Gravitational Search Algorithm

From the *no-free-lunch theorem*, one can conclude that fine-tuning an algorithm is essential to get the best performance out of a meta-heuristic algorithm. This would basically ensure that an algorithm is equipped to handle the unique characteristics of a given optimization problem. Of course, it is possible to use our intuition, experience, and default values suggested for an algorithm's parameters as a good

```
Begin
        Set the algorithm's parameter and input the data
        Generate the initial population
        Let M denote the population size
        Let N denote the number of decision variables
        While the termination criterion is not met
                Record the best and worst particles
                Evaluate the particles' mass
                For i in range 1 to M
                        For k in range 1 to M
                                Compute the distance between the ith and kth particles
                                Compute the force action imposed by ith particle on kth particle
                        Next k
                        Update the acceleration, velocity, and position of the ith solution
                Next i
                Update Newton's gravitational factor
        End while
        Report the best solution
End
```

Figure 15.3 Pseudocode for the gravitational search algorithm.

starting point, one should bear in mind that fine-tuning these parameters is more than anything a trial-and-error process. Thus, while it is possible to get a good enough result by having an educated guess for setting the parameters of these algorithms, to get the best possible performance, it is necessary to go through this fine-tuning process.

In the case of the gravitational search algorithm, these parameters are population size (M), the initiation value for the gravitational factor (Ω_0), controlling coefficient (C), and of course, opting for the termination criterion and all the parameters that are associated with these methods. For instance, if limiting the number of iterations has been selected as a termination criterion, the maximum iteration (T) is another parameter that needs to be defined by the user. As can be seen here, not only are there few parameters in the structure of the gravitational search algorithm, but for the most part, the role of these parameters on the final outcome can be deduced intuitively, especially for those with little experience with meta-heuristic optimization, which makes the fine-tuning process less challenging. That said, to get the absolute best results out of the gravitational search algorithm, it is best to dabble with these algorithms first to gain some experience and inside knowledge about such parameters. By doing so, you could better understand how to fine-tune these parameters as your initial guesses and parameter selection strategies become more educated. The pseudocode for the gravitational search algorithm is shown in Figure 15.3.

15.4 Python Codes

The code to implement the gravitational search algorithm can be found below:

```python
import numpy as np

def init_generator(pop_size, num_variabels, min_val, max_val):

    return np.random.uniform(min_val, max_val, (pop_size, num_
  variables))

def sorting_pop(pop, obj_func, minimizing):
    results = np.apply_along_axis(obj_func, 1, pop)
    indeces = np.argsort(results)
    if not minimizing:
        indeces = indeces[::-1]
    return pop[indeces]

def normalizing_ofs(sorted_pop):
    results = np.apply_along_axis(obj_func, 1, sorted_pop)
    best = results[0]
    worst = results[-1]
    return (results-worst)/(best-worst)

def particles_mass(sorted_pop):
    normalized_results = normalizing_ofs(sorted_pop)
    sum_values = np.sum(normalized_results)
    return normalized_results/sum_values

def distance(a, b):
    return np.sqrt(np.sum((a-b)**2))

def force(a, b, sorted_pop, mass_values, gamma, epsilon=1e-3):
    dist_val = distance(sorted_pop[a], sorted_pop[b])
    x = gamma*mass_values[a]*mass_values[b]*(sorted_pop[b]-sorted_
  pop[a])
    return x /(dist_val+epsilon)

def active_agents(iteration, agents_0, agents_last):
    x=(np.arange(iteration)*(agents_0-agents_last)/iteration)
    agents=agents_0-x
    return agents.astype(int)

def acceleration(sorted_pop, pop_size, num_variables, mass_values,
                 gamma, min_val,
                 max_val, agents_0, agents_last, epsilon=1e-3):
    acceleration_values = np.zeros_like(sorted_pop)
    agents = active_agents(iteration, agents_0, agents_last)
    for i in range(pop_size):
        a = np.zeros((1, num_variables))
        for j in range(agents[i]):
            rand_value = np.random.uniform()
            a+=(rand_value*force(i,j,sorted_pop,
```

```
                                mass_values,gamma,epsilon))
        if mass_values[i]!=0:
            acceleration_values[i] = (a)/mass_values[i]
        else:
            acceleration_values[i] = np.random.uniform(min_val,max_val,
                                                       (1,num_
                                                       variables))
    return acceleration_values
def gamma_values(gamma_0, alpha, iteration):
    x = np.arange(iteration)
    return gamma_0*np.exp(-alpha*x/iteration)
def gravitational_search_algorithm(num_variables, pop_size, min_val,
                        max_val, gamma_0, alpha,
                        iteration, obj_func, minimizing,
                        epsilon=1e-3, full_results=False):
    NFE_value = 0
    NFE = np.zeros(iteration)
    results = np.zeros(iteration)
    pop = init_generator(pop_size, num_variables, min_val, max_val)
    pop = sorting_pop(pop, obj_func, minimizing)
    NFE_value += pop_size
    gamma = gamma_values(gamma_0, alpha, iteration)
    velocity = np.zeros_like(pop)
    for k in range(iteration):
        mass_values = particles_mass(pop)
        acceleration_values = acceleration(pop, pop_size,
                                            num_variables, mass_
                                            values,
                                            gamma[k], min_val, max_
                                            val,
                                            agents_0, agents_last,
                                            epsilon)
        velocity = np.random.uniform()*velocity+acceleration_values
        pop += velocity
        pop = sorting_pop(pop, obj_func, minimizing)
        NFE_value += pop_size
        results[k] = obj_func(pop[0])
        NFE[k] = NFE_value
    if not full_results:
        return pop[0], obj_func(pop[0])
    else:
        return pop[0], obj_func(pop[0]), results, NFE
```

15.5 Concluding Remarks

Inspired by an idealized interpretation of Newton's law of universal gravitation and the law of motion, the gravitational search algorithm establishes itself as a formidable choice when handling real-world optimization problems. This algorithm can be categorized as a stochastic population-based meta-heuristic algorithm built on the swarm intelligence principle. Upon closer investigation of its algorithmic architecture, it can be seen that the gravitational search algorithm is, to some extent, an interpretation of a generalized particle swarm optimization algorithm. But the nuances and novelties of this algorithm certainly give it an edge over the more conventional swarm intelligent-based algorithm. One of the most notable characteristics of this algorithm is that it treats each of the particles in the population set as an independent local search engine that works in parallel with other agents. While this is naturally more computationally taxing than the typical approach, the overall effect is that it arms the algorithm with a more efficient search strategy.

The other distinctive characteristic of this algorithm is that it can smoothly transition between the exploration and exploitation phases. Two notable reasons for this effect are rooted in the algorithm's structure. Firstly, implementing excessively parallelized local search engines that work simultaneously yet somewhat independently from one another can create such an effect. The reason is that when the algorithm initiates, the particles far from each other will experience more pronounced motions. But as the algorithm progresses and the particles gravitate toward each other, the motions become less intense and, in effect, transit into a more localized searching process. The second reason is that the algorithm gradually reduces the motions' intensity by dropping the gravitational factor values in each iteration.

As a last note on this algorithm, it should be noted that not only the algorithm has a fairly straightforward computational structure that follows simple mathematics and easy-to-understand principles, but there are also a few parameters in the structure of the gravitational search algorithm, and for the most part, the role of these parameters on the final outcome can be deduced intuitively. All in all, one can safely state that the gravitational search algorithm can be considered a viable option when tackling real-world complex optimization problems.

References

Behrang, M.A., Assareh, E., Ghalambaz, M., Assari, M.R., & Noghrehabadi, A.R. (2011). Forecasting future oil demand in Iran using GSA (gravitational search algorithm). *Energy*, 36(9), 5649–5654.

Dowlatshahi M.B. & Nezamabadi-Pour, H. (2014). GGSA: A grouping gravitational search algorithm for data clustering. *Engineering Applications of Artificial Intelligence*, 36, 114–121.

Gauci, M., Dodd, T.J., & Groß, R. (2012). Why 'GSA: A gravitational search algorithm' is not genuinely based on the law of gravity. *Natural Computing*, 11(4), 719–720.

Gonzalez, B., Valdez, F., Melin, P., & Prado-Arechiga, G. (2015). Fuzzy logic in the gravitational search algorithm for the optimization of modular neural networks in pattern recognition. *Expert Systems with Applications*, 42(14), 5839–5847.

Jiang, J., Jiang, R., Meng, X., & Li, K. (2020). SCGSA: A sine chaotic gravitational search algorithm for continuous optimization problems. *Expert Systems with Applications*, 144, 113118.

Khajehzadeh, M., Taha, M.R., & Eslami, M. (2013). Efficient gravitational search algorithm for optimum design of retaining walls. *Structural Engineering and Mechanics*, 45(1), 111–127.

Lei, Z., Gao, S., Gupta, S., Cheng, J., & Yang, G. (2020). An aggregative learning gravitational search algorithm with self-adaptive gravitational constants. *Expert Systems with Applications*, 152, 113396.

Li, C. & Zhou, J. (2011). Parameters identification of hydraulic turbine governing system using improved gravitational search algorithm. *Energy Conversion and Management*, 52(1), 374–381.

Mittal, H., Pal, R., Kulhari, A., & Saraswat, M. (2016). Chaotic Kbest gravitational search algorithm (CKGSA). In *Proceeding 2016 9th International Conference on Contemporary Computing*, Noida, India.

Mittal, H. & Saraswat, M. (2018). An optimum multi-level image thresholding segmentation using non-local means 2D histogram and exponential Kbest gravitational search algorithm. *Engineering Applications of Artificial Intelligence*, 71, 226–235.

Moeini, R., Soltani-Nezhad, M., & Daei, M. (2017). Constrained gravitational search algorithm for large scale reservoir operation optimization problem. *Engineering Applications of Artificial Intelligence*, 62, 222–233.

Momeni, E., Yarivand, A., Dowlatshahi, M.B., & Armaghani, D.J. (2021). An efficient optimal neural network based on gravitational search algorithm in predicting the deformation of geogrid-reinforced soil structures. *Transportation Geotechnics*, 26, 100446.

Mosa, M.A. (2019). Real-time data text mining based on gravitational search algorithm. *Expert Systems with Applications*, 137, 117–129.

Niu, W.J., Feng, Z.K., Chen, Y.B., Zhang, H.R., & Cheng, C.T. (2020). Annual streamflow time series prediction using extreme learning machine based on gravitational search algorithm and variational mode decomposition. *Journal of Hydrologic Engineering*, 25(5), 04020008.

Niu, W.J., Feng, Z.K., Zeng, M., Feng, B.F., Min, Y.W., Cheng, C.T., & Zhou, J.Z. (2019). Forecasting reservoir monthly runoff via ensemble empirical mode decomposition and extreme learning machine optimized by an improved gravitational search algorithm. *Applied Soft Computing*, 82, 105589.

Rashedi, E., Nezamabadi-Pour, H., & Saryazdi, S. (2009). GSA: A gravitational search algorithm. *Information Sciences*, 179(13), 2232–2248.

Sun, G., Zhang, A., Wang, Z., Yao, Y., Ma, J., & Couples, G.D. (2016). Locally informed gravitational search algorithm. *Knowledge-Based Systems*, 104, 134–144.

Yildiz, B.S., Lekesiz, H., & Yildiz, A.R. (2016). Structural design of vehicle components using gravitational search and charged system search algorithms. *Materials Testing*, 58(1), 79–81.

Zahiri, S.H. (2012). Fuzzy gravitational search algorithm an approach for data mining. *Iranian Journal of Fuzzy Systems*, 9(1), 21–37.

Zibanezhad, B., Zamanifar, K., Nematbakhsh, N., & Mardukhi, F. (2009). An approach for web services composition based on QoS and gravitational search algorithm. In *Proceeding of 2009 International Conference on Innovations in Information Technology*, Al Ain, United Arab Emirates.

16 Plant Propagation Algorithm

Summary

Inspired by the life cycle of plants that use propagation to colonize a new habitat, the plant propagation algorithm established itself as a formidable swarm intelligence-based meta-heuristic optimization algorithm that can be quite efficient when it comes to handling real-world complex optimization problems. In this chapter, we will dig deep and explore the mechanisms used in this algorithm. We would get familiar with the plant propagation algorithm's terminology and see how one can implement this algorithm in the Python programming language. Finally, we will explore the potential merits and drawbacks of this algorithm.

16.1 Introduction

Plants, in general, have some of the most fascinating adaption mechanisms when it comes to colonizing a new habitat. In fact, one can rarely find a harsh environment wholly deserted of vegetation life and still house other living organisms. One way or another, a species of plants would be equipped with the proper adaptation feature that enables them to survive or even, in some cases, thrive in these barren environments. More impressively, such plants would always opt for a suitable reproduction strategy that ultimately increases their odds of survival as a community. This distinct feature of such plant species would ensure that the colony, as a whole, could withstand the harshest environments, as the previous generations would always actively attempt to locate a more suitable place for the offspring plants. As such, after a few generations, the plants would always be able to identify and, more importantly, colonize the most suitable places in a habitat that could meet their needs and requirements.

Of course, plant species have adopted various strategies to cope with such hardships as they get introduced to a new harsh and sterile environment. For instance, in previous chapters, we got familiar with how some invasive weeds used their seed dispensation mechanisms to colonize nearly any new habitat they were introduced to. The main idea was that weeds with access to more nutrition or are generally located in more suitable places would tend to create more seeds than other community members. The community would tend to spread these seeds to

DOI: 10.1201/9781003424765-16

colonize other parts of the currently barren habitat. These seeds would first be dispersed further from the parent plant to ensure that the colony has fully explored every corner of the said regions. But eventually, the later generations of the plants in these colonies would spread their seeds closer and closer to where the parents are located, creating denser weed-populated areas in places deemed more suitable to nourish the said species.

Other plants, such as most varieties of strawberries, use an asexual reproduction strategy called *propagation by runners*. A *runner* or a *stolon* is a shoot that grows out of the parent plant and along the ground and sends out roots from shoot nodes to other regions in the plants' vicinity. Eventually, such runners will develop their own roots, enabling them to extract water and other nourishment from the soil. Of course, as these adventitious roots get established in the soil, the runners begin to dry up and shrivel away. In effect, such propagation would create a new clone plant.

The governing principles of plant propagation by runners are somewhat similar to those that are associated with seed desertion. The general theme here is that plants exposed to more suitable environmental properties would tend to create more runners than plants located in spots that may not have access to enough water, sun exposure, or other nutrition materials. The former group also tends to issue shorter runners, given that the current location of the parent plant has already been deemed suitable to provide the plants' requirements. On the other hand, the latter group would attempt to issue longer runners to explore other potential areas in the habitat. If the plants would be able to identify a more suitable location in such situations, they tend to occupy the said spot, otherwise, they keep on exploring the area. The overall effect of these features is that the more suitable spots in the area would have a more dense population than less desirable areas where the plant population is more dispersed.

Inspired by an idealized interpretation of the above phenomenon, Salhi and Fraga (2011) theorized a novel meta-heuristic algorithm called the plant propagation algorithm. Built upon the principles of swarm intelligence, the plant propagation algorithm, also known as the *strawberry algorithm*, could be described as a stochastic population-based algorithm that emulates the procedure in which some plants use propagation by runners to colonize a new environment. Being based on the principles of the direct search would simply mean that, like other meta-heuristic algorithms, the plant propagation algorithm is also not bound by problems that are often associated with high dimensionality, multimodality, epistasis, non-differentiability, and discontinuous search space imposed by constraints (Du & Swamy, 2016; Bozorg-Haddad et al., 2017). The most notable distinguishing feature of this algorithm is that its structure of this algorithm enables it to conduct both local and exploratory searches simultaneously. Based on the architecture of this algorithm, the search agents would tend to resort to conducting local searches if they deem the position more desirable. If the position is not found to be quite suitable, on the other hand, the search agents will tend to resort to exploratory moves. Again it is important to note that, in opposition to conventional swarm intelligence-based algorithms, the search agents do not necessarily coordinate their moves based on direct communication with one another, but rather, their relative performance against their contemporary counterpart search agents would dictate how their reproduction mechanism would work. As such, having this relative dependency between the search agent's performances would give this algorithm a swarm of intelligence-based feelings.

The plant propagation algorithm is based on an abstract interpretation of the propagation by runner mechanism we have explored earlier. The idea is that the algorithm would initiate its search by randomly generating a series of search agents, referred to as plants, within the feasible area of the search space. Each plant would then continue to create a set of runners that would be issued from the parent plants within the vicinity of the parent weed. The number of issued runners is directly proportional to the suitability of the parent plant's position, while their length is inversely proportional to this measure. As such, the algorithm would conduct a local search within the vicinity of more suitable places while it resorts to exploratory moves if they are issued from less desirable plants. These runners would go on to grow and create a set of runners of their own and continue the colonization of the environment. When the number of plants exceeds the environment's capacity to host a new plant, the exceeding number of plants with inferior genes would be terminated as they cannot survive in the competing environment. This whole process would be repeated until a termination criterion is met, at which point the best plant encountered thus far would be returned as the solution to the optimization problem at hand. In this analogy, the plants represent the search agents that enumerate the search space, denoted as the environment that these plants inhabit. The desirability of a place within this space is measured with the quality of the objective function or the fitness function in constrained optimization problems.

As can be seen, from a computational standpoint, the architecture of the plant propagation algorithm closely resembles the computational structure of the invasive weed optimization algorithm. One of the main differences between these two algorithms is rooted in the way these algorithms compute the radius of the search premise for each search agent. In the invasive weed optimization algorithm, the dispersion length of these solutions is gradually reduced with each iteration, but it was not directly proportional to the solutions' relative performance. Here, however, the length of the runners is always directly proportional to the performance of the parent plant to the extent that they always seem to be a local search for more desirable solutions. At the same time, they become more exploratory in nature for bad solutions. Such a feature would enable the algorithm to conduct both exploratory and local searches simultaneously. However, coupling this with the mechanism by which the algorithm determines the number of runners for each plant would lead us to the inevitable conclusion that the algorithm favors local-based search more than exploratory moves. This could be a bit problematic without proper parameter tuning as the algorithm tends to rely on these local searches to identify the optimum solution. As such, there is the possibility that it could be trapped in local optima. Thus, being a meta-heuristic algorithm, the plant propagation algorithm also suffers from the same generic issues that are commonly associated with this branch of optimization, such as having no guarantee for reaching the optimum solution. That said, it should be noted that one of the most noteworthy characteristics of this algorithm is that it follows a series of arguably simple computational instructions, which ultimately makes the algorithm easy to understand and execute (Du & Swamy, 2016). The lack of complex mathematical formulas or having over-the-top interacted architecture makes it easy to understand and implement this algorithm to handle real-world optimization problems. More importantly, this algorithm has few parameters to begin with, and for the most part, understanding how these parameters would affect the final result should be fairly intuitive for those with a bit of background in meta-heuristic computation.

Over the years, many variants of the plant propagation algorithm have been proposed in the literature, some of which are discrete plant propagation algorithm (Selamoğlu & Salhi, 2016) and hybrid plant propagation algorithm (Sulaiman & Salhi, 2016), to name a few. That said, the standard plant propagation algorithm is still considered a viable option to handle real-world optimization problems. In fact, standard plant propagation optimization algorithm has been successfully used for architectural engineering (e.g., Fraga et al., 2015), educational research (e.g., Cheraitia et al., 2019), electrical engineering (e.g., Sulaiman et al., 2020), energy industry (e.g., Waqar et al., 2020), human resources management (e.g., Haddadi, 2020), image processing (e.g., Khirade & Patil, 2015), and water resources planning and management (e.g., Asvini & Amudha, 2016). In the following sections, we will explore the computational structure of the standard plant propagation algorithm.

16.2 Algorithmic Structure of the Plant Propagation Algorithm

In order to create a functional algorithm, we first need to extract and idealize some of governing principles of the plant propagation mechanism. The main idea here is to establish a simplified and abstract mathematical representation of the said procedure that would be utilized later on to assemble the search engine of the plant propagation algorithm. These simplifying assumptions are as follows:

I Plants that are located in more suitable habitats propagate by issuing a more significant number of runners, while poorly placed plants issue less number of runners.
II Plants that are located in more suitable habitats issue runners with shorter lengths, in opposition to poorly placed plants which would issue long runners.
III The suitability of a location is synonymous with the values associated with the objective or fitness functions.

Based on these simplifying principles, the plant propagation algorithm creates a search engine that governs the repositioning of the search agents so that they can ultimately locate what could potentially be the optimum solution.

The plant propagation algorithm's flowchart is depicted in Figure 16.1. A closer look at the architecture of the plant propagation algorithm would reveal that it

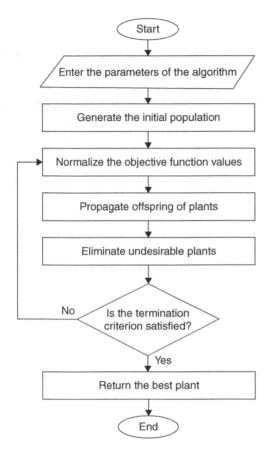

Figure 16.1 The computational flowchart of the plant propagation algorithm.

actually consists of three main stages that are the initiation, propagation, and termination stages. Using this structure, the algorithm would conduct a thorough search and locate what could be the optimum solution to the problem at hand. The following subsection will discuss each of these stages and their mathematical structures.

16.2.1 Initiation Stage

The plant propagation algorithm is a population-based meta-heuristic optimization algorithm, and as such, it works with multiple search agents, here called plants/runners, that would enumerate through the search space. As we have seen, in an optimization problem with N decision variables, an N-dimension coordination system could be used to represent the search space. In this case, any point within the search space, say X, can be represented mathematically as a $1 \times N$ array as follows:

$$X = \left(x_1, x_2, x_3, \ldots, x_j, \ldots, x_N\right) \tag{16.1}$$

where X represents a plant in the search space of an optimization problem with N decision variables, and x_j represents the value associated with the jth decision variable.

The plant propagation algorithm starts with randomly placing a series of plants within the feasible boundaries of the search space. This bundle of arrays, which in the plant propagation algorithm's terminology is referred to as the population, can be mathematically expressed as $M \times N$ matrix, where M denotes the number of plants or what is technically referred to as the population size. In such a structure, each row represents a single search agent. A population, denoted by *pop*, can be represented as follows:

$$pop = \begin{bmatrix} X_1 \\ X_2 \\ \vdots \\ X_i \\ \vdots \\ X_M \end{bmatrix} = \begin{bmatrix} x_{1,1} & x_{1,2} & \cdots & x_{1,j} & \cdots & x_{1,N} \\ x_{2,1} & x_{1,2} & \cdots & x_{2,j} & \cdots & x_{2,N} \\ & \vdots & & & & \\ x_{i,1} & x_{i,2} & \cdots & x_{i,j} & \cdots & x_{i,N} \\ & \vdots & & & & \\ x_{M,1} & x_{M,2} & \cdots & x_{M,j} & \cdots & x_{M,N} \end{bmatrix} \tag{16.2}$$

where X_i represents the ith plant in the population, and $x_{i,j}$ denotes the jth decision variable of the ith plant.

The initially generated population represents the first generation of plants. As we progress, the values stored in the *pop* matrix will be altered according to the computational structure of the plant propagation algorithm's propagation stage. By the end of this iterative computation process, when the termination criterion is met, one or possibly multiple plants will converge to the optimum solution.

16.2.2 Propagation Stage

By mimicking the propagation by the runner, the algorithm would tend to create a new set of search agents in this stage. The propagation stage consists of two main pillars: *Creating and issuing runners* and *competitive exclusion*. These mechanisms are used as alteration tools to enhance the population's good qualities and eradicate undesirable properties.

The first step in the propagation stage is to emulate the asexual reproduction procedure of plant propagation. The idea is that each virtual plant can issue a number of runners directly proportional to its environmental condition, which in the mathematical context is expressed as the objective function associated with the said plant. Naturally, the more desirable the environmental conditions are, the more productive a plant could get.

In order to compute the number of runners for each plant, the algorithm would first need to bring the performance of all the tentative solutions to the same numeric scale. As such, the algorithm would first need to normalize the values for the objective functions for each plant in the population set. This could be mathematically expressed as follows:

$$\sigma_i = \frac{f(X_i) - f(Xworst)}{f(Xbest) - f(Xworst)} \qquad \forall i \tag{16.3}$$

in which σ_i denotes the normalized suitability of the ith plant; *Xbest* and *Xworst* represent the best and worst plants in the current population, respectively; and lastly, $f()$ represents the objective function. Note that normalized suitability values always range somewhere between 0 and 1, where the worst plant has the normalized suitability of 0, while the best plant has a normalized suitability equal to 1. Of course, it should go without saying that desirable conditions (i.e., identifying the best and worst plants in the population set) have a different interpretation for maximization and minimization problems. For instance, higher objective functions are considered more desirable in a maximization problem, while we are looking for lower values for the objective function in a minimization problem.

The normalized suitability values would then be used to determine the number of runners that each plant issues. The general idea is that the number of runners should be directly proportional to the desirability of a plant's position within the search space. The better these properties, the more runners can be issued by the said plant. These runners would then establish plants of their own, which could be interpreted as a set of new tentative solutions. Mathematically speaking, the number of runners for each plant can be computed as follows:

$$\mu_i = \left\lceil \lambda_{max} \times Rand \times \sigma_i \right\rceil \qquad \forall i \tag{16.4}$$

where μ_i represents the number of runners issued by the ith plant; λ_{max} denotes the maximum number of runners that can be issued by a plant, which happens to be a user-defined parameter of the algorithm; *Rand* is a randomly generated value within the range 0–1; and $\lceil \; \rceil$ denotes the *ceiling function*, which retunes the smallest nearest integer to the specified value in a number line (i.e., rounds up the specified number to the closest integer number).

As stated earlier, in opposition to the number of issued runners, the length of these runners is inversely proportional to the normalized suitability value of the plants. The idea here is to create the effect that the plants located in more suitable spots should issue runners within the close approximate of their current position, while those located in poor environments must issue longer runners to give their offspring better odds of survival. The overall effect of this would be that the algorithm would be conducting a local search in areas that are deemed more suitable while resorting to exploratory moves in less suitable positions. These could be mathematically expressed as follows:

$$d_{i,j} = 2 \times (Rand - 0.5) \times (1 - \sigma_i) \qquad \forall i, j \tag{16.5}$$

in which $d_{i,j}$ denotes the length of the runner issued from the ith plant in the jth dimension.

Given that the runner is, in fact, issued from the parent plant before transforming into a new plant itself, the position of the said plant could be computed as follows:

$$x'_{r,j} = x_{i,j} + (U_j - L_j) \times d_{i,j} \qquad \forall i, j, r \tag{16.6}$$

where $x'_{r,j}$ is the position of the rth new plant in the jth dimension; and U_j and L_j represent the upper and lower feasible boundaries of the jth decision variables, respectively. These procedures would be repeated until the algorithm computes the new position for all runners of all plants.

The most significant problem here is that as these new plants get cumulatively staged on top of the original population, the number of generated plants could easily get out of hand, given that the plant population is increasing exponentially. As such, the algorithm is equipped with a competitive exclusion mechanism to thin the population by removing undesirable plants. This process would emulate selective competition, where the fittest members of the species would survive the battle over limited natural resources. To implant this, the algorithm would check the population size at the end of each iteration. If the population surpasses a predefined population size, the algorithm will simply eliminate the exceeding weeds by removing the least fit values. Given that all plants are actively participating in the composition of the new generation, it is crucial to have this mechanism in place to eliminate undesirable tentative solutions and keep things in check. More importantly, this would ensure that the computation task would not get out of hand, as without a competitive exclusion procedure, the number of computations in each iteration could quickly get overwhelmingly taxing within a few iterations.

The propagation mechanism described here gives the plant propagation algorithm a slight edge over the more traditional stochastic-based meta-heuristic algorithms. The significance of this mechanism is that while, on paper, all participating are, in fact, participating in composing the next generation of plants, in practice, some of the more elite members of the said community have a more active role than, say, the more inferior plants. This ensures that the population is always moving toward what could potentially be the right direction. This was not always the case in a more conventional stochastic-based meta-heuristic algorithm, as in most cases, the architecture of these algorithms is based on selecting the participating member in a probabilistic manner. While the algorithm was explicitly designed so that the better members would have a better chance of being selected in such a process, the bottom line was that there was always a chance, as slim as it was, that the worst members would be selected and divert the search from the more promising areas. Of course, abandoning this probabilistic section procedure in such structures could lead to premature convergence or being trapped in local optima. All in all, at least in theory, the reproduction mechanism embedded in this algorithm seems like an elegant solution to address these issues.

16.2.3 Termination Stage

Based on the propagation stage described above, in each iteration, the plant propagation algorithm would create a new set of runners that would be dispersed across the search space and even continue reproducing their own runners. At any given time, if the population of the plants surpasses the defined limits, the algorithm would eliminate the exceeding undesirable plants. Through this stochastic process, in each given iteration, the algorithm conducts a series of simultaneous searches that are more local-based for more desirable solutions and more exploratory in nature for less desirable solutions. This process would be repeated until the algorithm eventually locates what could potentially be the optimum solution.

Like other meta-heuristic algorithms, the sequence of operational structures of this algorithm needs to be executed iteratively until a certain termination criterion is met, at which point the execution of the algorithm would be terminated, and the best plant recorded in the memory would be reported as the solution to the optimization problem. Note that without such a termination stage, the algorithm would potentially be executed in an infinite loop. The termination stage would, in effect, determine whether the algorithm has reached what could be the optimum solution.

As the plant propagation algorithm is not equipped with an explicitly defined, unique termination mechanism, one could implement the commonly available options, most notably limiting the number of iterations, run time, or perhaps monitoring the improvement made to the best solution in consecutive iterations. Among these options, limiting the number of iterations is arguably the most cited mechanism to create a termination stage for the plant propagation algorithm. The idea being the process would be executed only for a specified number of times, a parameter known as the maximum iteration. In any case, it should be noted that the selection of a termination mechanism is also considered one of the algorithm's parameters. Bear in mind that in most cases, these termination mechanisms may require setting up additional parameters.

16.3 Parameter Selection and Fine-Tuning the Plant Propagation Algorithm

From the *no-free-lunch theorem*, one can conclude that fine-tuning an algorithm is essential to get the best performance out of a meta-heuristic algorithm. This would basically ensure that an algorithm is equipped to handle the unique characteristics of a given optimization problem. Of course, it is possible to use our intuition, experience, and default values suggested for an algorithm's parameters as a good

```
Begin
        Set the algorithm's parameter and input the data
        Generate the initial population
        Let M denote the population size
        Let N denote the number of decision variables
        While the termination criterion is not met
                Evaluate the fitness of all solutions
                Normalize the fitness values
                For i in range 1 to M
                        Compute the number of runners denoted by μ_i
                        For j in range 1 to μ_i
                                Compute the position of the runner
                        Next j
                Next i
                Add the new tentative solutions to the main population
                Remove inferior solutions from the population
        End while
        Report the best solution
End
```

Figure 16.2 Pseudocode for the plant propagation algorithm.

starting point, one should bear in mind that fine-tuning these parameters is, more than anything, a trial-and-error process. Thus, while it is possible to get a good enough result by having an educated guess for setting the parameters of these algorithms, to get the best possible performance, it is necessary to go through this fine-tuning process.

In the case of the plant propagation algorithm, these parameters are population size (M), the maximum number of runners that can be issued by a plant (λ_{max}), and of course, opting for the termination criterion, and all the parameters that are associated with these methods. For instance, if limiting the number of iterations has been selected as a termination criterion, the maximum iteration (T) is another parameter that needs to be defined by the user. As can be seen here, not only the plant propagation algorithm has few parameters to begin with, but for the most part, the role of these parameters on the final outcome is easy to deduce intuitively, even for those with little experience with meta-heuristic optimization, which makes the fine-tuning process a bit more manageable than other population-based optimization algorithms we have seen thus far. That said, to get the best results from the plant propagation algorithm, it is best to dabble with these algorithms first to gain some experience and inside knowledge about such parameters. By doing so, you could better understand how to fine-tune these parameters as your initial guesses and parameter selection strategies become more educated. The pseudocode for the plant propagation algorithm is shown in Figure 16.2.

16.4 Python Codes

The code to implement the plant propagation algorithm can be found below:

```python
import numpy as np

def init_generator(pop_size, num_variables, min_val, max_val):
    return np.random.uniform(min_val, max_val, (pop_size, num_
variables))

def sorting_pop(pop, obj_func, minimizing):
    ofs = np.apply_along_axis(obj_func, 1, pop)
    indeces = np.argsort(ofs)
    if not minimizing:
        indeces = indeces[::-1]
    ofs = ofs[indeces]
    best_of = ofs[0]
    worst_of = ofs[-1]
    normailized_ofs = (ofs-worst_of)/(best_of-worst_of)
    return pop[indeces], normailized_ofs

def num_runners(normalized_ofs, max_runners):
    rand = np.random.uniform(0, 1, size=normalized_ofs.shape)
    runners_val = np.ceil(normalized_ofs*rand*max_runners)
    return runners_val.astype(int)

def dist_runners(normailized_of, num_variables, seed):
    rand = np.random.uniform(-1, 1, size=(seed, num_variables))
    dists = (1-normailized_of)
    dist_vals = dists*rand
    return dist_vals

def relocating(x, dist_val, min_val, max_val):
    new_x = x+((max_val-min_val)*dist_val)
    new_x= np.where(new_x>max_val, max_val, new_x)
    new_x = np.where(new_x<min_val, min_val, new_x)
    return new_x

def generate_new_pop(pop, pop_size, num_variables, normailized_ofs,
                     runners_val, min_val, max_val):
    new_pop = pop
    for i in range(1, pop_size):
        seed = runners_val[i]
        x = pop[i]
        dist_val = dist_runners(normailized_ofs[i], num_
variables, seed)
        new_x = relocating(x, dist_val, min_val, max_val)
        new_pop = np.concatenate((new_pop, new_x), 0)
    return new_pop

def select(new_pop, pop_size, obj_func, minimizing):
    ofs = np.apply_along_axis(obj_func, 1, new_pop)
```

```
    indeces = np.argsort(ofs)
    if not minimizing:
        indeces = indeces[::-1]
    new_pop = new_pop[indeces]
    return new_pop[:pop_size]

def plant_propagation_algorithm(pop_size, num_variables, min_val,
                                max_val, obj_func, iteration,
                                max_runners, minimizing=True,
                                full_results=False):
    NFE_value = 0
    NFE = np.zeros(iteration)
    results = np.zeros(iteration)
    pop = init_generator(pop_size, num_variables, min_val, max_val)
    NFE_value += pop_size
    for i in range(iteration):
        pop, normailized_ofs = sorting_pop(pop, obj_func,
    minimizing)
        runners_val = num_runners(normailized_ofs, max_runners)
        new_pop = generate_new_pop(pop, pop_size, num_variables,
                                   normailized_ofs, runners_val,
    min_val,
                                        max_val)
        pop = select(new_pop, pop_size, obj_func, minimizing)
        NFE_value += pop_size
        NFE[i] = NFE_value
        results[i] = obj_func(pop[0])
    if not full_results:
        return pop[0], obj_func(pop[0])
    else:
        return pop[0], obj_func(pop[0]), results, NFE
```

16.5 Concluding Remarks

Inspired by the generic mechanism by which plants would spread into a new environment, the plant propagation algorithm presents itself as a formidable meta-heuristic algorithm when it comes to tackling complex real-world problems. Though the algorithm is heavily influenced by the computational structure of the invasive weed optimization algorithm, the nuances and subtle improving features used in this algorithm made it an interesting stochastic population-based meta-heuristic optimization method from a computational standpoint. First and foremost, this algorithm has found an elegant way to carry out both local and exploratory searches simultaneously. The idea is that the search for plants that are deemed more suitable resembles the characteristics of an intensive local search. For the plants that are found to be not that suitable, these searches would take the form of exploratory moves. Building upon such architecture prevents the algorithm from being equipped with an intricate mechanism to smoothly transition from the exploration to the exploitation phase. The most notable outcomes of this type of computational structure are having a relatively straightforward to implement computational structure, lack of an overwhelming number of complicated parameters, and an efficient searching strategy. However, this would also mean that there are no guarantees that the algorithm could always identify the optimum solution, and if the algorithm is not fine-tuned properly, there is a possibility it gets trapped in local optima. All in all, being a simplified take on the invasive weed optimization algorithm, the plant propagation algorithm can certainly be considered a viable option when tackling real-world complex optimization problems.

References

Asvini, M.S. & Amudha, T. (2016). An efficient methodology for reservoir release optimization using plant propagation algorithm. *Procedia Computer Science*, 93, 1061–1069.

Bozorg-Haddad, O., Solgi, M., & Loáiciga, H.A. (2017). *Meta-heuristic and evolutionary algorithms for engineering optimization*. John Wiley & Sons. ISBN: 9781119386995

Cheraitia, M., Haddadi, S., & Salhi, A. (2019). Hybridising plant propagation and local search for uncapacitated exam scheduling problems. *International Journal of Services and Operations Management*, 32(4), 450–467.

Du, K.L. & Swamy, M.N.S. (2016). *Search and optimization by metaheuristics: Techniques and algorithms inspired by nature*. Springer International Publishing Switzerland. ISBN: 9783319411910

Fraga, E.S., Salhi, A., Zhang, D., & Papageorgiou, L.G. (2015). Optimisation as a tool for gaining insight: An application to the built environment. *Journal of Algorithms & Computational Technology*, 9(1), 13–26.

Haddadi, S. (2020). Plant propagation algorithm for nurse rostering. *International Journal of Innovative Computing and Applications*, 11(4), 204–215.

Khirade, S.D. & Patil, A. B. (2015). Plant disease detection using image processing. In Proceeding of 2015 International Conference on Computing, Communication Control, and Automation, Pune, India.

Salhi, A. & Fraga, E.S. (2011). Nature-inspired optimisation approaches and the new plant propagation algorithm. In *Proceedings of t*he International Conference on Numerical Analysis and Optimization, Yogyakarta, Indonesia.

Selamoğlu, B.İ. & Salhi, A. (2016). The plant propagation algorithm for discrete optimisation: The case of the travelling salesman problem. In Yang, X.S. ed. *Nature-inspired computation in engineering*. Springer, 43–61.

Sulaiman, M. & Salhi, A. (2016). A hybridisation of runner-based and seed-based plant propagation algorithms. In Yang, X.S. ed. *Nature-inspired computation in engineering*. Springer, 195–215.

Sulaiman, M., Sulaman, M., Hamdi, A., & Hussain, Z.H. (2020). The plant propagation algorithm for the optimal operation of directional over-current relays in electrical engineering. *Mehran University Research Journal of Engineering and Technology*, 39(2), 223–236.

Waqar, A., Subramaniam, U., Farzana, K., Elavarasan, R.M., Habib, H.U.R., Zahid, M., & Hossain, E. (2020). Analysis of optimal deployment of several DGs in distribution networks using plant propagation algorithm. *IEEE Access*, 8, 175546–175562.

17 Teaching-Learning-Based Optimization Algorithm

Summary

Inspired by an idealized representation of how a group of individuals would teach and learn a given subject, the teaching-learning-based optimization algorithm is considered one of the most notable cases of meta-heuristic algorithms that actually uses social structures and human interaction as a source of inspiration to create a search engine for an optimization algorithm. A fairly straightforward computational structure and, more importantly, few parameters make it rather easy to implement this algorithm. These features have made this a popular choice for handling complex real-world optimization problems. In this chapter, we will dig deep and explore the mechanisms used in this algorithm. We would get familiar with the teaching-learning-based algorithm's terminology and see how one can implement this algorithm in the Python programming language. Finally, we will explore the potential merits and drawbacks of this algorithm.

17.1 Introduction

For most of us, our college years are some of the fondest memories of our life. It is a time when you experience independency, engage in thought-provoking conversations, expand your inner circle with interesting individuals, have fun, and learn a thing or two about subjects that interest you. In a sense, college prepares us by teaching us the way of life. But our learning did not start from college, and for that matter, nor should it stop there. As a society, we have collectively decided that the youngsters of our community should spend most of their early adolescents learning things we deem crucial for them to survive and thrive in the world. But the desire to learn and imitate comes instinctively to us even when we are toddlers. We have learned to walk or speak simply because we have observed these behaviors from the outside world. In fact, arguably, the very reason that we as a species were able to thrive the way we did could be attributed to the fact that we were able to pass the gained knowledge and experience to the other members of our species and, by the same token, to the next generation. Little by little, these collective pearls of wisdom have shaped our past and present and continue to build our future.

DOI: 10.1201/9781003424765-17

Learning from the surrounding environment is, by no means, limited to humankind, as other species have learned to communicate with one another in various ways to pass helpful information. But the level of sophistication in these communications and exchanging of information with other species cannot even come close to what we have perfected over the years. Learning makes individuals adapt better to their surroundings by acquiring new skill sets and additional information about a subject (Du & Swamy, 2016).

There are undoubtedly different ways to acquire a new skill set. For instance, one can learn about something through observing, experiencing, and mimicking a given behavior. But perhaps, one of the most common approaches is teaching and learning. This is the primary mechanism we choose to preserve our academic body of knowledge by passing it to the next generation. The gist of this process is based on an experienced and knowledgeable individual who knows the ins and outs of a subject, from here onward, called the *teacher*, to share the information with others who want to learn about the said subject. These students, or *learners*, would soak up this information to enhance their skill sets. Naturally, learners would have different capacities to understand and absorb this new information. While some may be more talented and pick up these skill sets rather effortlessly, others may struggle and accrue only part of this new information. By the same token, teaching is also something that is natural for some and challenging for others. As such, while some are more successful in making others grasp the fundamentals of a study subject, others may fail to present and share their knowledge in the best way possible. What is interesting about this process is that the learners could also accrue some information or improve themselves, in a sense, by observing and mimicking other learners in their group. In fact, these are why most education systems encourage students to engage in teamwork activities. This way, students accrue the critical skill of social interaction and can enhance their learning experience by sharing and interacting with their peers and friends.

Inspired by an idealized interpretation of the above phenomenon, Rao et al. (2011) theorized a novel meta-heuristic algorithm called the teaching-learning-based optimization algorithm. Built upon the principles of swarm intelligence, the teaching-learning-based optimization algorithm could be described as a stochastic population-based algorithm that emulates the procedure in which information is passed between individuals. Being based on the principles of the direct search would simply mean that, like other meta-heuristic algorithms, the teaching-learning-based optimization algorithm is also not bound by problems that are often associated with high dimensionality, multimodality, epistasis, non-differentiability, and discontinuous search space imposed by constraints (Du & Swamy, 2016; Sarzaeim et al., 2018). One of the exciting things about this algorithm is that it actually uses social structures and human interaction as a source of inspiration to power its search engine. But from a computational perspective, perhaps the most distinctive feature of this algorithm is that it has the bare minimum number of parameters that we came to expect from meta-heuristic algorithms. Coupling this with a fairly straightforward computational structure of this algorithm makes this algorithm a formidable choice to tackle optimization problems, even for those with little experience with meta-heuristic optimization algorithms.

The teaching-learning-based optimization algorithm is based on an abstract interpretation of the philosophy of teaching and learning within the context of a group of individuals. From a computational standpoint, the architecture of this algorithm is based on two main phases that are the *teacher phase* and the *learner phase*. The idea is that the algorithm would initiate its search by randomly generating a series of search agents called learners. The best individual group would be elected as the teacher. Using the computational principles of the teacher phase, the algorithm would use the selected teacher as a guiding point to adjust the properties of the learners in the group. Note that the algorithm is heavily based on the *greedy strategy*. This means that the algorithm would only permit the improving moves to take effect and ignore any other suggested adjustments. After the teacher phase, the individuals would be paired randomly so that the learners would help enhance the properties of their partners. As such, in each pair, the algorithm would use the properties of the superior learner to enhance the quality of one of the learners. Again, it is essential to note that the greedy strategy used in this algorithm would only accept the improving adjustments. This whole process would be repeated until a termination criterion is met, at which point the best teacher/learner encountered thus far would be returned as the solution to the optimization problem at hand.

Over the years, many variants of the teaching-learning-based optimization algorithm have been proposed in the literature, some of which are the chaotic teaching-learning-based optimization algorithm (Farah et al., 2106), the elitism-based teaching-learning optimization algorithm (Bhadoria et al., 2106), and the binary teaching-learning-based optimization algorithm (Khuat & Le, 2019), to name a few. That said, the standard teaching-learning-based optimization algorithm is still considered a viable option for handling real-world optimization problems. In fact, standard teaching-learning-based optimization algorithm has been successfully used for computer science (e.g., Mousavi et al., 2017), civil engineering (e.g., Shahrouzi & Sabzi, 2018), energy industry (e.g., Bhattacharyya & Babu, 2016), environmental engineer (e.g., Bayram et al., 2015), financial market (e.g., Nayak et al., 2020), hydrology (e.g., Bozorg-Haddad et al., 2021), image processing (e.g., Jin & Wang, 2014), land-use planning and management (e.g., Deb et al., 2021), material engineering (e.g., Suresh et al., 2020), medical science (e.g., Balakrishnan, 2020), nuclear engineer (e.g., Sahu et al., 2015), project management (e.g., Abirami et al., 2014), robotic science (e.g., Majumder et al., 2021), structural engineering (e.g., Degertekin & Hayalioglu, 2013), and water resources planning and management (e.g., Sarzaeim et al., 2018; Patel & Bhavsar, 2021; Yaghoubzadeh-Bavandpour et al., 2022). In the following sections, we will explore the computational structure of the standard teaching-learning-based optimization algorithm.

17.2 Algorithmic Structure of the Teaching-Learning-Based Optimization Algorithm

This population-based meta-heuristic algorithm enumerates through the search space through a set of search agents. But what distinguishes this algorithm from the most meta-heuristic algorithm we have seen thus far is that there is no explicit transformation from the exploration to the exploitation phase, as there are few parameters to control this transition. However, this algorithm works because the greedy strategy's core principle is embedded within this algorithm's main computational structure. The idea is that the algorithm, for the most part, prevents any non-improving move. So, in a sense, it is a controlled randomized local search that utilizes the idea of the greedy search so that it could constantly improve the properties of the population set.

The teaching-learning-based optimization algorithm's flowchart is depicted in Figure 17.1. A closer look at the architecture of the teaching-learning-based optimization algorithm would reveal that it actually consists of three main stages that are the initiation, teaching/learning, and termination stages. The teaching/learning stage itself consists of two main pillars that are the teacher phase and the learner phase. Using this structure, the algorithm would conduct a thorough search and locate what could be the optimum solution to the problem at hand. The following subsection will discuss each of these stages and their mathematical structures.

17.2.1 Initiation Stage

The teaching-learning-based optimizing algorithm is a population-based meta-heuristic optimization algorithm, and as such, it works with multiple search agents, here called teacher/learner, that would enumerate through the search space. As we have seen, in an optimization problem with N decision variables, an N-dimension coordination system could be used to represent the search space. In this case, any point within the search space, say X, can be represented mathematically as a $1 \times N$ array as follows:

$$X = \left(x_1, x_2, x_3, \ldots, x_j, \ldots, x_N \right) \tag{17.1}$$

where X represents a search agent in the search space of an optimization problem with N decision variables, and x_j represents the value associated with the jth decision variable. Note that while these search agents would be labeled as teacher and learner further down the line, they both have the same mathematical structure.

The teaching-learning-based optimizing algorithm starts with randomly placing a series of search agents within the feasible boundaries of the search space. This bundle of arrays, which in the teaching-learning-based optimizing algorithm's terminology is referred to as the population, can be mathematically expressed as $M \times N$ matrix, where M denotes the number of particles or what is technically referred to as the population size. In such a structure, each row represents a single search agent. A population, denoted by *pop*, can be represented as follows:

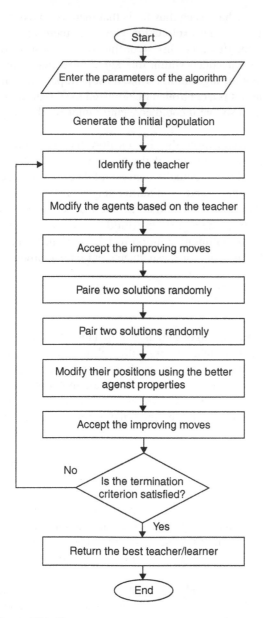

Figure 17.1 The computational flowchart of the teaching-learning-based optimization algorithm.

$$pop = \begin{bmatrix} X_1 \\ X_2 \\ \vdots \\ X_i \\ \vdots \\ X_M \end{bmatrix} = \begin{bmatrix} x_{1,1} & x_{1,2} & \cdots & x_{1,j} & \cdots & x_{1,N} \\ x_{2,1} & x_{1,2} & \cdots & x_{2,j} & \cdots & x_{2,N} \\ & & & \vdots & & \\ x_{i,1} & x_{i,2} & \cdots & x_{i,j} & \cdots & x_{i,N} \\ & & & \vdots & & \\ x_{M,1} & x_{M,2} & \cdots & x_{M,j} & \cdots & x_{M,N} \end{bmatrix} \tag{17.2}$$

where X_i represents the ith agent in the population, and $x_{i,j}$ denotes the jth decision variable of the ith agent.

The initially generated population represents the initial position of search agents. As we progress, the values stored in the *pop* matrix would be altered according to the computational structure of the teaching-learning-based optimizing algorithm's teaching/learning stage. By the end of this iterative computation process, when the termination criterion is met, one or possibly multiple agents could converge to the optimum solution.

17.2.2 *Teaching/Learning Stage*

Through mimicking an abstract representation of how knowledge is passed among a group of individuals, in this stage, the algorithm would tend to improve the properties of the population set. The teaching/learning stage itself consists of two main pillars that are the teacher phase and the learner phase. These mechanisms are used as alteration tools to ultimately enhance the population set. It should be noted that the idea of the greedy strategy is integrated into both these phases. Using the idea of the greedy strategy, the algorithm would compare these solutions with their original counterparts and replace them in case they show better performance. As such, the algorithm ensures that it always preserves the best properties of the population set.

As stated, the repositioning of search agents in this algorithm is done via two main mechanisms, which are the teacher phase and the learner phase. The former phase is based on the notion that the primary way to educate a group of individuals is to assign a teacher to the said group to absorb the tutor's knowledge. From a computational perspective, this could mean that the algorithm tends to encourage the community's learners to adapt to the properties of the teacher the best way they can. As such, the first step is to assign one of the population members set as the teacher. The teacher should be the best member of the population set. As such, the algorithm would need to evaluate and identify the best search agent in the population set, denoted by *Xbest*. The rest of the members would then be labeled as learners. It goes without saying that the most desirable solutions (i.e., identifying the best organism in the population set) have a different interpretation for maximization and minimization problems. For instance, higher objective functions are considered more desirable in a maximization problem, while we are looking for lower values for the objective function in a minimization problem. The said search agent could be mathematically represented as follows:

$$Xbest = \left(x_{best,1}, x_{best,2}, x_{best,3}, \ldots, x_{best,j}, \ldots, x_{best,N}\right) \tag{17.3}$$

where $x_{best,j}$ denotes the position of the best-identified search agent in the population in the jth dimension.

Again the main idea of this phase is to create a generic motion that pushes the entire set toward the teacher's position. For this, the teaching-learning-based optimization algorithm suggests finding the vector that connects the center of gravity of the learners to the teacher's position. The center of gravity of the learners, here denoted by *Xmean*, can be computed as follows:

$$Xmean = \left(x_{mean,1}, x_{mean,2}, x_{mean,3}, \ldots, x_{mean,j}, \ldots, x_{mean,N}\right) \tag{17.4}$$

$$x_{mean,j} = \frac{\sum_{i=1}^{M} x_{i,j}}{M} \quad \forall j \tag{17.5}$$

in which $x_{mean,j}$ denotes the learner's center of gravity position in the population in the jth dimension.

The algorithm would then use these points to adjust the population set as follows:

$$diff_{i,j} = Rand \times \left(x_{best,j} - T_F \times x_{mean,j}\right) \quad \forall j, T_F \in \{1,2\} \tag{17.6}$$

$$x'_{i,j} = x_{i,j} + diff_{i,j} \quad \forall j \tag{17.7}$$

where $x'_{i,j}$ denotes the new position in the jth dimension for the tentative solution associated with the ith search agent, $diff_{i,j}$ denotes the value of the repositioning vector for the ith learner in the jth dimension, *Rand* is a randomly generated value within the range 0–1, and T_F denotes the teaching factor which is randomly assuming the value 1 or 2.

Given that the greedy strategy is embedded in this algorithm, it is essential that the repositioned agents are checked against their previous positions. The algorithm would only permit moves that improve the search agents' properties while disregarding the rest. As such, the repositioned agents are compared against their previous positions. If the new position is more desirable in terms of the objective function value, the agent would be moved to the new position, otherwise, it stays where it was before.

After the teacher phase comes the learner phase, which basically emulates the process where learners help improve the properties of other community members.

The algorithm would continue by randomly coupling two agents from the population set to emulate this sort of interaction to implement this idea. Let us assume that the *r*th search agent, denoted by X_r, has been randomly selected from the population set to be coupled with the *i*th learner. As one could imagine, in any paired agent, the learner with the better properties would assume the role of the guide, the local teacher, for this interaction. As such, the algorithm would evaluate both agents and identify the more suitable agent in terms of the objective function value. Let us assume that the local teacher and the local learner are denoted by *Xlocbest* and *Xlocworst*, respectively. The algorithm would then attempt to update the position of the *i*th learner as follows:

$$x'_{i,j} = x_{i,j} + Rand \times \left(x_{locbest,j} - x_{locworst,j} \right) \qquad \forall j \tag{17.9}$$

where $x_{locbest,j}$ denotes the position of the *Xlocbest* in the *j*th dimension, and $x_{locworst,j}$ denotes the position of the *Xlocworst* in the *j*th dimension. Again it is important to note that given that the greedy strategy is embedded in this algorithm, the repositioned agents are checked against their previous positions. The algorithm would only permit those moves that improve the properties of the search agents while disregarding the rest. As such, the repositioned agents are compared against their previous positions. If the new position is more desirable in terms of the objective function value, the agent would be moved to the new position, otherwise, it stays where it was before.

17.2.3 Termination Stage

Based on the above procedure, in each given iteration, after an overall generic repositioning of the entire set based on the teacher's properties, the algorithm would go over the agents in the population set one by one and apply the learning phase. The idea here is that two given agents would be selected randomly, and the superior learner would attempt to improve the properties of its partner. It is important to note here that based on the greedy strategy, the algorithm is designed so that it can only accept improving moves. Thus, in any stage of the search, anytime the algorithm detects an improving move, the position of the search agents would be updated within the search space, otherwise, the agent's position would remain unchanged. Like other meta-heuristic algorithms, the sequence of operational structures of this algorithm needs to be executed iteratively until a certain termination criterion is met, at which point the execution of the algorithm would be terminated, and the best agent recorded in the memory would be reported as the solution to the optimization problem. Note that without such a termination stage, the algorithm would potentially be executed in an infinite loop. The termination stage would, in effect, determine whether the algorithm has reached what could be the optimum solution.

As the teaching-learning-based optimization algorithm is not equipped with an explicitly defined, unique termination mechanism, one could implement the commonly available options, most notably limiting the number of iterations, run time, or perhaps monitoring the improvement made to the best solution in consecutive iterations. Among these options, limiting the number of iterations is arguably the most cited mechanism to create a termination stage for the teaching-learning-based optimization algorithm. The idea being the process would be executed only for a specified number of times, a parameter known as the maximum iteration. In any case, it should be noted that the selection of the termination mechanism is also considered one of the algorithm's parameters. Bear in mind that in most cases, these termination mechanisms may require setting up additional parameters.

17.3 Parameter Selection and Fine-Tuning the Teaching-Learning-Based Optimization Algorithm

From the *no-free-lunch theorem*, one can conclude that fine-tuning an algorithm is essential to get the best performance out of a meta-heuristic algorithm. This would basically ensure that an algorithm is equipped to handle the unique characteristics of a given optimization problem. Of course, it is possible to use our intuition, experience, and default values suggested for an algorithm's parameters as a good starting point, one should bear in mind that fine-tuning these parameters is more than anything a trial-and-error process. Thus, while it is possible to get a good enough result by having an educated guess for setting the parameters of these algorithms, to get the best possible performance, it is necessary to go through this fine-tuning process.

In the case of the teaching-learning-based optimization algorithm, it should be noted that the algorithm has been specifically designed to have the minimum number of parameters. In this case, the only parameters used in the structure of the algorithm are the population size (M) and, of course, opting for the termination criterion and all the parameters that are associated with these methods. For instance, if limiting the number of iterations has been selected as a termination criterion, the maximum iteration (T) is another parameter that needs to be defined by the user. As can be seen here, not only does the teaching-learning-based optimization algorithm have very few parameters to begin with, but for the most part, the role of these parameters on the final outcome is easy to deduce intuitively. This makes the fine-tuning process less challenging for those less experienced. That said, to get the absolute best results out of the teaching-learning-based optimization algorithm, it is best to dabble with these algorithms first to gain some experience and inside knowledge about such parameters. By doing so, you could better understand how to fine-tune these parameters as your initial guesses and parameter selection strategies become more educated. The pseudocode for the teaching-learning-based optimization algorithm is shown in Figure 17.2.

```
Begin
        Set the algorithm's parameter and input the data
        Generate the initial population
        Let M denote the population size
        While the termination criterion is not met
                Evaluate the generated agents
                Let Xbest denote the best agent in the set
                Compute the population mean, denoted by Xmean
                Let diff denote the distance between Xbest and Xmean
                Use diff to update the position of the population
                Accept improving moves and reject the non-improving ones
                For i in range 1 to M
                        Select the rth agent denoted by Xr randomly
                        Use the better agent to improve the Xr position
                        Accept an improving move and reject a non-improving one
                Next i
        End while
        Report the best solution
End
```

Figure 17.2 Pseudocode for the teaching-learning-based optimization algorithm.

17.4 Python Codes

The code to implement the teaching-learning-based optimization algorithm can be found below:

```python
import numpy as np

def init_genrator(min_val, max_val, num_variables, pop_size):
    return np.random.uniform(min_val, max_val, (pop_size, num_
    variables))

def sorting_pop(pop, minimizing, obj_func):
    results = np.apply_along_axis(obj_func, 1, pop)
    indeces = np.argsort(results)
    if not minimizing:
        indeces = indeces[::-1]
    return pop[indeces]

def teacher_phase(pop, pop_size, obj_func, minimizing):
    teacher = pop[0]
    center = np.mean(pop, axis=0)
    rand = np.random.uniform(0,1,size=(pop_size,1))
    teacher_factor = np.random.choice(np.arange(1,3), size=(pop_size,1))
    diff = rand*(teacher-center*teacher_factor)
    pop_new = pop + diff
    of = np.apply_along_axis(obj_func, 1, pop).reshape(-1,1)
    of_new = np.apply_along_axis(obj_func, 1, pop_new).reshape(-1,1)
    if minimizing:
        return np.where(of_new<of, pop_new, pop)
    else:
        return np.where(of_new>of, pop_new, pop)
```

```
def learner_phase(pop, pop_size, obj_func, minimizing):
    rand = np.random.uniform(0,1,size=(pop_size,1))
    pop_new = np.zeros_like(pop)
    for i in range(pop_size):
        x, y = np.sort(np.random.choice(np.arange(pop_size),
                                        size=2, replace=False))
        pop_new[i] = pop[i] + rand[i]*(pop[x]-pop[y])
    of = np.apply_along_axis(obj_func, 1, pop).reshape(-1,1)
    of_new = np.apply_along_axis(obj_func, 1, pop_new).reshape(-1,1)
    if minimizing:
        return np.where(of_new<of, pop_new, pop)
    else:
        return np.where(of_new>of, pop_new, pop)

def teaching_learning_based_optimization(num_variables, min_val,
                                         max_val, pop_size,
                                         iteration,
                                         minimizing, obj_func,
                                         full_results=False):
    NFE_value = 0
    NFE = np.zeros(iteration)
    results = np.zeros(iteration)
    pop = init_genrator(min_val, max_val, num_variables, pop_size)
    pop = sorting_pop(pop, minimizing, obj_func)
    for i in range(iteration):
        pop = teacher_phase(pop, pop_size, obj_func, minimizing)
        pop = sorting_pop(pop, minimizing, obj_func)
        pop = learner_phase(pop, pop_size, obj_func, minimizing)
        pop = sorting_pop(pop, minimizing, obj_func)
        NFE_value += (3*pop_size)
        NFE[i] = NFE_value
        results[i] = obj_func(pop[0])
    if not full_results:
        return pop[0], obj_func(pop[0])
    else:
        return pop[0], obj_func(pop[0]), results, NFE
```

17.5 Concluding Remarks

Inspired by an idealized representation of how a group of individuals would teach and learn a given subject, the teaching-learning-based optimization algorithm presents itself as a formidable meta-heuristic algorithm when it comes to tackling complex real-world problems. A straightforward computational structure and few parameters are perhaps one of the most notable features of this algorithm. However, there are two major problems with this algorithm. Firstly, the teaching-learning-based optimization algorithm uses the greedy strategy rather excessively. The point is that the algorithm would only accept improving moves at any stage. This generally ensures that the algorithm is either improving its properties or at least holding its current relatively superior properties. As such, the solutions obtained in each iteration are either improving or at least staying the same as the previous iterations. However, there are a few pitfalls when an algorithm relies heavily on this strategy, for instance, to the same extent that we have seen in the teaching-learning-based algorithm.

First and foremost, it can be rather computationally taxing, especially in real-world problems when the simulation process is a bit more intricate and requires significant computational power. The point is any update would require at least two simulations to determine whether the move could be accepted or not. And for this algorithm, for each search agent in any given iteration, you may need to call the simulation part multiple times, given that the algorithm actually has multiple phases. As a result, this is not particularly the most efficient way to solve problems from a computational standpoint. The other notable pitfall here is that by resorting to this strategy, the algorithm set up the search agent to move toward local optima. If a search agent moves toward a local optimum in such a situation, there are no ways that it could release itself from this trap. So the only way that the user could effectively address this issue is by increasing the number of search agents, which could be computationally taxing. The problem gets amplified with this algorithm, given that there is no meaningful way for the user to adjust the algorithm to the problem at hand as an alternative approach.

This brings us to the second major drawback of this algorithm which is the lack of adjusting parameters. Of course, not being riddled with parameters would simply mean that the fine-tuning of the algorithm is much more manageable, which, in turn, facilitates the algorithm's implementation even for those with little to no experience working with meta-heuristic optimization algorithms. But this seemingly huge advantage comes at a price, lacking enough controlling parameters. This means that there are not many ways to control the properties of the search engine of the algorithm. In other words, the user does not have many options to tailor the algorithm to a specified optimization problem. As a result, the user, for instance, cannot control the transition between the exploration and exploitation phase during the search. This is not to say there is no exploration or exploitation phase during the search procedure. In fact, as the algorithm initiates, the movement is more exploratory, but as the search progresses and the solutions get closer together, the search is more in line with the idea of the exploitation phase. But the critical thing to note here is that the user has no control over the pace of this transition, given that there are no significant parameters to control the algorithm's search properties. As such, the only way to avoid being trapped in local optima or premature convergence is to increase the population size or the number of iterations. Even with these changes, we cannot guarantee we will avoid these pitfalls altogether. With that said, all in all, from a practical standpoint, the teaching-learning-based optimization algorithm can undoubtedly be considered a viable option when it comes to tackling real-world complex optimization problems.

References

Abirami, M., Ganesan, S., Subramanian, S., & Anandhakumar, R. (2014). Source and transmission line maintenance outage scheduling in a power system using teaching learning based optimization algorithm. *Applied Soft Computing*, 21, 72–83.

Balakrishnan, S. (2020). Feature selection using improved teaching learning based algorithm on chronic kidney disease dataset. *Procedia Computer Science*, 171, 1660–1669.

Bayram, A., Uzlu, E., Kankal, M., & Dede, T. (2015). Modeling stream dissolved oxygen concentration using teaching–learning based optimization algorithm. *Environmental Earth Sciences*, 73(10), 6565–6576.

Bhadoria, A., Singh, M., & Gupta, M. (2016). An improved elitism based teaching-learning optimization algorithm. In *Proceeding of 2016 International Conference on Electrical, Electronics, and Optimization Techniques*, Chennai, India.

Bhattacharyya, B. & Babu, R. (2016). Teaching learning based optimization algorithm for reactive power planning. *International Journal of Electrical Power & Energy Systems*, 81, 248–253.

Bozorg-Haddad, O., Sarzaeim, P., & Loáiciga, H.A. (2021). Developing a novel parameter-free optimization framework for flood routing. *Scientific Reports*, 11(1), 1–14.

Deb, S., Gao, X.Z., Tammi, K., Kalita, K., & Mahanta, P. (2021). A novel chicken swarm and teaching learning based algorithm for electric vehicle charging station placement problem. *Energy*, 220, 119645.

Degertekin, S.O., & Hayalioglu, M.S. (2013). Sizing truss structures using teaching-learning-based optimization. *Computers & Structures*, 119, 177–188.

Du, K.L. & Swamy, M.N.S. (2016). *Search and optimization by metaheuristics: Techniques and algorithms inspired by nature.* Springer International Publishing Switzerland. ISBN: 9783319411910

Farah, A., Guesmi, T., Abdallah, H.H., & Ouali, A. (2016). A novel chaotic teaching–learning-based optimization algorithm for multi-machine power system stabilizers design problem. *International Journal of Electrical Power & Energy Systems*, 77, 197–209.

Jin, H. & Wang, Y. (2014). A fusion method for visible and infrared images based on contrast pyramid with teaching learning based optimization. *Infrared Physics & Technology*, 64, 134–142.

Khuat, T.T. & Le, M.H. (2019). Binary teaching–learning-based optimization algorithm with a new update mechanism for sample subset optimization in software defect prediction. *Soft Computing*, 23(20), 9919–9935.

Majumder, A., Majumder, A., & Bhaumik, R. (2021). Teaching–learning-based optimization algorithm for path planning and task allocation in multi-robot plant inspection system. *Arabian Journal for Science and Engineering*, 46, 1–23.

Mousavi, S., Mosavi, A., & Varkonyi-Koczy, A.R. (2017). A load balancing algorithm for resource allocation in cloud computing. In *Proceeding of International Conference on Global Research and Education*, Iaşi, Romania.

Nayak, S.C., Das, S., & Ansari, M.D. (2020). Tlbo-fln: Teaching-learning based optimization of functional link neural networks for stock closing price prediction. *International Journal of Sensors Wireless Communications and Control*, 10(4), 522–532.

Patel, J.N. & Bhavsar, P.N. (2021). Optimal distribution of water resources for long-term agricultural sustainability and maximization recompense from agriculture. *ISH Journal of Hydraulic Engineering*, 27(1), 110–116.

Rao, R.V., Savsani, V.J., & Vakharia, D.P. (2011). Teaching–learning-based optimization: A novel method for constrained mechanical design optimization problems. *Computer-Aided Design*, 43(3), 303–315.

Sahu, B.K., Pati, S., Mohanty, P.K., & Panda, S. (2015). Teaching–learning based optimization algorithm based fuzzy-PID controller for automatic generation control of multi-area power system. *Applied Soft Computing*, 27, 240–249.

Sarzaeim, P., Bozorg-Haddad, O., & Chu, X. (2018). Teaching-learning-based optimization (TLBO) algorithm. In *Advanced optimization by nature-inspired algorithms*. Springer, 51–58.

Shahrouzi, M. & Sabzi, A.H. (2018). Damage detection of truss structures by hybrid immune system and teaching–learning-based optimization. *Asian Journal of Civil Engineering*, 19(7), 811–825.

Suresh, S., Elango, N., Venkatesan, K., Lim, W.H., Palanikumar, K., & Rajesh, S. (2020). Sustainable friction stir spot welding of 6061-T6 aluminium alloy using improved non-dominated sorting teaching learning algorithm. *Journal of Materials Research and Technology*, 9(5), 11650–11674.

Yaghoubzadeh-Bavandpour, A., Bozorg-Haddad, O., Rajabi, M., Zolghadr-Asli, B., & Chu, X. (2022). Application of swarm intelligence and evolutionary computation algorithms for optimal reservoir operation. *Water Resources Management*, 36(7), 2275–2292.

18   Bat Algorithm

Summary

Inspired by an idealized interpretation of the echolocation characteristics of microbats, the bat algorithm holds itself as a formidable swarm intelligence-based meta-heuristic optimization algorithm that can be quite efficient when it comes to handling real-world complex optimization problems. In this chapter, we will dig deep and explore the mechanisms used in this algorithm. We would get familiar with the bat algorithm's terminology and see how one can implement this algorithm in the Python programming language. Finally, we will explore the potential merits and drawbacks of this algorithm.

18.1 Introduction

As the sun goes down every evening in southern Pacific regions, say northern Australia, millions of giant winged creatures dominate the night skies. Though such imposing sights might not be everyone's cup of tea, the reality is that these are just harmless species of bats called giant bats, the biggest of their kind in the world. On average, their wingspread could span a whopping 1.7 m [≈6 ft] and weigh up to 1.5 kg (Jones & Purvis, 1997; Anand & Sripathi, 2004). But bats, the only known winged mammals, are some of the most diverse species on the planet. For instance, there are other species of bat in western Thailand and southeast Myanmar commonly known as the bumblebee bat, the smallest mammal in the world that weighs less than a penny and is smaller than a thumbnail (Jones & Purvis, 1997). It is estimated that there are more than 1,300 identified species of bats, nearly a quarter of the world's mammal species (Aizpurua & Alberdi, 2018).

Bat species have picked up different biological evolutionary mechanisms and habits to adapt to their habitats by living in various habitats and facing different environmental challenges. For instance, while some bat species, such as giant bats, are primarily frugivorous, others, such as microbat species, are insectivores. And while this is not always the case, most bats are considered nocturnal animals that use *echolocation*, or what is also known as biosonar, for navigation, hunting, and food foraging purposes.

DOI: 10.1201/9781003424765-18

Echolocating is an adaptation ability in some animals, including bats, where the said animal emits sonar signals and use the returned bounced echoes to get a sense of their surrounding environment. Through echolocation, the animal could locate and often even identify the objects within its vicinity. This is an evolutionary coping mechanism in bat species that can perform echolocating, compensating for the challenges that these species should face as they tend to prey or forage for food during nighttime when eyesight vision cannot be as useful.

Microbats are one bat species that use their echolocation capabilities to navigate, detect, and hunt prey. While emitted pulse properties, including their frequency, loudness, or the number of pulses emitted per second, may vary depending on the circumstance, the activity in mind, and the bat species themselves, the gist of them are quite similar. The idea is that bat would often emit a loud and relatively short frequency pulse that would bounce back from the environment and be picked up by the bat. This ultrasonic sonar signal that usually has a frequency that ranges somewhere between 25 and 150 kHz is, in fact, so high-pitched that it goes undetected by human ears. The bats would process these returned sonar signals to get a sense of their surroundings. As these sound pulses last for a brief time span that could range between 8 and 10 ms, the bats need to constantly emit these sonar signals to improve their information about their surrounding environment. In some cases where the microbats need more frequent updates on their situation, say as they get closer to the prey, they would pick the pace of these sonar signal bursts. The impressive thing about this process is that the microbats' capacity to analyze the obtained information is efficient and fast.

Inspired by an idealized interpretation of the echolocation characteristics of microbats, Yang and Gandomi (2012) theorized a novel meta-heuristic algorithm called the bat algorithm. Built upon the principles of swarm intelligence, the bat algorithm could be described as a stochastic population-based algorithm that emulates the preying of microbats to create its search engine. One of the most noteworthy characteristics of this algorithm is that it follows a series of arguably simple computational instructions, which ultimately makes the algorithm easy to understand and execute. The other notable feature of this algorithm is the dynamic structure of the algorithm, where the majority of search agents' properties would be adjusted in each iteration. The idea is that through this feature algorithm would gradually decrease the searching domain of each agent. This, in turn, ensures that there is a smooth transition from the exploitation phase to the exploration phase.

The bat algorithm starts its search by randomly placing a set of search agents, here called bats, within the feasible span of the search space. The bats would then be stochastically assigned to one of the two movement patterns defined in this algorithm, one of which is to coordinate the search agent's move with the best position in the search engine's memory, and the other one is a random fly where the algorithm would select a trajectory based on pure randomness. The idea of the last move is to converge the solutions to what could be the optimum solution, while the latter move is designed to expand the algorithm's exploration capacities. While assigning these movements to each search agent is stochastic by nature, one of the brilliancies of this algorithm is that it tilts the odds of selecting for the converging moves as the algorithm progresses to create a smooth transition from the exploration phase to the exploitation phase. The described process would be repeated until the algorithm reaches a termination point, where the best-encountered position through the searching process would be returned as the optimum solution.

A closer look at the architecture of the bat algorithm could simply reveal that this algorithm is a clever mixture of a simplified and trimmed version of the particle swarm optimization algorithms and a purely random search. What is interesting here is that the algorithm builds a stochastic mechanism to transit from a phase where the pure random search is more pronounced to another phase where the search agents' realignment is more in line with the idea presented in the simplified particle swarm optimization algorithm. As such, the search agents are provided with more freedom to explore the search space at the beginning of the search till they are gradually forced to converge to the optimum solution. That said, being a meta-heuristic algorithm, the bat algorithm still suffers from the same generic issues that are commonly associated with this branch of optimization, such as having no guarantee to reach the optimum solution or being trapped in a local optimum solution. Of course, on the other hand, this could also mean that the algorithm that, like other meta-heuristic algorithms, the bat algorithm is not bound by the problems that are often associated with high dimensionality, multimodality, epistasis, non-differentiability, and discontinuous search space imposed by constraints (Yang, 2010; Bozorg-Haddad et al., 2017).

Over the years, many variants of the bat algorithm have been proposed in the literature, some of which are the adaptive bat algorithm (Wang et al., 2013), binary bat algorithm (Mirjalili et al., 2014), chaotic bat algorithm (Gandomi & Yang, 2014), complex-valued bat algorithm (Li & Zhou, 2014), and island bat algorithm (Al-Betar & Awadallah, 2018), to name a few. That said, the standard bat algorithm is still considered a viable option to handle real-world optimization problems. In fact, the standard bat algorithm has been successfully used for climatology (e.g., Dong et al., 2021), the energy industry (e.g., Niknam et al., 2013), economy (e.g., Dehghani & Bogdanovic, 2018), hydrology (e.g., Farzin et al., 2018), medical science (e.g., Lu et al., 2017), nuclear engineering (e.g., Kashi et al., 2014), remote sensing (e.g., Senthilnath et al., 2016), robotic science (e.g., Rahmani et al., 2016), structural engineering (e.g., Zenzen et al., 2018), and water resources planning and management (e.g., Bozorg-Haddad et al., 2015; Ahmadianfar et al., 2016). In the following sections, we will explore the computational structure of the standard bat algorithm.

18.2 Algorithmic Structure of the Bat Algorithm

From a practical point of view, one needs to capture the essence of the bat's echolocation procedure in a mathematical context in order to build the computational structure of a meta-heuristic optimization algorithm. To create a functional algorithm, we first need to extract and idealize some of governing principles of the bat's preying mechanism. The main idea here is to establish a simplified and abstract mathematical representation of the said procedure that would be utilized later on to assemble the search engine of the bat algorithm. These simplifying assumptions are as follows:

I   We assume that all bats can use echolocation to sense the distance between their current location and their prey's location.
II It is assumed that simulated bats fly randomly with a velocity v_i, at position x_i with a fixed frequency λ_i, varying wavelength W_i, and loudness A_i to search for prey. Furthermore, it is also assumed that bats are capable of automatically adjusting their echolocation capabilities by dynamically modifying the wavelength of the emitted sonar signals and rate of pulse emission, $r \in [0, 1]$, depending on how close they are estimated to be to the approximated prey.[1]
III For simplicity, we assume that all the properties of the echolocation process, including loudness, frequency, and pulsation rate, always range within two predefined upper and lower boundaries.
IV As the bats approach their prey, as a general strategy, they tend to decrease the loudness of the emitted signals but increase the pulsation rate.

Based on these principles, the bat algorithm creates a search engine that coordinates the movement of the search agents and navigates them to the point that could potentially be the optimum solution. It should be noted that the self-organizing nature of the search agents' movement in this algorithm qualifies it as a swarm intelligence-based optimization algorithm.

The bat algorithm's flowchart is depicted in Figure 18.1. A closer look at the architecture of the bat algorithm would reveal that it actually consists of three main stages that are the initiation, repositioning, and termination stages. Using this structure, the algorithm would conduct a thorough search and locate what could be the optimum solution to the problem at hand. The following subsection will discuss each of these stages and their mathematical structures.

18.2.1 Initiation Stage

The bat algorithm is a population-based meta-heuristic optimization algorithm, and as such, it works with multiple search agents, here called bats, that would enumerate through the search space. As we have seen, in an optimization problem with N decision variables, an N-dimension coordination system could be used to represent the search space. In this case, any point within the search space, say X, can be represented mathematically as a 1×N array as follows:

$$X = \left(x_1, x_2, x_3, \ldots, x_j, \ldots, x_N\right) \tag{18.1}$$

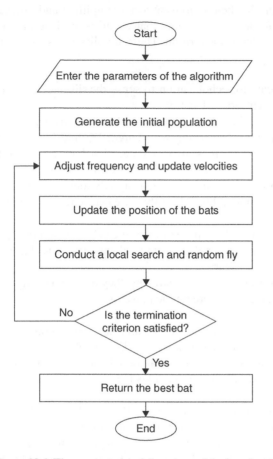

Figure 18.1 The computational flowchart of the bat algorithm.

where X represents a bat in the search space of an optimization problem with N decision variables, and x_j represents the value associated with the jth decision variable.

The bat algorithm starts with randomly placing a series of bats within the feasible boundaries of the search space. This bundle of arrays, which in the bat algorithm's terminology is referred to as the population, can be mathematically expressed as $M \times N$ matrix, where M denotes the number of particles or what is technically referred to as the population size. In such a structure, each row represents a single search agent. A population, denoted by *pop*, can be represented as follows:

$$pop = \begin{bmatrix} X_1 \\ X_2 \\ \vdots \\ X_i \\ \vdots \\ X_M \end{bmatrix} = \begin{bmatrix} x_{1,1} & x_{1,2} & \cdots & x_{1,j} & \cdots & x_{1,N} \\ x_{2,1} & x_{1,2} & \cdots & x_{2,j} & \cdots & x_{2,N} \\ & & & \vdots & & \\ x_{i,1} & x_{i,2} & \cdots & x_{i,j} & \cdots & x_{i,N} \\ & & & \vdots & & \\ x_{M,1} & x_{M,2} & \cdots & x_{M,j} & \cdots & x_{M,N} \end{bmatrix} \qquad (18.2)$$

where X_i represents the ith bat in the population, and $x_{i,j}$ denotes the jth decision variable of the ith bat.

The initially generated population represents the initial position of bats. As we progress, the values stored in the *pop* matrix will be altered according to the computational structure of the bat algorithm's repositioning stage. By the end of this iterative computation process, when the termination criterion is met, one or possibly multiple bats could converge to the optimum solution.

18.2.2 Repositioning Stage

By emulating the fundamental principle that governs the preying process of microbats, the bat algorithm attempts to alter the components of the search agents within each iteration. Here, like many other population-based meta-heuristic algorithms, each individual search agent is, in fact, conducting a series of parallelized local searches. What distinguishes this algorithm from other population-based algorithms we know thus far is that here the bat algorithm has built a structure that stochastically assigns one of two predefined position-updating procedures to each search agent. One of these movement patterns resembles a simplified interpretation of the particle swarm optimization algorithm, while the others more or less follow the principles of a random search paradigm. While the former pattern is more useful when the algorithm tends to converge the solutions into a local spot and focuses the search in more restricted regions of the search space, the latter pattern is more applicable when the goal is to promote more exploratory moves for the search agents. Thus, the former move is more in line with the idea of exploitation, while the latter is better used when the algorithm prioritizes the exploiting phase. What is interesting about this algorithm is that it creates a stochastic structure that dynamically adjusts the probabilities of selecting these patterns so that the random moves are more likely to be selected at the beginning of the search when it makes more sense to promote exploratory moves. But as the algorithm progresses, the odds would be adjusted so that it becomes more likely for the coordinating moving pattern to be selected so that the algorithm could converge the solutions into the optimum solution.

The algorithm would first need to define the computational structure of both these moving patterns to implement this idea. A simplified interpretation of the particle swarm optimization algorithm inspires the first moving pattern. The general theme of this move is for the search agents to coordinate their position with the best-encountered point by the entire population thus far. With this in mind, the algorithm would first need to identify the best position encountered by the search agents. The better the objective function of these search agents, the more desirable the corresponding bats' positions would be. Needless to say, the interpretation of being better depends on the type of optimization problem we are dealing with. For the maximization problem, for instance, the higher values for the objective function are considered better. Conversely, in a minimization problem, the lower values for objective function are deemed more desirable. In this case, let *Xbest* denote the best-encountered position.

The mathematical structure of this array can be represented as follows:

$$Xbest = \left(x_{best,1}, x_{best,2}, x_{best,3}, \ldots, x_{best,j}, \ldots, x_{best,N} \right) \tag{18.3}$$

in which *Xbest* denotes the best position encountered by the population thus far, and $x_{best,j}$ is the *j*th decision variable of the current global best solution.

The above-described process of updating the bats' positions can be mathematically expressed as follows:

$$X_i^{new} = \left(x_{i,1}', x_{i,2}', x_{i,3}', \ldots, x_{i,j}', \ldots, x_{i,N}' \right) \quad \forall i \tag{18.4}$$

$$x_{i,j}' = x_{i,j} + v_{i,j}' \quad \forall i,j \tag{18.5}$$

$$v_{i,j}' = v_{i,j} + \left[\lambda_{i,j} \times \left(x_{best,j} - x_{i,j} \right) \right] \quad \forall i,j \tag{18.6}$$

$$\lambda_{i,j} = \lambda_{min} + Rand \times \left(\lambda_{max} - \lambda_{min} \right) \quad \forall i,j \tag{18.7}$$

where X_i^{new} denotes the new position for the *i*th bat, $x_{i,j}'$ represents the upgraded value for the *j*th decision variable of the *i*th bat, *Rand* is a randomly generated value within the range 0–1, $v_{i,j}$ denotes the velocity of the *i*th bat in the *j*th dimension in the previous iteration, $v_{i,j}'$ is the velocity of the *i*th bat in the *j*th dimension in the current iteration, $\lambda_{i,j}$ denotes the frequency value of the *i*th bat in the *j*th dimension, and λ_{min} and λ_{max} are lower and upper boundaries of frequency, respectively, which are two of the user-defined parameters of the algorithm. What is happening here is that these boundaries are, in effect, working as control parameters to keep the boundaries of the movement premise in check.

As stated, the other movement pattern is based on the idea of a pure random move, which in this case is referred to as a random fly. The idea of introducing the random fly in the structure of this algorithm was to expand its exploration capabilities (Yaghoubzadeh-Bavandpour et al., 2022). Whenever the random fly is called upon in this algorithm, the following formulation could be used to update the position of the bats:

$$x'_{i,j} = x_{i,j} + \left[\left(2 \times Rand - 1 \right) \times A_t \right] \qquad \forall i, j \tag{18.8}$$

in which A_t denotes the loudness of emitted sonar signals at the tth iteration.

As stated, one of the main features of the bat algorithm was its capacity to create a smooth transition between the exploration and exploitation phase. This idea is achieved by creating a dynamic way to transit between the ways the algorithm updates the positions of virtual bats. As we have seen earlier, one of the movement patterns was more in line with the idea of random search, which enables the algorithm to explore the search space's landscape thoroughly. The other movement is more applicable when the algorithm tends to converge the solutions to a focal point set to be the best global solution encountered thus far. While the former is more applicable at the beginning of the search process, the second pattern becomes more advantageous near the end.

In order to establish a dynamic way to switch between these two methods, the bat algorithm creates a stochastic structure that is controlled via two controlling mechanisms, namely, the loudness and pulsation rate. These controlling factors basically determine which moving pattern should be used to update each search agent at any given time.

After the algorithm updates the position of all bats in the population set using the repositioning mechanisms introduced in Equations (18.4)–(18.7), it conducts two tests to see whether additional repositioning would be needed for each of these individual bats. These tests are of stochastic nature, and whenever the algorithm deems that a bat needs additional alteration, a random fly would be used to update the said bat's position.

The first test is based on the pulsation rate. As we have seen in the previous sections, bats tend to increase the number of signal bursts as they approach their prey to have a better sense of their surroundings. To emulate this idea, the algorithm uses a numeric parameter to track the pulsation rate and increase its value in each iteration. This could be mathematically expressed as follows:

$$\delta_t = \delta_{final} \times \left(1 - e^{-\gamma t} \right) \qquad \forall t \tag{18.9}$$

where δ_t denotes the pulsation rate at the tth iteration; γ is the pulsation intensification factor, another parameter of the algorithm which should be selected from the range $(0, \infty)$; and δ_{final} denotes the pulsation rate at the last iteration, which is again another parameter of the bat algorithm.

In order to conduct the first test, in each iteration, the algorithm generates a random value, denoted by *Rand*, that ranges between 0 and 1. Suppose the randomly generated value exceeds the pulsation rate of the population at that iteration. In that case, the algorithm randomly selects a bat from the population set and updates its position using a random fly operator. What is important to note here is that as the algorithm increases the pulsation rate, the odds of conducting such a random fly would reduce gradually as the algorithm progresses.

The second test is based on the loudness of the emitted sonar signals. As we have stated earlier, the loudness of sonar signals is a dynamic property that would decrease as bats approach their prey. To create such effects, the algorithm would attempt to decrease the value of this property in each iteration. This could be mathematically expressed as follows:

$$A_t = \alpha \times A_{t-1} \quad \forall t \tag{18.10}$$

in which α is the loudness cooling factor parameter, another user-defined parameter of the algorithm that ranges between 0 and 1. Note that to compute this parameter, you need to have the loudness value in iteration zero, denoted by A_0, which happens to be one of the algorithm's parameters.

To conduct this test, the algorithm would first randomly select a bat from the set and apply the random fly operator on the said bat to create a tentative solution. The algorithm would continue to generate another random value, denoted by *Rand*, that ranges between 0 and 1. If the said value is smaller than the loudness at that given iteration and the tentative solution is deemed better than the current solution, the algorithm would accept this repositioning. Note that here the algorithm follows the principles of the *greedy strategy*, as only the improving moves get to be accepted, while the non-improving repositionings would be tossed aside.[2] Again, as the algorithm gradually decreases the loudness values in each iteration, the odds of conducting such random fly would gradually reduce as the algorithm progresses.

18.2.3 Termination Stage

Based on the repositioning stage described above, in each iteration, the bat algorithm would stochastically update the position of each individual bat in the population set. As the algorithm progresses, the searches tend to be less random by nature and gravitate toward what could be the optimum solution.

Like other meta-heuristic algorithms, the sequence of operational structures of this algorithm needs to be executed iteratively until a certain termination criterion is met, at which point the execution of the algorithm would be terminated, and the best bat recorded in the memory would be reported as the solution to the optimization problem. Note that without such a termination stage, the algorithm would

potentially be executed in an infinite loop. The termination stage would, in effect, determine whether the algorithm has reached what could be the optimum solution.

As the bat algorithm is not equipped with an explicitly defined, unique termination mechanism, one could implement the commonly available options, most notably limiting the number of iterations, run time, or perhaps monitoring the improvement made to the best solution in consecutive iterations. Among these options, limiting the number of iterations is arguably the most cited mechanism to create a termination stage for the bat algorithm. The idea being the process would be executed only for a specified number of times, a parameter known as the maximum iteration. In any case, it should be noted that the selection of a termination mechanism is also considered one of the algorithm's parameters. Bear in mind that in most cases, these termination mechanisms may require setting up additional parameters.

18.3 Parameter Selection and Fine-Tuning the Bat Algorithm

From the *no-free-lunch theorem*, one can conclude that fine-tuning an algorithm is essential to get the best performance out of a meta-heuristic algorithm. This would basically ensure that an algorithm is equipped to handle the unique characteristics of a given optimization problem. Of course, it is possible to use our intuition, experience, and default values suggested for an algorithm's parameters as a good starting point, one should bear in mind that fine-tuning these parameters is more than anything a trial-and-error process. Thus, while it is possible to get a good enough result by having an educated guess for setting the parameters of these algorithms, to get the best possible performance, it is necessary to go through this fine-tuning process.

In the case of the bat algorithm, these parameters are population size (M), lower (λ_{min}) and upper (λ_{maz}) boundaries of frequency, pulsation intensification factor (γ), pulsation rate at the last iteration (δ_{final}), loudness cooling factor parameter (α), loudness in iteration zero (A_0), and of course, opting for the termination criterion, and all the parameters that are associated with these methods. For instance, if limiting the number of iterations has been selected as a termination criterion, the maximum iteration (T) is another parameter that needs to be defined by the user. As can be seen here, not only the bat algorithm is riddled with a relatively high number of parameters, but for the most part, the role of these parameters on the final outcome is not easy to deduce intuitively, especially for those with little experience with meta-heuristic optimization, which makes the fine-tuning process a bit more challenging. That said, to get the absolute best results out of the bat algorithm, it is best to dabble with these algorithms first to gain some experience and inside knowledge about such parameters. By doing so, you could better understand how to fine-tune these parameters as your initial guesses and parameter selection strategies become more educated. The pseudocode for the bat algorithm is shown in Figure 18.2.

```
Begin
        Set the algorithm's parameter and input the data
        Generate the initial population
        Let M denote the population size
        While the termination criterion is not met
                Evaluate the fitness of all bats
                Rank all solutions and record the best solution
                For i in range 1 to M
                        Adjusting the frequency and update velocity of all solutions
                        Relocate the bats
                        Let Rand denote a random number between 0 and 1
                        Let δᵢ denote the pulsation rate of the ith bat
                        If Rand> δᵢ
                                Select a solution among the best solutions
                                Generate a local solution among the selected bat using random fly
                        End if
                        Let Rand denote a random number between 0 and 1
                        Let Aᵢ denote the loudness of the ith bat
                        If (Rand< Aᵢ) & (the new solution is better than the old one)
                                Accept new solution
                        End if
                Next i
                Increase δᵢ value
                Reduce Aᵢ value
        End while
        Report the best solution
End
```

Figure 18.2 Pseudocode for the bat algorithm.

18.4   Python Codes

The code to implement the bat algorithm can be found below:

```python
import numpy as np

def init_generator(pop_size, num_variables, min_val, max_val):
    return np.random.uniform(min_val, max_val, (pop_size, num_
    variables))

def frequancy(frequancy_min, frequancy_max):
    return np.random.uniform(frequancy_min, frequancy_max)

def sort_pop(pop, obj_func, minimizing):
    results = np.apply_along_axis(obj_func, 1, pop)
    index = np.argsort(results)
    if not minimizing:
        index = index[::-1]
    return pop[index]

def velocity_calculator(pop, velocity_values, frequancy_val,
  xbest):
    return velocity_values + frequancy_val*(xbest-pop)
```

```python
def pulsation_rate(final_pulse, C1, iteration):
    return final_pulse*(1-np.exp(-C1*np.arange(iteration)))

def loudness(initial_loudness, C2, iteration):
    return initial_loudness*C2**np.arange(iteration)

def random_fly(pop, pop_size, num_variables, obj_func,
               minimizing, loudness_value):
    index = np.random.randint(pop_size)
    rand = np.random.uniform(-1,1,num_variables)
    new_x = pop[index] + (rand*loudness_value)
    if minimizing:
        if obj_func(new_x)<obj_func(pop[index]):
            pop[index]=new_x
    else:
        if obj_func(new_x)>obj_func(pop[index]):
            pop[index]=new_x
    return pop

def bat_algorithm(pop_size, num_variables, min_val, max_val,
                  frequency_min, frequancy_max, obj_func,
                  final_pulse, initial_loudness, C1, C2, iteration,
                  minimizing, full_result=False):
    NFE_value = 0
    NFE = np.zeros(iteration)
    results = np.zeros(iteration)
    pop = init_generator(pop_size, num_variables, min_val, max_val)
    pop = sort_pop(pop, obj_func, minimizing)
    xbest = pop[0]
    velocity_values = np.zeros_like(pop)
    pulse_values = pulsation_rate(final_pulse, C1, iteration)
    loudness_value = loudness(initial_loudness, C2, iteration)
    for i in range(iteration):
        frequancy_val = frequancy(frequency_min, frequancy_max)
        velocity_values = velocity_calculator(pop, velocity_values,
                                              frequancy_val, xbest)
        pop += velocity_values
        if np.random.uniform(0,1)>pulse_values[i]:
            pop = random_fly(pop, pop_size, num_variables,
                             obj_func, minimizing, loudness_value[i])
        if np.random.uniform(0,1)<loudness_value[i]:
            pop = random_fly(pop, pop_size, num_variables,
                             obj_func, minimizing, loudness_value[i])
        pop = sort_pop(pop, obj_func, minimizing)
        xbest = pop[0]
        NFE_value += pop_size
        NFE[i] = NFE_value
        results[i] = obj_func(pop[0])
    if not full_result:
        return pop[0], obj_func(pop[0])
    else:
        return pop[0], obj_func(pop[0]), results, NFE
```

18.5 Concluding Remarks

Inspired by the echolocating capabilities of microbats, the bat algorithm is considered a stochastic population-based meta-heuristic optimization algorithm that is built upon the main principles of swarm intelligence. Upon closer investigation of its algorithmic architecture, it can be deduced that the bat algorithm is, in fact, a clever combination of a pure randomized search and a simplified interpretation of the particle swarm optimization algorithm. One of the most notable features of this algorithm is that it found a clever way to dynamically switch between these two vastly different approaches on the fly. As a result, the algorithm would have a smooth transition between the exploration and exploitation phases. At the beginning of the search process, the random search component is more pronounced, as such, the algorithm would be more likely to demonstrate its exploration capabilities. But as the algorithm progress with its search, the movements become more coordinated toward the global optimum solution located thus far, searching to exhibit its exploitation phase. That said, being a meta-heuristic algorithm, it can be seen that the bat algorithm still suffers from the same generic issues that are commonly associated with this branch of optimization, such as having no guarantee to reach the optimum solution or being trapped in a local optimum solution. The other notable problem is that the algorithm is practically riddled with many parameters that are often not intuitively clear on fine-tuning them for those with less experience with meta-heuristic optimization algorithms. All in all, however, one can safely state that the bat algorithm can be considered a viable option when it comes to tackling real-world complex optimization problems.

Notes

1 Given that these sonar signals are, in fact, sound waves, we should note that the product of wavelength and frequency is always a constant value as $v_s = \lambda \times W$; where v_s denotes the speed of sound, W is the wavelength, and f represents the frequency of the sound signal. As such, rather than using the frequency, we could have also rewritten this structure based on the fixed wavelength and varying frequency.
2 It could be argued that using the greedy strategy makes the parameter tuning procedure for the loudness operator much more manageable, as we are ensuring that no non-improving move could jeopardize the quality of the emerging solution. By the same token, one could argue that applying the greedy strategy to any solution that random flies generated could benefit from the same logic. This is why, in this chapter, we have used an embedded greedy search whenever the random_fly function was called. While this helps the algorithm's convergence, it should be noted that this idea was not used in the standard variation of the bat algorithm.

References

Ahmadianfar, I., Adib, A., & Salarijazi, M. (2016). Optimizing multireservoir operation: Hybrid of bat algorithm and differential evolution. *Journal of Water Resources Planning and Management*, 142(2), 05015010.

Aizpurua, O. & Alberdi, A. (2018). Ecology and evolutionary biology of fishing bats. *Mammal Review*, 48(4), 284–297.

Al-Betar, M.A. & Awadallah, M.A. (2018). Island bat algorithm for optimization. *Expert Systems with Applications*, 107, 126–145.

Anand, A.A.P. & Sripathi, K. (2004). Digestion of cellulose and xylan by symbiotic bacteria in the intestine of the Indian flying fox (Pteropus giganteus). *Comparative Biochemistry and Physiology Part A: Molecular & Integrative Physiology*, 139(1), 65–69.

Bozorg-Haddad, O., Karimirad, I., Seifollahi-Aghmiuni, S., & Loáiciga, H.A. (2015). Development and application of the bat algorithm for optimizing the operation of reservoir systems. *Journal of Water Resources Planning and Management*, 141(8), 04014097.

Bozorg-Haddad, O., Solgi, M., & Loáiciga, H.A. (2017). *Meta-heuristic and evolutionary algorithms for engineering optimization*. John Wiley & Sons. ISBN: 9781119386995

Dehghani, H. & Bogdanovic, D. (2018). Copper price estimation using bat algorithm. *Resources Policy*, 55, 55–61.

Dong, L., Zeng, W., Wu, L., Lei, G., Chen, H., Srivastava, A.K., & Gaiser, T. (2021). Estimating the pan evaporation in Northwest China by coupling CatBoost with bat algorithm. *Water*, 13(3), 256.

Farzin, S., Singh, V.P., Karami, H., Farahani, N., Ehteram, M., Kisi, O., ... & El-Shafie, A. (2018). Flood routing in river reaches using a three-parameter Muskingum model coupled with an improved bat algorithm. *Water*, 10(9), 1130.

Gandomi, A.H. & Yang, X.S. (2014). Chaotic bat algorithm. *Journal of Computational Science*, 5(2), 224–232.

Jones, K.E. & Purvis, A. (1997). An optimum body size for mammals? Comparative evidence from bats. *Functional Ecology*, 11(6), 751–756.

Kashi, S., Minuchehr, A., Poursalehi, N., & Zolfaghari, A. (2014). Bat algorithm for the fuel arrangement optimization of reactor core. *Annals of Nuclear Energy*, 64, 144–151.

Li, L. & Zhou, Y. (2014). A novel complex-valued bat algorithm. *Neural Computing and Applications*, 25(6), 1369–1381.

Lu, S., Qiu, X., Shi, J., Li, N., Lu, Z.H., Chen, P., ... & Zhang, Y. (2017). A pathological brain detection system based on extreme learning machine optimized by bat algorithm. *CNS & Neurological Disorders-Drug Targets (Formerly Current Drug Targets-CNS & Neurological Disorders)*, 16(1), 23–29.

Mirjalili, S., Mirjalili, S.M., & Yang, X.S. (2014). Binary bat algorithm. *Neural Computing and Applications*, 25(3), 663–681.

Niknam, T., Azizipanah-Abarghooee, R., Zare, M., & Bahmani-Firouzi, B. (2013). Reserve constrained dynamic environmental/economic dispatch: A new multiobjective self-adaptive learning bat algorithm. *IEEE Systems Journal*, 7(4), 763–776.

Rahmani, M., Ghanbari, A., & Ettefagh, M.M. (2016). Robust adaptive control of a bio-inspired robot manipulator using bat algorithm. *Expert Systems with Applications*, 56, 164–176.

Senthilnath, J., Kulkarni, S., Benediktsson, J.A., & Yang, X.S. (2016). A novel approach for multispectral satellite image classification based on the bat algorithm. *IEEE Geoscience and Remote Sensing Letters*, 13(4), 599–603.

Wang, X., Wang, W., & Wang, Y. (2013). An adaptive bat algorithm. In *Proceeding of International Conference on Intelligent Computing*, Berlin, Heidelberg.

Yaghoubzadeh-Bavandpour, A., Bozorg-Haddad, O., Zolghadr-Asli, B., & Gandomi, A.H. (2022). Improving approaches for meta-heuristic algorithms: A brief overview. In Bozorg-Haddad, O., Zolghadr-Asli, B. eds. *Computational intelligence for water and environmental sciences*. Springer Singapore, 35–61.

Yang, X. & Gandomi, A.H. (2012). Bat algorithm: A novel approach for global engineering optimization. *Engineering Computations*, 29(5), 464–483.

Yang, X.S. (2010). *Nature-inspired metaheuristic algorithms*. Luniver Press. ISBN: 9781905986286

Zenzen, R., Belaidi, I., Khatir, S., & Wahab, M.A. (2018). A damage identification technique for beam-like and truss structures based on FRF and bat algorithm. *Comptes Rendus Mécanique*, 346(12), 1253–1266.

19 Flower Pollination Algorithm

Summary

Inspired by the life cycle of plants that use pollination as a reproduction mechanism, the flower pollination algorithm established itself as a formidable swarm intelligence-based meta-heuristic optimization algorithm. A fairly straightforward computational structure and, more importantly, few parameters make it rather easy to implement and, in turn, a popular choice when handling real-world complex optimization problems. In this chapter, we will dig deep and explore the mechanisms used in this algorithm. We would get familiar with the flower pollination algorithm's terminology and see how one can implement this algorithm in the Python programming language. Finally, we will explore the potential merits and drawbacks of this algorithm.

19.1 Introduction

Flowering plants, technically known as *Angiospermae*, which literally translates to "hidden seeds" in Greek, are known to be one of the most diverse groups of land plants with 64 orders, 416 families, approximately 13,000 known genera, and nearly 300,000 known species (Cantino et al., 2007; Christenhusz & Byng, 2016). A 125-million-year evolutionary process has equipped these branches of land plants with one of the most fascinating reproduction mechanisms, *pollination* (Yang, 2012). The ultimate goal of this mechanism is to produce and spread the pollens for reproduction purposes.

There are two known pollination mechanisms that are *biotic* and a*biotic* pollination. While the former form accounts for nearly 90% of the identified flower pollination-based plants, the latter only represents less than 10% of the said group (Yang, 2012). In biotic pollination, the pollen is transferred via pollinator agents, which could refer to a range of insects, birds, or other species of animals that help transfer the pollen from one location to another. On the other hand, some external agents are tasked explicitly with dispersing the pollens in abiotic pollination. Here, the winds or water helps carry the pollens to a new location, where the plant could start growing.

DOI: 10.1201/9781003424765-19

From the reproduction mechanism point of view, the pollination process has two primary forms: *self-pollination* and *cross-pollination*. In cross-pollination, or what is technically known as allogamy, the pollen is transferred from one flower's anther to another flower's stigma on a different individual of the same plant species. Some of the most well-known cross-pollinating plants are grass and maple trees. However, in self-pollination, pollen from the anther is deposited on the stigma of the same flower or another flower on the same plant. Some of the most known self-pollinating plants are wheat, tomatoes, potatoes, and peaches. It is important to note that cross-pollinating plants would require external agents such as biotic pollinators to help spread their pollens. As such, the pollens of such species could spread over a long distance, as the agents such as honeybees or birds are known to travel a long distance to extract the flower nectars that such plants offer to lure these pollinators.

Inspired by an idealized representation of pollination, Yang (2012) theorized a novel meta-heuristic algorithm called the flower pollination algorithm. Built upon the principles of swarm intelligence, the flower pollination algorithm could be described as a stochastic population-based algorithm that emulates the flower pollination of some plants for reproduction purposes. Being based on the principles of the direct search would simply mean that, like other meta-heuristic algorithms, the flower pollination algorithm is also not bound by problems that are often associated with high dimensionality, multimodality, epistasis, non-differentiability, and discontinuous search space imposed by constraints (Du & Swamy, 2016). From a computational perspective, perhaps the most distinctive feature of this algorithm is its bare minimum number of parameters, which is not common for typical meta-heuristic algorithms. Coupling this with a fairly straightforward computational structure of this algorithm makes the flower pollination algorithm a formidable choice to tackle optimization problems, even for those with little experience with meta-heuristic optimization algorithms.

The flower pollination algorithm is based on an abstract interpretation of the procedure in which some plants use flower pollination for reproduction. From a computational standpoint, the architecture of this algorithm is based on two main pillars, which are *global* and *local pollination*. The idea is that the algorithm would initiate its search by randomly generating a series of search agents called pollens. The algorithm would then continue by identifying the best search agent in the set. After this, the algorithm would randomly assign each pollen to one of the two pollination mechanisms introduced earlier. Using the computational principles of these mechanisms, the algorithm would adjust the properties of the pollen in the population set. Note that the algorithm is heavily based on the *greedy strategy*. This means that the algorithm would only permit the improving moves to take effect and ignore any other suggested adjustments. This whole process would be repeated until a termination criterion is met, at which point the best pollen encountered thus far would be returned as the solution to the optimization problem at hand.

Over the years, many variants of the flower pollination algorithm have been proposed in the literature, some of which are binary flower pollination algorithm (Rodrigues et al., 2015), elite opposition-based flower pollination algorithm (Zhou et al., 2016), chaotic flower pollination algorithm (Yousri et al., 2019), discrete greedy flower pollination algorithm (Zhou et al., 2019), and island flower pollination algorithm (Al-Betar et al., 2019), to name a few. That said, the standard flower pollination algorithm is still a viable option for real-world optimization problems. In fact, the standard flower pollination algorithm has been successfully used for civil engineering (e.g., Bekdaş et al., 2015), computer science (e.g., Kabir et al., 2017), energy industry (e.g., Xu & Wang, 2017; Priya & Rajasekar, 2019), environmental engineering (e.g., Pham et al., 2021), geohydrology (e.g., Akram, 2020), hydrology (e.g., Chatterjee et al., 2018), image processing (e.g., Shen et al., 2018), manufacturing management (e.g., Guan et al., 2019), medical science (e.g., Gupta et al., 2021), robotic science (e.g., Mehta et al., 2020), structural engineering (e.g., Nigdeli et al., 2016; Mergos & Mantoglou, 2020), and water resources planning and management (e.g., Akinsunmade & Aina, 2021). In the following sections, we will explore the computational structure of the standard flower pollination algorithm.

19.2 Algorithmic Structure of the Flower Pollination Algorithm

In order to create a functional algorithm, we first need to extract and idealize some of governing principles of the flower pollination procedure. The main idea here is to establish a simplified and abstract mathematical representation of the said procedure that would be utilized later on to assemble the search engine of the flower pollination algorithm. These simplifying assumptions are as follows:

I   Global pollination is the representation of biotic cross-pollination. A Lévy flight is used to capture this process mathematically.

II   Local pollination is the representation of abiotic self-pollination.

III   For simplicity, it is assumed here that any flower can pollinate via either of global or local pollination. We assume this is basically a random procedure in which there is a fixed possibility, denoted by p, which determines which type of pollination would occur for any given pollen.

IV   For simplicity's sake, we assume there is no distinction between pollen, gamete, flower, and the plant itself.

V   To make things more manageable from a computational standpoint, it is assumed that each plant has one flower and can only produce one pollen gamete.

Based on these principles, the flower pollination algorithm creates a search engine that coordinates the movement of the search agents and navigates them to the point that could potentially be the optimum solution. It should be noted that the self-organizing nature of the search agents' movement in this algorithm qualifies the algorithm to be categorized as a swarm intelligence-based optimization algorithm.

The flower pollination algorithm's flowchart is depicted in Figure 19.1. A closer look at the architecture of the flower pollination algorithm would reveal that it actually consists of three main stages that are the initiation, pollination, and termination stages. The pollination stage consists of two main pillars that are global

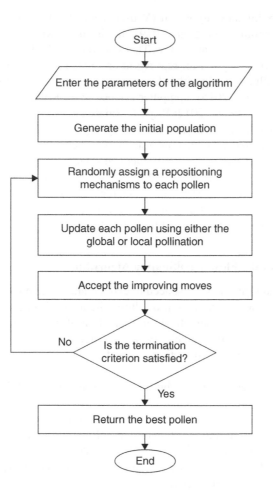

Figure 19.1 The computational flowchart of the flower pollination algorithm.

and local pollination. Using this structure, the algorithm would conduct a thorough search and locate what could be the optimum solution to the problem at hand. The following subsection will discuss each of these stages and their mathematical structures.

19.2.1 Initiation Stage

The flower pollination algorithm is a population-based meta-heuristic optimization algorithm, and as such, it works with multiple search agents, here called pollen, that would enumerate through the search space. As we have seen, in an optimization problem with N decision variables, an N-dimension coordination system could be used to represent the search space. In this case, any point within the search space, say X, can be represented mathematically as a $1 \times N$ array as follows:

$$X = \left(x_1, x_2, x_3, \ldots, x_j, \ldots, x_N\right) \tag{19.1}$$

where X represents a pollen in the search space of an optimization problem with N decision variables, and x_j represents the value associated with the jth decision variable.

The flower pollen algorithm starts with randomly placing a series of search agents within the feasible boundaries of the search space. This bundle of arrays, which in the flower pollination algorithm's terminology is referred to as the population, can be mathematically expressed as $M \times N$ matrix, where M denotes the number of particles or what is technically referred to as the population size. In such a structure, each row represents a single search agent. A population, denoted by *pop*, can be represented as follows:

$$pop = \begin{bmatrix} X_1 \\ X_2 \\ \vdots \\ X_i \\ \vdots \\ X_M \end{bmatrix} = \begin{bmatrix} x_{1,1} & x_{1,2} & \cdots & x_{1,j} & \cdots & x_{1,N} \\ x_{2,1} & x_{1,2} & \cdots & x_{2,j} & \cdots & x_{2,N} \\ & & & \vdots & & \\ x_{i,1} & x_{i,2} & \cdots & x_{i,j} & \cdots & x_{i,N} \\ & & & \vdots & & \\ x_{M,1} & x_{M,2} & \cdots & x_{M,j} & \cdots & x_{M,N} \end{bmatrix} \tag{19.2}$$

where X_i represents the ith pollen in the population, and $x_{i,j}$ denotes the jth decision variable of the ith pollen.

The initially generated population represents the initial position of search agents. As we progress, the values stored in the *pop* matrix will be altered according to the computational structure of the flower pollination algorithm's pollination stage. By the end of this iterative computation process, when the termination criterion is met, one or possibly multiple pollens will converge to the optimum solution.

19.2.2 Pollination Stage

By mimicking a generic representation of the flower pollination process, the algorithm would tend to update the population in this stage. The pollination stage consists of two main pillars that are global and local pollination. The former tends to represent the general theme of a biotic cross-pollination, which from a mathematical standpoint tends to gravitate a given pollen toward the best-identified point thus far. The other option is local pollination which is a representation of abiotic self-pollination. The algorithm would randomly call up these mechanisms to adjust the pollens in the population set. As such, these mechanisms are used in this context as alteration tools to create a new set of tentative solutions. Using the greedy strategy, the algorithm would compare these solutions with their original counterparts and replace them in case they show better performance.

As can be seen here, the first step here is to determine which of the pollination mechanism should be used for given pollen in the population set. The flower pollination algorithm uses a pure random procedure to make this call. Using a parameter called switch or proximity probability, denoted by p, the algorithm would determine how likely each of these mechanisms will be selected in this process. Switch probability is, in fact, the probability of selecting the global pollination mechanism as an alternation toll to create a new tentative solution for given pollen in the population set. Naturally, if the global pollination approach is not assigned to a pollen, the algorithm would resort to local pollination to carry out this task. In practice, to create this effect, for any pollen in the set, a random value, denoted by *Rand*, would be randomly generated within the range 0–1. If the said value is less than the switch probability threshold, the algorithm will activate the global pollination mechanism for the said pollen, otherwise, the local pollination mechanism will be used instead. It should be noted while here a parameter fine-tuning is needed, it is suggested that a good general starting point for p could be 0.8 (Yang, 2012).

As we have seen, the repositioning of search agents in this algorithm is done via two main mechanisms that are global and local pollination. The former phase is based on the notion that the pollens would be carried by biological pollinators, which, in turn, indicates that they could potentially travel a long distance before fertilizing another flower. One way to capture the motion of biological pollinators in a mathematical context is to implement the idea of Lévy flight. In fact, numerous studies have pointed out that the flying patterns of many bird species and insects, say the fruit flies, resemble the characteristics of a Lévy flight (Yang, 2010). As such, this procedure can be used as the core idea for repositioning the search agent within the search space.

A Lévy flight is a special form of the *random walk* where the step lengths are drawn from a Lévy distribution (Yaghoubzadeh-Bavandpour et al., 2022). The proponents of implementing the Lévy flight instead of the usual random walk argue that this is an efficient way to explore the search space as the step lengths drawn via this procedure are more heavy-tailed (Yang, 2010). As a result, it is more likely to generate longer steps and cover more ground with this mechanism. A special case of Lévy distribution can be defined as follows:

$$L(s, \gamma, \mu) \approx \sqrt{\frac{\gamma}{2\pi}} \frac{1}{s^{3/2}} \tag{19.3}$$

in which s, γ, and μ are all parameters of the Lévy distribution.

From an implementation standpoint, however, generating a random step from the Lévy distribution consists of two main steps that are opting for the direction and selecting the step size. The Mantegna algorithm is one of the most straightforward ways to generate a step size that bears the stochastic properties of a Lévy distribution (Mantegna, 1994). Based on this algorithm, the step size, here denoted by l, can be computed as follows:

$$l = \frac{u}{|v|^{1/\beta}} \tag{19.4}$$

in which β is a parameter of the Lévy distribution, and both u and v are randomly generated values that are drawn from normal distributions that are defined as follows:

$$u \sim Norm\left(0, \sigma_u^2\right) \tag{19.5}$$

$$v \sim Norm\left(0, \sigma_v^2\right) \tag{19.6}$$

$$\sigma_u = \left[\frac{\Gamma(1+\beta) \times \sin\left(\frac{\pi\beta}{2}\right)}{\Gamma\left(\frac{1+\beta}{2}\right) \times \beta \times 2^{\frac{\beta-1}{2}}}\right]^{1/\beta} \tag{19.7}$$

$$\sigma_v = 1 \tag{19.8}$$

where *Norm* denotes the normal distribution.

Using the above-described procedure, the length of flight can be computed. The global pollination thus can be formulated as follows:

$$x'_{i,j} = x_{i,j} + l \times \left(x_{i,j} - x_{best,j}\right) \qquad \forall j \tag{19.9}$$

in which $x'_{i,j}$ denotes the new position in the jth dimension for the tentative solution associated with the ith search agent, and $x_{best,j}$ denotes the position of the best-identified search agent in the population in the jth dimension. As can be seen, the global pollination mechanism tends to push a pollen toward the best-identified pollen in the set, denoted by *Xbest*. It goes without saying that the most desirable solutions (i.e., identifying the best organism in the population set) have a different interpretation for maximization and minimization problems. For instance, higher objective functions are considered more desirable in a maximization problem, while we are looking for lower values for the objective function in a minimization problem. The said search agent could be mathematically represented as follows:

$$Xbest = \left(x_{best,1}, x_{best,2}, x_{best,3}, \ldots, x_{best,j}, \ldots, x_{best,N}\right) \tag{19.10}$$

Given that the greedy strategy is embedded in this algorithm, it is crucial that the repositioned agents are checked against their previous positions. The algorithm would only permit those moves that improve the properties of the search agents while disregarding the rest. As such, the repositioned agents are compared against their previous positions. If the new position is more desirable in terms of the objective function value, the agent would be moved to the new position, otherwise it stays where it was before.

Alternatively, a pollen may be assigned to be repositioned via the local pollination mechanism. This represents that some pollens are spread via abiotic agents in the self-pollination process. From the computational standpoint, this procedure is more or less in line with the idea of local search and enables the pollens to search its vicinity. The algorithms would first select two random pollens from the population set to implement this idea. Let us assume that the rth and kth search agent, denoted, respectively, by X_r and X_k, has been randomly selected from the population set for the ith pollen. The algorithm would use the properties of these pollens to alter the ith search agent. The local pollination mechanism is formulated as follows:

$$x'_{i,j} = x_{i,j} + Rand \times \left(x_{r,j} - x_{k,j} \right) \qquad \forall j \qquad (19.11)$$

Again it is important to note that given that the greedy strategy is embedded in this algorithm, the repositioned agents are checked against their previous positions. The algorithm would only permit those moves that improve the properties of the search agents while disregarding the rest. As such, the repositioned agents are compared against their previous positions. If the new position is more desirable in terms of the objective function value, the agent would be moved to the new position, otherwise it stays where it was before.

19.2.3 Termination Stage

Based on the above procedure, in each given iteration, the algorithm would simultaneously switch between a global-oriented and local-based search to update the position of the pollens within the search space. As such, exploration and exploitation phases are carried out simultaneously in this algorithm. It is important to note here that based on the greedy strategy, the algorithm is designed so that it can only accept improving moves. Thus, in any stage of the search, anytime the algorithm detects an improving move, the position of the search agents would be updated within the search space, otherwise, the agent's position would remain unchanged. Like other meta-heuristic algorithms, the sequence of operational structures of this algorithm needs to be executed iteratively until a certain termination criterion is met, at which point the execution of the algorithm would be terminated, and the best agent recorded in the memory would be reported as the solution to the optimization problem. Note that without such a termination stage, the algorithm would

potentially be executed in an infinite loop. The termination stage would, in effect, determine whether the algorithm has reached what could be the optimum solution.

As the flower pollination algorithm is not equipped with an explicitly defined, unique termination mechanism, one could implement the commonly available options, most notably limiting the number of iterations, run time, or perhaps monitoring the improvement made to the best solution in consecutive iterations. Among these options, limiting the number of iterations is arguably the most cited mechanism to create a termination stage for the flower pollination algorithm. The idea being the process would be executed only for a specified number of times, a parameter known as the maximum iteration. In any case, it should be noted that the selection of the termination mechanism is also considered one of the algorithm's parameters. Bear in mind that in most cases, these termination mechanisms may require setting up additional parameters.

19.3 Parameter Selection and Fine-Tuning the Flower Pollination Algorithm

From the *no-free-lunch theorem*, one can conclude that fine-tuning an algorithm is essential to get the best performance out of a meta-heuristic algorithm. This would basically ensure that an algorithm is equipped to handle the unique characteristics of a given optimization problem. Of course, it is possible to use our intuition, experience, and default values suggested for an algorithm's parameters as a good starting point, one should bear in mind that fine-tuning these parameters is more than anything a trial-and-error process. Thus, while it is possible to get a good enough result by having an educated guess for setting the parameters of these algorithms, to get the best possible performance, it is necessary to go through this fine-tuning process.

In the case of the flower pollination algorithm, it should be noted that the algorithm has been specifically designed to have the minimum number of parameters. In this case, the only parameters used in the structure of the algorithm are the population size (M), switch or proximity probability (p), and of course, opting for the termination criterion and all the parameters that are associated with these methods. For instance, if limiting the number of iterations has been selected as a termination criterion, the maximum iteration (T) is another parameter that needs to be defined by the user. As can be seen here, not only does the flower pollination algorithm have very few parameters to begin with, but for the most part, the role of these parameters on the final outcome is easy to deduce intuitively. This makes the fine-tuning process less challenging for those less experienced. That said, to get the absolute best results out of the flower pollination algorithm, it is best to dabble with these algorithms first to gain some experience and inside knowledge about such parameters. By doing so, you could better understand how to fine-tune these parameters as your initial guesses and parameter selection strategies become more educated. The pseudocode for the flower pollination algorithm is shown in Figure 19.2.

```
Begin
        Set the algorithm's parameter and input the data
        Generate the initial population
        Let M denote the population size
        Let p denote the switch probability
        While the termination criterion is not met
                For i in range 1 to M
                        Let Xbest denote the best solution in the population
                        Let Rand denote a random number between 0 and 1
                        If Rand < p
                                Update the ith pollen using the global pollination mechanism
                        Else
                                Update the ith pollen using the local pollination mechanism
                        End if
                        Accept an improving move and reject a non-improving one
                Next i
        End while
        Report the best solution
End
```

Figure 19.2 Pseudocode for the flower pollination algorithm.

19.4 Python Codes

The code to implement the flower pollination algorithm can be found below:

```python
import numpy as np

from scipy import special
def init_genrator(min_val, max_val, pop_size, num_variables):
  return np.random.uniform(min_val, max_val, (pop_size, num_variables))

def levy_flight(x, beta, step_size):
    sigma_v = 1
    num_u = (special.gamma(beta+1)*np.sin(np.pi*beta*.5)
    denom_u = special.gamma((1+beta)*.5)*beta*2**((beta-1)*.5))
    sigma_u = (num_u)/( denom_u)**(1/beta)
    v = np.random.normal(0,sigma_v,size=x.shape)
    u = np.random.normal(0,sigma_u,size=x.shape)
    s = u/(np.abs(v)**(1/beta))
    return x + step_size*s

def best_solution(pop, obj_func, minimizing):
    results = np.apply_along_axis(obj_func, 1, pop)
    if minimizing:
        index = np.argmin(results)
    else:
        index = np.argmax(results)
    return pop[index]

def global_pollination(pop, pop_size, xbest, beta):
    L = levy_flight(pop_size, beta)
    return pop + L*(pop-xbest)
```

```
def local_pollination(pop, pop_size):
    index_1 = np.random.choice(np.arange(pop_size),
                               size=pop_size,replace=False)
    index_2 = np.random.choice(np.arange(pop_size),
                               size=pop_size,replace=False)
    pop_1 = pop[index_1]
    pop_2 = pop[index_2]
    rand = np.random.uniform(0,1,size=(pop_size,1))
    return pop + rand*(pop_1-pop_2)
def merge(pop, pop_new, obj_func, minimizing):
    of = np.apply_along_axis(obj_func, 1, pop)
    of_new = np.apply_along_axis(obj_func, 1, pop_new)
    if minimizing:
        indeces = (of<of_new).reshape(-1,1)
    else:
        indeces = (of>of_new).reshape(-1,1)
    return np.where(indeces, pop, pop_new)
def flower_pollination_algorithm(min_val, max_val, num_variables,
                                 pop_size, beta, switch_prob,
                                 iteration, obj_func, minimizing,
                                 full_results=False):
    NFE_value = 0
    NFE = np.zeros(iteration)
    results = np.zeros(iteration)
    pop = init_genrator(min_val, max_val, pop_size, num_variables)
    xbest = best_solution(pop, obj_func, minimizing)
    for i in range(iteration):
        rand = np.random.uniform()
        if rand<switch_prob:
            pop_new = global_pollination(pop, pop_size, xbest, beta)
        else:
            pop_new = local_pollination(pop, pop_size)
        pop = merge(pop, pop_new, obj_func, minimizing)
        xbest = best_solution(pop, obj_func, minimizing)
        NFE_value += 4*pop_size
        results[i] = obj_func(xbest)
        NFE[i] = NFE_value
    if not full_results:
        return xbest, obj_func(xbest)
    else:
        return xbest, obj_func(xbest), results, NFE
```

19.5 Concluding Remarks

Inspired by an idealized representation of how some plants use pollination for reproduction purposes, the flower pollination algorithm presents itself as a formidable meta-heuristic algorithm when it comes to tackling complex real-world problems. A straightforward computational structure and few parameters are perhaps one of the most notable features of this algorithm. However, there are two major problems with this algorithm. Firstly, the flower pollination algorithm is heavily based on the greedy strategy. The point is that the algorithm would only accept improving moves at any stage. This generally ensures that the algorithm always improves or holds relatively superior properties. As such, the solutions obtained in each iteration are either improving or at least staying the same as the previous iterations. However, a few pitfalls exist when an algorithm with such computational structures excessively exploits this strategy. First and foremost, it can be relatively computationally taxing, especially in real-world problems when the simulation process is a bit more intricate and require significant computational power. The point is any update would require at least two simulations to determine whether the move could be accepted or not. And for this algorithm, even with the optimal computational structure for each search agent in any given iteration, you may need to call the simulation part multiple times, given that you are using more than one agent in each operator. As a result, this is not particularly the most efficient way to solve a problem from a computational standpoint.

The other notable pitfall here is that by resorting to this strategy, the algorithm set up the search agent to move toward local optima. If a search agent moves toward local optima in such a situation, there are no ways that it could release itself from this trap. So the only way that the user could effectively address this issue is by increasing the number of search agents, which could be computationally taxing. The problem gets amplified with this algorithm, given that there is no meaningful way for the user to adjust the algorithm to the problem. This brings us to the second major drawback of this algorithm which is the limited number of adjusting parameters.

Of course, not being riddled with parameters would simply mean that the fine-tuning of the algorithm is much more manageable, which, in turn, facilitates the algorithm's implementation even for those with little to no experience working with meta-heuristic optimization algorithms. But this seemingly huge advantage comes at a price, and lacking enough controlling parameters means that there are not many ways to control the algorithm's properties of the search engine. In other words, the user does not have many options to tailor the algorithm to a specified optimization problem. As a result, the user, for instance, has limited control over the transition between the exploration and exploitation phase during the search. In fact, based on this algorithm, this is a purely random procedure to switch between two types of motions for search agents. As such, in a sense, both exploration and exploitation phases are co-occurring in the algorithm. Of course, you could put an emphasis on one aspect of the search by adjusting the switch or proximity probability parameter. But this stochastic ratio is more or less constant during the entire search, and the user nor the algorithm would not attempt to adjust this value dynamically. As such, the only way to avoid being trapped in local optima or premature convergence is to increase the population size or the number of iterations. Even with these changes, we cannot guarantee we will avoid these pitfalls altogether. With that said, all in all, from a practical standpoint, the flower pollination algorithm can undoubtedly be considered a viable option when it comes to tackling real-world complex optimization problems.

References

Akinsunmade, A.E. & Aina, I.I. (2021). Water distribution network design using hybrid self-adaptive multi-population elitist pollination intelligence (HSAMPEPI) jaya algorithm. *Earthline Journal of Mathematical Sciences*, 5(2), 329–343.

Akram, S. (2020). Improved flower pollination algorithm for optimal groundwater management. *International Journal of Computational Intelligence and Applications*, 19(03), 2050022.

Al-Betar, M.A., Awadallah, M.A., Doush, I.A., Hammouri, A.I., Mafarja, M., & Alyasseri, Z.A.A. (2019). Island flower pollination algorithm for global optimization. *The Journal of Supercomputing*, 75(8), 5280–5323.

Bekdaş, G., Nigdeli, S.M., & Yang, X.S. (2015). Sizing optimization of truss structures using flower pollination algorithm. *Applied Soft Computing*, 37, 322–331.

Cantino, P.D., Doyle, J.A., Graham, S.W., Judd, W.S., Olmstead, R.G., Soltis, D.E., ... & Donoghue, M.J. (2007). Towards a phylogenetic nomenclature of Tracheophyta. *Taxon*, 56(3), E1–E44.

Chatterjee, S., Datta, B., & Dey, N. (2018). Hybrid neural network based rainfall prediction supported by flower pollination algorithm. *Neural Network World*, 28(6), 497–510.

Christenhusz, M.J. & Byng, J.W. (2016). The number of known plants species in the world and its annual increase. *Phytotaxa*, 261(3), 201–217.

Du, K.L. & Swamy, M.N.S. (2016). *Search and optimization by metaheuristics: Techniques and algorithms inspired by nature*. Springer International Publishing Switzerland. ISBN: 9783319411910

Guan, C., Zhang, Z., & Li, Y. (2019). A flower pollination algorithm for the double-floor corridor allocation problem. *International Journal of Production Research*, 57(20), 6506–6527.

Gupta, K.D., Dwivedi, R., & Sharma, D.K. (2021). Prediction of Covid-19 trends in Europe using generalized regression neural network optimized by flower pollination algorithm. *Journal of Interdisciplinary Mathematics*, 24(1), 33–51.

Kabir, M.N., Ali, J., Alsewari, A.A., & Zamli, K.Z. (2017). An adaptive flower pollination algorithm for software test suite minimization. In *Proceeding of 2017 3rd International Conference on Electrical Information and Communication Technology*, Khulna, Bangladesh.

Mantegna, R.N. (1994). Fast, accurate algorithm for numerical simulation of Levy stable stochastic processes. *Physical Review E*, 49(5), 4677.

Mehta, I., Singh, G., Gigras, Y., Dhull, A., & Rastogi, P. (2020). Robotic Path Planning Using Flower Pollination Algorithm. *Recent Advances in Computer Science and Communications (Formerly: Recent Patents on Computer Science)*, 13(2), 191–199.

Mergos, P.E. & Mantoglou, F. (2020). Optimum design of reinforced concrete retaining walls with the flower pollination algorithm. *Structural and Multidisciplinary Optimization*, 61(2), 575–585.

Nigdeli, S.M., Bekdaş, G., & Yang, X.S. (2016). Application of the flower pollination algorithm in structural engineering. In Yang, X.S., Bekdaş, G., & Nigdeli, S. M. eds. *Metaheuristics and optimization in civil engineering*. Springer, 25–42.

Pham, Q.B., Sammen, S.S., Abba, S.I., Mohammadi, B., Shahid, S., & Abdulkadir, R.A. (2021). A new hybrid model based on relevance vector machine with flower pollination algorithm for phycocyanin pigment concentration estimation. *Environmental Science and Pollution Research*, 28, 1–16.

Priya, K. & Rajasekar, N. (2019). Application of flower pollination algorithm for enhanced proton exchange membrane fuel cell modelling. *International Journal of Hydrogen Energy*, 44(33), 18438–18449.

Rodrigues, D., Yang, X.S., De Souza, A.N., & Papa, J.P. (2015). Binary flower pollination algorithm and its application to feature selection. In Yang, X.S. ed. *Recent advances in swarm intelligence and evolutionary computation*. Springer, 85–100.

Shen, L., Fan, C., & Huang, X. (2018). Multi-level image thresholding using modified flower pollination algorithm. *IEEE Access*, 6, 30508–30519.

Xu, S. & Wang, Y. (2017). Parameter estimation of photovoltaic modules using a hybrid flower pollination algorithm. *Energy Conversion and Management*, 144, 53–68.

Yaghoubzadeh-Bavandpour, A., Bozorg-Haddad, O., Zolghadr-Asli, B., & Gandomi, A. H. (2022). Improving approaches for meta-heuristic algorithms: A brief overview. In Bozorg-Haddad, O., Zolghadr-Asli, B. eds. *Computational intelligence for water and environmental sciences*. Springer Singapore, 35–61.

Yang, X.S. (2010). *Nature-inspired metaheuristic algorithms*. Luniver Press. ISBN: 9781905986286

Yang, X.S. (2012). Flower pollination algorithm for global optimization. In *Proceeding of the 11th International Conference on Unconventional Computing and Natural Computation*, Orléan, France.

Yousri, D., Babu, T.S., Allam, D., Ramachandaramurthy, V.K., & Etiba, M.B. (2019). A novel chaotic flower pollination algorithm for global maximum power point tracking for photovoltaic system under partial shading conditions. *IEEE Access*, 7, 121432–121445.

Zhou, Y., Wang, R., & Luo, Q. (2016). Elite opposition-based flower pollination algorithm. *Neurocomputing*, 188, 294–310.

Zhou, Y., Wang, R., Zhao, C., Luo, Q., & Metwally, M.A. (2019). Discrete greedy flower pollination algorithm for spherical traveling salesman problem. *Neural Computing and Applications*, 31(7), 2155–2170.

20 Water Cycle Algorithm

Summary

Inspired by the Earth's hydrological cycle, the water cycle algorithm is a formidable swarm intelligence-based meta-heuristic optimization algorithm that can be quite efficient when handling real-world complex optimization problems. In this chapter, we will dig deep and explore the mechanisms used in this algorithm. We would get familiar with the water cycle algorithm's terminology and see how one can implement this algorithm in the Python programming language. Finally, we will explore the potential merits and drawbacks of this algorithm.

20.1 Introduction

Have you ever wondered about how water resources get distributed around the world? Though many of us who have the privilege of readily available access to freshwater resources may take this basic human right for granted, unfortunately, this is not the case for the entire population of planet Earth. Although this blue planet of ours is seemingly covered with oceans and seas across its surface, approximately only 3% of available water resources throughout the entire world, called freshwater resources, have the vital characteristics to meet the needs of humans (Gleick, 1993; Oki & Kanae, 2006). What is even more shocking is that nearly 2 billion people worldwide inhabit highly water-stressed regions because of the uneven temporal and spatial distribution of freshwater resources. Matters such as climate change and rapid population growth are posing some new challenges in arid and semi-arid regions, including the Middle East and North African nations, which, needless to say, are already experiencing mild to severe water stress due to the limited water availability and growing water demands (Zolghadr-Asli et al., 2017; Bozorg-Haddad et al., 2020, 2021). In an attempt to emphasize the importance of water resources as a human right, the World Health Organization (WHO) declared 2005–2015 the *decade of water*, with the goal of establishing the framework of ultimately providing full access to water supply and sanitation worldwide (Rietveld et al., 2009). In a similar gesture, as part of their *millennium development goals*, the United Nations (UN) set a target of halving the number of individuals

DOI: 10.1201/9781003424765-20

who may not have access to safe and hygienic drinking water by 2015 (Binagwaho & Sachs, 2005).

One must first understand the hydrological or water cycle concept to understand the true nature of water problems and the pattern of freshwater distribution. Driven by solar energy, this natural mechanism ensures that the water constantly cycles throughout the planet. This process basically redistributes freshwater within the hydrological boundaries. Of course, being a cycle simply means that there cannot be a beginning or an end to the water cycle per se, but for the sake of argument, let us start exploring this cycle from the moisture stored in the atmosphere. The condensed moisture in the air would then precipitate in different forms, such as rainfalls or snowfalls. Some of these waters would be directly returned to surface water bodies such as lakes, ponds, seas, or oceans. In colder regions, such as mountaintops or polar regions, these precipitations may transform into glaciers and dense snow that may stay in that form for millions of years. In warmer regions, the water may evaporate immediately and return back to the atmosphere. More often than not, however, precipitated water forms a series of streams and found its paths to more significant streams and rivers. While some of the water would eventually infiltrate the ground and form groundwater streams and aquifers, the remaining water would reach surface water bodies. Of course, it is possible for the water to return back to the atmosphere through evaporation or transpiration. And this cycle would continue forever. As can be seen here, water is not being created nor destroyed in this process; rather, it changes forms and location; hence, the said cycle is a *closed system*. It is important to note here that all the components in the said cycle are occurring simultaneously, making it a remarkably simple yet intricate phenomenon nonetheless.

Inspired by an idealized interpretation of the hydrological cycle, Eskandar et al. (2012) theorized a novel meta-heuristic algorithm called the water cycle algorithm. Built upon the principles of swarm intelligence, the water cycle algorithm could be described as a stochastic population-based algorithm that emulates an idealized representation of how water circulates through the Earth. Being based on the principles of the direct search would simply mean that, like other meta-heuristic algorithms, the water cycle algorithm is also not bound by problems that are often associated with high dimensionality, multimodality, epistasis, non-differentiability, and discontinuous search space imposed by constraints (Du & Swamy, 2016; Bozorg-Haddad et al., 2017).

The water cycle algorithm is based on an abstract interpretation of the hydrological cycle, which we have explored earlier. The idea is that the algorithm would initiate its search by randomly generating a series of search agents, referred to as raindrops, within the feasible area of the search space. Based on their performance, the algorithm would continue evaluating these raindrops and labeling them as streams, rivers, or seas. Based on what we have in the hydrological cycle, the streams would tend to move toward rivers, and the rivers would, in turn, flow to the sea. By the same token, these search agents would use a corresponding guiding point to navigate their movement. The water cycle algorithm proposes an evaporation/precipitation mechanism to introduce some exploratory components to this process. The idea is that the agents close enough to the globally identified ideal point here, referred to as the sea, would get evaporated and regenerated randomly. The entire procedure would be repeated, and in each iteration, the algorithm would allow the search agents to get closer to the sea without being evaporated. The point is that by doing so, the algorithm is making a smooth transition from the exploratory phase to the exploitation phase. After a termination criterion is met, the algorithm would terminate the search process and return the best-encountered position in the search space as the optimums solution.

As can be seen here, from a computational standpoint, the architecture of the water cycle algorithm closely resembles the structural computational of the particle swarm optimization algorithm. But this algorithm has certain novelties and nuances that distinguish it from other swarm intelligence-based meta-heuristic optimization algorithms. The most notable difference is that the algorithm would dynamically set up a series of guiding points to navigate the repositioning procedure of agents that are selected from the same generation of search agents. As such, the algorithm ensures that in each given iteration, there are a series of simultaneous local searches to enumerate the search space rather thoroughly. Through the evaporation/precipitation operator, the algorithm finds an elegant mechanism to create a smooth transition from the exploration phase to the exploitation phase. That said, this is not particularly the most straightforward computational structure. As such, it could be challenging for those less experienced to understand and execute this algorithm in practice. Being a meta-heuristic algorithm, the water cycle algorithm also suffers from the same generic issues that are commonly associated with this branch of optimization, such as having no guarantee to reach the optimum solution or being trapped in local optima. That said, it should be noted that the algorithm has shown promising results when tackling real-world, complex optimization problems.

Over the years, many variants of the water cycle algorithm were proposed in the literature, some of which are dual-system water cycle algorithm (Luo et al., 2016), efficient chaotic water cycle algorithm (Heidari et al., 2017), and discrete water cycle algorithm (Osaba et al., 2018), to name a few. That said, the standard water cycle algorithm is still considered a viable option to handle real-world optimization problems. In fact, the standard water cycle optimization algorithm has been successfully used for civil engineering (e.g., Zhang et al., 2021), climatology (e.g., Sadollah et al., 2015a,b), data science (e.g., Taib & Bahreininejad, 2021), energy industry (e.g., Khodabakhshian et al., 2016), financial market (e.g., Moradi, 2017), hydrology (e.g., Akbarifard et al., 2018), image processing (e.g., Kandhway & Bhandari, 2019), robotic science (e.g., Tuba et al., 2018), structural engineering (e.g., Eskandar et al., 2013; Sadollah et al., 2015a,b), and water resources planning and management (e.g., Bozorg-Haddad et al., 2015). In the following sections, we will explore the computational structure of the standard water cycle algorithm.

20.2 Algorithmic Structure of the Water Cycle Algorithm

In order to create a functional algorithm, we first need to extract and idealize some of the governing principles of the hydrological cycle. The main idea here is to

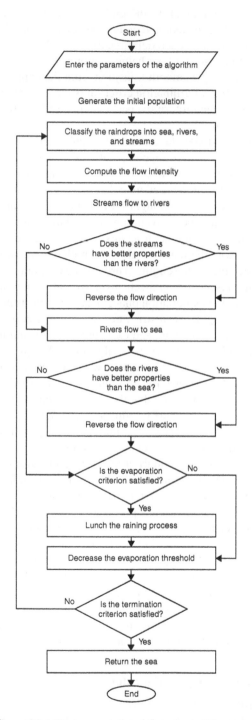

Figure 20.1 The computational flowchart of the water cycle algorithm.

establish a simplified and abstract mathematical representation of the said procedure that would be utilized later on to assemble the search engine of the water cycle algorithm. These simplifying assumptions are as follows:

I In order to simplify the hydrological cycle, here we are simply ignoring the geohydrology-related components of the said process. This means that we are assuming that groundwater flows and infiltration are negligible. Transpiration is also used interchangeably with evaporation. We also assume that the water is constantly in motion; as such, water would not be transformed into ice or glacier form during this idealized representation of the hydrological cycle. All in all, in this simplified interpretation of the water cycle, we are only dealing with raindrops, streams, rivers, and the sea.

II While a given source can have multiple charging inputs, it cannot have multiple outflows. From the computational standpoint, this means that while a search agent can serve as a guide point for multiple other agents, the repositioning of a single agent cannot be influenced by multiple guide points.

Based on these simplifying principles, the water cycle algorithm creates a search engine that governs the repositioning of the search agents so that they can ultimately locate what could potentially be the optimum solution.

The water cycle algorithm's flowchart is depicted in Figure 20.1. A closer look at the architecture of the water cycle algorithm would reveal that it actually consists of three main stages that are the initiation, hydrological simulation, and termination stages. Using this structure, the algorithm would conduct a thorough search and locate what could be the optimum solution to the problem at hand. The following subsection will discuss each of these stages and their mathematical structures.

20.2.1 Initiation Stage

The water cycle algorithm is a population-based meta-heuristic optimization algorithm, and as such, it works with multiple search agents, here called raindrops, that would enumerate through the search space. As we have seen, in an optimization problem with N decision variables, an N-dimension coordination system could be used to represent the search space. In this case, any point within the search space, say X, can be represented mathematically as a $1{\times}N$ array as follows:

$$X = \left(x_1, x_2, x_3, \ldots, x_j, \ldots, x_N\right) \tag{20.1}$$

where X represents a raindrop in the search space of an optimization problem with N decision variables, and x_j represents the value associated with the jth decision variable. Note that while these raindrops would be labeled as streams, rivers, or the sea further down the line, they still have the same mathematical structure.

The water cycle algorithm starts with randomly placing a series of raindrops within the feasible boundaries of the search space. This bundle of arrays, which in the water cycle algorithm's terminology is referred to as the population, can be mathematically expressed as $M \times N$ matrix, where M denotes the number of particles or what is technically referred to as the population size. In such a structure, each row represents a single search agent. A population, denoted by *pop*, can be represented as follows:

$$
pop = \begin{bmatrix} X_1 \\ X_2 \\ \vdots \\ X_i \\ \vdots \\ X_M \end{bmatrix} = \begin{bmatrix} x_{1,1} & x_{1,2} & \cdots & x_{1,j} & \cdots & x_{1,N} \\ x_{2,1} & x_{1,2} & \cdots & x_{2,j} & \cdots & x_{2,N} \\ & & & \vdots & & \\ x_{i,1} & x_{i,2} & \cdots & x_{i,j} & \cdots & x_{i,N} \\ & & & \vdots & & \\ x_{M,1} & x_{M,2} & \cdots & x_{M,j} & \cdots & x_{M,N} \end{bmatrix} \tag{20.2}
$$

where X_i represents the ith raindrop in the population, and $x_{i,j}$ denotes the jth decision variable of the ith raindrop.

The initially generated population represents the initial position of raindrops. As we progress, the values stored in the *pop* matrix will be altered according to the computational structure of the water cycle algorithm's hydrological simulation stage. By the end of this iterative computation process, one or possibly multiple raindrops could converge to the optimum solution when the termination criterion is met.

20.2.2 Hydrological Simulation Stage

As stated earlier, the main idea behind this algorithm is to assign each search agent with a guiding point so that the agent would coordinate its move to flow toward said aspiration point. What is important here is that these guiding points would dynamically change through the search. To create this structure, the water cycle algorithm takes inspiration from a basic hydrological principle that water tends to move toward places with lower altitudes constantly. This can be interpreted that the water particles, called raindrops, tend to move toward better positions. As such, small streams would connect together and form a more significant stream, here called rivers. By the same token, the rivers would merge with one another and eventually flow to the sea, which in this context refers to the best-identified point in the set.

In order to create the virtual hydrological cycle, the algorithm's first step in this stage is to label the generated raindrops as streams, rivers, and the sea. From a mathematical standpoint, this means that the algorithm needs to sort the raindrops of the population based on their performance from the best down to the worst. This, of course, would have a different interpretation in maximization and minimization problems. For instance, in a maximization problem, the raindrops with the greater objective functions would assume a higher position than those with lower objective function values. On the contrary, in a minimization problem, the raindrop with the lowest objective function would assume the highest position in the population, and the raindrop with the greatest objective function value would be placed at the bottom of the list. As such, the raindrop on the top of the sorted list is labeled as the sea. The schematic procedure in which the raindrops are labeled in this algorithm is illustrated in Figure 20.2. The algorithm would then label a number of raindrops, here denoted by R, as the rivers. Note that R is a user-defined parameter of the algorithm. With that in mind, the number of streams, denoted by S, can be computed as follows:

$$S = M - R - 1 \tag{20.3}$$

As shown in Figure 20.2, the algorithm needs to designate a guiding point for the streams and rivers. As such, the algorithm needs to compute the number of search agents that use the rivers and the sea as a guiding point. This process is

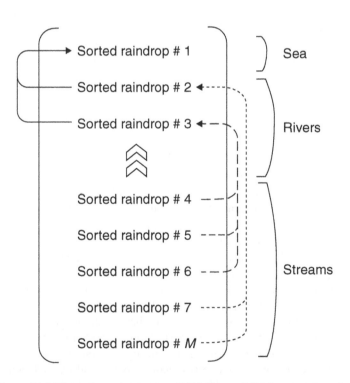

Figure 20.2 The schematic theme of labeling raindrops procedure in the water cycle algorithm.

based on the objective function values of the raindrops. The following equations could be used to determine how many raindrops would look at the sea or a given river as a guiding point.

$$\lambda_{sea} = round\left(\left|\frac{f(sea)}{f(sea)+\sum_{r=1}^{R}f(river_r)}\right| \times S\right) \tag{20.4}$$

$$\lambda_i = round\left(\left|\frac{f(river_i)}{f(sea)+\sum_{r=1}^{R}f(river_r)}\right| \times S\right) \quad i = 1,2,\ldots,R \tag{20.5}$$

in which *sea* denotes the best raindrop identified in the population, *river_r* is the *r*th river, λ_{sea} denotes the number of rivers or streams that flow directly to the sea, λ_i denotes the number of rivers or streams that flow directly to the *i*th river, *round*() is a mathematical function that returns the round off the value passed to the function to the closest integer number, and *f*() represents the objective function.

After determining what agents would serve as guiding points for the other members, the algorithm would continue by repositioning the location of the search agents using these guiding points. The main idea here is that a given stream or river tends to move toward a more suitable place, that is, its guiding point. This procedure can be mathematically expressed as follows:

$$x_{i,j}^{new} = x_{i,j} + Rand \times C \times \left(guide_{i,j} - x_{i,j}\right) \tag{20.6}$$

in which $x_{i,j}^{new}$ denotes the relocated value for the *i*th raindrop in the *j*th dimension,

$guide_{i,j}$ is the position of the guiding point for the *i*th raindrop in the *j*th dimension, *Rand* is a randomly generated value within the range 0–1, and *C* is the hydrological step length which is a user-defined parameter of the algorithm that ranges between 1 and 2. It is recommended that values closer to the upper limit should be opted for this parameter (Bozorg-Haddad et al., 2017).

It is important to note that both streams and rivers would coordinate their motion based on the corresponding guiding points in this system. Rivers, for instance, would look up to the sea as the guiding point, while streams would select rivers to guide their movements. What is important to note here is that such assignments are deliberately designed to be dynamic. The idea is that the algorithm may use different guiding points for a given search agent. More importantly, it is possible that the role of the search agents may change during the search process. For instance, the position of a stream and its corresponding river could be exchanged if the properties of the stream are deemed better than the river to which the stream is discharging. In such a situation, the flow direction would be reversed. A similar situation could happen between the river and the sea.

In the actual hydrological cycle, there is always the possibility that water gets evaporated and returns back into the atmosphere. The moisture in the air would get condensed and released back to the earth through precipitation.[1] The water cycle algorithm uses this process as an inspiration to add an additional mechanism called the evaporation/precipitation operator that is there basically to ensure that the algorithm would not have a premature convergence. From a computational standpoint, this operator is, for the most part, a safety mechanism that adds a hint of exploration to the algorithm's search engine. The central theme is that the algorithm would eradicate those search agents it deems too close to the sea and regenerate them from scratch through this mechanism. The criterion for identifying the search agents that need to go through this process is dynamically adjusted through the search in a way that it gets more challenging for the algorithm to resort to this procedure as the search progresses. As a result, the algorithm makes a conscious attempt to make a smooth transition from the exploration phase to the exploitation one.

The first step to conduct the evaporation/precipitation mechanism is to measure the distance between a given search agent and the sea. Often, measuring the *Euclidean distance* between two search agents is the most practical approach to computing this distance. For a given search agent, this distance can be mathematically measured as follows:

$$r_i = \|X_i - sea\| = \sqrt{\sum_{j=1}^{N}\left(x_{i,j} - sea_j\right)^2} \qquad \forall i \tag{20.7}$$

where r_i denotes the Euclidean distance between the ith search agent and the sea. If the distance between a given search agent and the sea falls below a certain threshold, here denoted by δ, based on the water cycle algorithm analogy, it is assumed that the said river has reached the sea. As such, the algorithm can initiate the evaporation/precipitation procedure for this search agent, that is, the water gets evaporated and joins back the cycle through precipitation. From a computational standpoint, the search agent would be swapped with a newly generated value. As can be seen, this process basically serves as a controlling measure to ensure that the algorithm does not suffer from a premature convergence or that search agents do not get trapped in local optima. As stated, as the search progresses, the algorithm would dynamically adjust the value of this threshold so that it permits the search agents to get close to the sea without being evaporated. As such, the algorithm makes a smooth transition between the exploration phase and the exploitation phase. In order to create this effect, the algorithm would gradually decrease the value of the said threshold in each iteration. This process can be mathematically expressed as follows:

$$\delta_t = \delta_{t-1} - \frac{\delta_{t-1}}{T} \qquad \forall t \tag{20.8}$$

in which δ_t denotes the evaporation threshold at the *t*th iteration; T denotes the maximum number of iterations, a user-defined parameter for this algorithm; and *t* represents the counter for the current iteration. Note that to use this mechanism, you must first establish an initial value for the evaporation threshold, denoted by δ_0, making it another algorithm parameter that needs to be set up by the user. It should be noted that the above formula can be altered in a way that would also apply to other termination criteria. For instance, if the idea is to run the algorithm within a specific time span, the maximum time limit would replace the maximum number of iterations, while the current iteration count would be swapped with the current run time value.

By passing the evaporation threshold at any given point, the search agent triggers the evaporation/precipitation procedure. The idea is that the said agent would be swapped with a newly generated value. For non-constrained optimization problems, the algorithm follows the below mathematic formula to generate a new search agent:

$$x_{i,j}^{new} = L_j + Rand \times \left(U_j - L_j\right) \tag{20.9}$$

in which $x_{i,j}^{new}$ represents the *j*th decision variable of the *i*th raindrop that has triggered the evaporation/precipitation procedure, U_j and L_j represent the upper and lower feasible boundaries of the *j*th decision variable, respectively.

It was argued that using the above procedure could potentially have an adverse effect on the convergence rate of constrained optimization problems (Eskandar et al., 2012, 2013). As such, the following equation could be used to regenerate the new agent:

$$x_{i,j}^{new} = sea_j + \sqrt{\eta} \times Norm\left(0,1\right) \tag{20.9}$$

where *Norm*(0, 1) denotes a function that generates a random value that is drawn from a normal distribution function with a mean of 0 and a standard deviation of 1; and η is the searching premise coefficient, which is a user-defined parameter of the algorithm. Assuming larger values for these parameters makes it possible for the search agent to be placed further from the sea; however, this could also mean that the convergence rate of the algorithm could suffer a bit. However, it should be noted that selecting small values for this parameter would also mean that there is the risk of premature convergence or being trapped in local optima.

20.2.3 Termination Stage

Based on the hydrological simulation stage described above, the water cycle algorithm constantly adjusts the search agent's positions within the search space in each iteration. These agents would continually move toward their corresponding local optima, and once in a while, if the evaporation criterion is met, the algorithm would reposition these agents randomly to make sure that the algorithm retains enough exploration capacity during the searching process. Of course, by adjusting the evaporation threshold in each iteration, the algorithm ensures that there is a smooth transition between the exploration and the exploitation phases, given that it would gradually decrease the odds of initiating the raining procedure, through which the raindrops would be randomly regenerated. Through this stochastic process, in each given iteration, the algorithm conducts a series of simultaneous searches that tends to gravitate more around a focal point as the algorithm progress. This process would be repeated until the algorithm eventually locates what could potentially be the optimum solution.

Like other meta-heuristic algorithms, the sequence of operational structures of this algorithm needs to be executed iteratively until a certain termination criterion is met, at which point the execution of the algorithm would be terminated, and the best raindrop recorded in the memory would be reported as the solution to the optimization problem. Note that without such a termination stage, the algorithm would potentially be executed in an infinite loop. The termination stage would, in effect, determine whether the algorithm has reached what could be the optimum solution.

As the water cycle algorithm is not equipped with an explicitly defined, unique termination mechanism, one could implement the commonly available options, most notably limiting the number of iterations, run time, or perhaps monitoring the improvement made to the best solution in consecutive iterations. Among these options, limiting the number of iterations is arguably the most cited mechanism to create a termination stage for the water cycle algorithm. The idea being the process would be executed only for a specified number of times, a parameter known as the maximum iteration. In any case, it should be noted that the selection of a termination mechanism is also considered one of the algorithm's parameters. Bear in mind that in most cases, these termination mechanisms may require setting up additional parameters.

20.3 Parameter Selection and Fine-Tuning the Water Cycle Algorithm

From the *no-free-lunch theorem*, one can conclude that fine-tuning an algorithm is essential to get the best performance out of a meta-heuristic algorithm. This would basically ensure that an algorithm is equipped to handle the unique characteristics of a given optimization problem. Of course, it is possible to use our intuition, experience, and default values suggested for an algorithm's parameters as a good starting point, one should bear in mind that fine-tuning these parameters is more than anything a trial-and-error process. Thus, while it is possible to get a good-enough result by having an educated guess for setting the parameters of these algorithms, to get the best possible performance, it is necessary to go through this fine-tuning process.

In the case of the water cycle algorithm, these parameters are population size (M), the number of rivers (R), hydrological step length (C), the initial value for the

```
Begin
        Set the algorithm's parameter and input the data
        Generate the initial population
        Let M denote the population size
        Let S denote the total number of streams
        Let R denote the total number of rivers
        Evaluate the generated raindrops
        Classify the solutions into streams, rivers, and seas based on their performance
        While the termination criterion is not met
                For i in range 1 to S
                        Flow the ith stream toward the corresponding river or sea
                        If the newly generated stream is better than the said river
                                Reverse the flow direction
                        End if
                Next i
                For i in range 1 to R
                        Flow the ith river toward the sea
                        If the newly generated river is better than the sea
                                Reverse the flow direction
                        End if
                Next i
                If the evaporation criterion is satisfied
                        Initiate the rain process
                End if
                Decrease the evaporation threshold
        End while
        Report the best solution
End
```

Figure 20.3 Pseudocode for the water cycle algorithm.

evaporation threshold (δ_0), the searching premise coefficient (η) in case of non-constrained optimization problems, and of course, opting for the termination criterion, and all the parameters that are associated with these methods. For instance, if limiting the number of iterations has been selected as a termination criterion, the maximum iteration (T) is another parameter that needs to be defined by the user. As can be seen here, not only is the water cycle algorithm riddled with parameters to begin with, but for the most part, the role of these parameters on the final outcome is not easy to deduce intuitively. This makes the fine-tuning process a bit challenging for those less experienced. That said, to get the absolute best results out of the water cycle algorithm, it is best to dabble with these algorithms first to gain some experience and inside knowledge about such parameters. By doing so, you could better understand how to fine-tune these parameters as your initial guesses and parameter selection strategies become more educated. The pseudocode for the water cycle algorithm is shown in Figure 20.3.

20.4 Python Codes

The code to implement the water cycle algorithm can be found below:

```python
import numpy as np

def init_generator(pop_size, num_variables, min_val, max_val):
    return np.random.uniform(min_val, max_val, (pop_size, num_
    variables))

def sorting_pop(pop, obj_func, minimizing):
    results = np.apply_along_axis(obj_func, 1, pop)
    indeces = np.argsort(results)
    if not minimizing:
        indeces = indeces[::-1]
    return pop[indeces]

def involved_ranindrop_counter(pop, obj_func, R, S):
    costs = np.apply_along_axis(obj_func,1,pop[:R+1])
    sum_costs = np.sum(costs)
    involved_raindrops = np.round(S*np.abs(costs/sum_costs)).
  astype(int)
    involved_raindrops = np.insert(involved_raindrops,0,0)
    return np.cumsum(involved_raindrops)
def update_positions(pop, obj_func, R, S, C):

    involved_raindrops = involved_ranindrop_counter(pop, obj_
    func, R, S)
    for i in range(R+1):
        pop_reverse = pop[::-1]
        low = involved_raindrops[i]
        up = involved_raindrops[i+1]
        rand = np.random.uniform(0,1,(up-low)).reshape(-1,1)
        x = rand*C*(pop[i]-pop_reverse[low:up])
        pop_reverse[low:up]=(pop_reverse[low:up]+x)
    return pop_reverse[::-1]

def evaporation(pop, sea_threshold, min_val, max_val,
                search_coef, constrained):
    dist_values = np.sqrt(np.sum((pop[0]-pop)**2,1))
    where = (dist_values<sea_threshold).flat
    replacing_index = np.array(where)[1:]
    if constrained:
        replacing_values=pop[0]+np.random.normal(0,np.sqrt(search_
        coef),
                                                 size=pop.shape)
    else:
        replacing_values = np.random.uniform(min_val, max_val, pop.
        shape)
    pop[replacing_index] = replacing_values[replacing_index]
    return pop
```

```
def water_cycle_algorithm(min_val, max_val, num_variables, pop_size,
                          R, C, sea_threshold,
                          obj_func, iteration, minimizing,
                          search_coef=.1, constrained=False,
                          full_results=False):
    NFE = np.zeros(iteration)
    results = np.zeros(iteration)
    NFE_values = 0
    pop = init_generator(pop_size, num_variables, min_val, max_val)
    pop = sorting_pop(pop, obj_func, minimizing)
    NFE_values += pop_size
    S = pop_size-R-1
    for i in range(iteration):
        pop = update_positions(pop, obj_func, R, S, C)
        pop = sorting_pop(pop, obj_func, minimizing)
        pop = evaporation(pop, sea_threshold, min_val, max_val,
                          search_coef, constrained)
        pop = sorting_pop(pop, obj_func, minimizing)
        sea_threshold *= (1-1/iteration)
        NFE_values += pop_size
        NFE[i] = NFE_values
        results[i] = obj_func(pop[0])
    if not full_results:
        return pop[0], obj_func(pop[0])
    else:
        return pop[0], obj_func(pop[0]), results, NFE
```

20.5 Concluding Remarks

Inspired by an idealized representation of how water gets redistributed throughout the plant via the hydrological cycle, the water cycle algorithm is a formidable meta-heuristic algorithm for tackling complex real-world problems. Though it could be argued that the computational structure of this algorithm is influenced to some degree by other trajectory-based swarm intelligence-based algorithms, such as the particle swarm optimization algorithms, the nuances and subtle improving features used in this algorithm made it an interesting stochastic population-based meta-heuristic optimization method from a computational standpoint. First and fore-most, the water cycle algorithm is built on a structure that allows multiple search agents to conduct their own local search as they tend to move toward their guiding point. What is important here is the dynamic structure of the algorithm that permits updating the guiding point, which in turn helps improve the efficiency of the search process. Through the evaporation/precipitation procedure, the algorithm would find a way to reduce the odds of being trapped in local optima or experiencing premature convergence. Again, it is important to note that the algorithm dynamic-ally adjusts the computational premise of the evaporation/precipitation procedure throughout the search so that it experiences a smooth transition between the explor-ation and exploitation phases. The algorithm does not particularly benefit from the most straightforward computational structure. It could be argued that the algorithm has a lot of parameters, for most of which it is hard to grasp their impact on the final outcome of the search intuitively. These could make the fine-tuning process a bit more challenging for those with less experience with meta-heuristic optimization algorithms. All in all, however, from a practical standpoint, the water cycle algorithm can certainly be considered a viable option when it comes to tackling real-world complex optimization problems.

Note

1 In reality, the process is a bit more complicated, as other components play a role. For instance, water can be stored in plant life, or a proportion of water resources get infiltrated into the ground. However, as discussed in the previous section, the water cycle algorithm emulates a simplified version of the hydrological cycle that is mainly concerned with a limited number of components of this cycle.

References

Akbarifard, S., Qaderi, K., & Alinnejad, M. (2018). Parameter estimation of the nonlinear muskingum flood-routing model using water cycle algorithm. *Journal of Watershed Management Research*, 8(16), 34–43.

Binagwaho, A. & Sachs, J.D. (2005). *Investing in development: A practical plan to achieve the millennium development goals*. Earthscan Publication.

Bozorg-Haddad, O., Moravej, M., & Loáiciga, H.A. (2015). Application of the water cycle algorithm to the optimal operation of reservoir systems. *Journal of Irrigation and Drainage Engineering*, 141(5), 04014064.

Bozorg-Haddad, O., Solgi, M., & Loáiciga, H.A. (2017). *Meta-heuristic and evolutionary algorithms for engineering optimization*. John Wiley & Sons. ISBN: 9781119386995

Bozorg-Haddad, O., Zolghadr-Asli, B., Chu, X., & Loáiciga, H.A. (2021). Intense extreme hydro-climatic events take a toll on society. *Natural Hazards*, 108, 1–7.

Bozorg-Haddad, O., Zolghadr-Asli, B., Sarzaeim, P., Aboutalebi, M., Chu, X., & Loáiciga, H.A. (2020). Evaluation of water shortage crisis in the Middle East and possible remedies. *Journal of Water Supply: Research and Technology-AQUA*, 69(1), 85–98.

Du, K.L. & Swamy, M.N.S. (2016). *Search and optimization by metaheuristics: Techniques and algorithms inspired by nature*. Springer International Publishing Switzerland. ISBN: 9783319411910

Eskandar, H., Sadollah, A., & Bahreininejad, A. (2013). Weight optimization of truss structures using water cycle algorithm. *Iran University of Science & Technology*, 3(1), 115–129.

Eskandar, H., Sadollah, A., Bahreininejad, A., & Hamdi, M. (2012). Water cycle algorithm – A novel metaheuristic optimization method for solving constrained engineering optimization problems. *Computers & Structures*, 110, 151–166.

Gleick, P.H. (1993). *Water in crisis: A guide to the world's fresh water resources*. Oxford University Press.

Heidari, A.A., Abbaspour, R.A., & Jordehi, A.R. (2017). An efficient chaotic water cycle algorithm for optimization tasks. *Neural Computing and Applications*, 28(1), 57–85.

Kandhway, P. & Bhandari, A.K. (2019). A water cycle algorithm-based multilevel thresholding system for color image segmentation using masi entropy. *Circuits, Systems, and Signal Processing*, 38(7), 3058–3106.

Khodabakhshian, A., Esmaili, M.R., & Bornapour, M. (2016). Optimal coordinated design of UPFC and PSS for improving power system performance by using multi-objective water cycle algorithm. *International Journal of Electrical Power & Energy Systems*, 83, 124–133.

Luo, Q., Wen, C., Qiao, S., & Zhou, Y. (2016). Dual-system water cycle algorithm for constrained engineering optimization problems. In Huang, D.S., Bevilacqua, V., and Premaratne, P. eds. *International Conference on Intelligent Computing*. Springer, 730–741.

Moradi, M. (2017). Portfolio optimization in Tehran stock exchange by water cycle algorithm. *Journal of Financial Management Perspective*, 7(20), 9–32.

Oki, T. & Kanae, S. (2006). Global hydrological cycles and world water resources. *Science*, 313(5790), 1068–1072.

Osaba, E., Del Ser, J., Sadollah, A., Bilbao, M.N., & Camacho, D. (2018). A discrete water cycle algorithm for solving the symmetric and asymmetric traveling salesman problem. *Applied Soft Computing*, 71, 277–290.

Rietveld, L.C., Harrhoff, J., & Jagals, P. (2009). A tool for technical assessment of rural water supply systems in South Africa. *Physics and Chemistry of the Earth*, 34(1–2), 43–49.

Sadollah, A., Eskandar, H., Bahreininejad, A., & Kim, J.H. (2015a). Water cycle algorithm with evaporation rate for solving constrained and unconstrained optimization problems. *Applied Soft Computing*, 30, 58–71.

Sadollah, A., Eskandar, H., Bahreininejad, A., & Kim, J.H. (2015b). Water cycle, mine blast and improved mine blast algorithms for discrete sizing optimization of truss structures. *Computers & Structures*, 149, 1–16.

Taib, H. & Bahreininejad, A. (2021). Data clustering using hybrid water cycle algorithm and a local pattern search method. *Advances in Engineering Software*, 153, 102961.

Tuba, E., Dolicanin, E., & Tuba, M. (2018). Water cycle algorithm for robot path planning. In *Proceeding of 2018 10th International Conference on Electronics, Computers and Artificial Intelligence*, Iasi, Romania.

Zhang, Y.G., Tang, J., Liao, R.P., Zhang, M.F., Zhang, Y., Wang, X.M., & Su, Z.Y. (2021). Application of an enhanced BP neural network model with water cycle algorithm on landslide prediction. *Stochastic Environmental Research and Risk Assessment*, 35(6), 1273–1291.

Zolghadr-Asli, B., Bozorg-Haddad, O., & Chu, X. (2017). Strategic importance and safety of water resources. *Journal of Irrigation and Drainage Engineering*, 143(7), 02517001.

21 Symbiotic Organisms Search Algorithm

Summary

Inspired by generic biological interactions found in nature, the symbiotic organisms search algorithm is a formidable swarm intelligence-based meta-heuristic optimization algorithm that can be quite efficient when handling real-world complex optimization problems. The symbiotic organisms search algorithm is known to have very few parameters, which is one of the distinctive features of this meta-heuristic algorithm. In this chapter, we will dig deep and explore the mechanisms used in this algorithm. We would get familiar with the symbiotic organisms search algorithm's terminology and see how one can implement this algorithm in the Python programming language. Finally, we will explore the potential merits and drawbacks of this algorithm.

21.1 Introduction

One of the most fascinating things about nature is how various species in an ecosystem establish a harmonious lifestyle as they coexist. In fact, it would not be an exaggeration to assume that, one way or another, every life form on this planet of ours, regardless of its size, shape, or form, has benefited, or still benefits, from such coexistence somewhere down their evolutionary line. For years, understanding these sorts of interactions between different species has become a topic of interest for researchers and scholars. The term *Symbiosis*, a Greek term that literally translates to *living together*, has been used as an umbrella term for these sorts of behaviors in nature. In short, symbiosis is a technical term to identify two or more species living together in close association (Margulis, 1971).

In 1879, Heinrich Anton de Bary coined the phrase symbiosis to identify the concept of "the living together of unlike organisms" (Martin & Schwab, 2012). In this context, symbiosis was used as an umbrella term to identify any close and long-term biological interaction between two different biological organisms. The application of this term to identify different interactions that can be found in nature has been, and to some extent, still is, the subject of some debates among biologists. That said, one of the more known schools of thought in this regard, which happens to be the theoretical foundation upon which the subject matter of this chapter was

DOI: 10.1201/9781003424765-21

built, states that there are three distinctive symbiosis interactions in nature, namely, *mutualism, commensalism*, and *parasitism*.

A symbiosis relationship can be categorized as mutualism or *interspecies reciprocal altruism* when the long-term interaction between individual organisms of different species is considered beneficial for all involved parties (Paracer & Ahmadjian, 2000). Mutualistic relationships may be obligatory for both species, such as the interaction between the goby fish and the shrimp. The shrimp, which is practically almost blind, is responsible for digging and cleaning up a burrow in the sand in which both species coexist. In case of a threat, the goby fish warns the shrimp by touching it so they can both hide inside the burrow. Mutualistic relationships may also be obligatory for one species and facultative for the other, or in some cases, the relationship is purely facultative for both species.

Another form of symbiosis relationship is commensalism. As the name suggests, in a commensalism relationship, while one involved party receives all the benefits from this long-term interaction, the other species are not significantly helped or harmed by this exchange (Paracer & Ahmadjian, 2000). The term itself is derived from the Medieval Latin term *commensa*, which meant sharing food. In this variation of symbiosis relationship, one organism may rely on the other species for transportation, housing, or even salvaging what the others have created after their death for their own survival. A simple example of this would be those species of spiders that build their webs in plants to lure other insects.

Finally, a symbiosis relationship could be parasitism of nature. The idea is that while one involved party is defiantly benefiting from this interaction, the other species or the *host* is being harmed during this long-term coexistence (Paracer & Ahmadjian, 2000). It is estimated that a whopping 40% of all identified living creatures are considered to have a form of parasitic life (Dobson et al., 2008). There are different forms of parasitism relationship. For instance, like most bacteria, the parasite may nest within the host's body. In other cases, such as mosquitoes or leeches, the parasite may hatch on the surface of the host.

Inspired by different forms of the symbiosis relationship that can be found in nature, Cheng and Prayogo (2014) theorized a novel meta-heuristic algorithm called the symbiotic organisms search algorithm. Built upon the principles of swarm intelligence, the symbiotic organisms search algorithm could be described as a stochastic population-based algorithm that emulates the procedure in which organisms of different species would establish some form of symbiosis relationship to improve their odds of survival. In this analogy, each search agent acts as an individual organism that tries to survive in a hypothetical ecosystem. Being based on the principles of the direct search would simply mean that, like other meta-heuristic algorithms, the symbiotic organisms search algorithm is also not bound by problems that are often associated with high dimensionality, multimodality, epistasis, non-differentiability, and discontinuous search space imposed by constraints (Bozorg-Haddad et al., 2017a,b). Perhaps the most distinctive feature of this algorithm is that it has the bare minimum number of parameters that we came to expect from meta-heuristic algorithms. Coupling this with a fairly straightforward computational structure of this algorithm makes the symbiotic organisms search algorithm a formidable choice to tackle optimization problems, even for those with little experience with meta-heuristic optimization algorithms.

The symbiotic organisms search algorithm is based on an abstract interpretation of the primary known symbiosis relationships that can be observed in nature. The idea is that the algorithm would initiate its search by randomly generating a series of search agents, referred to as organisms, within the feasible area of the search space. The algorithm would continue by randomly pairing the agents and applying three main operators on the selected organisms that basically emulate an idealized representation of the primary symbiosis interaction: Mutualism, commensalism, and parasitism. Each operator would create a new set of tentative solutions, which the algorithm evaluates and replaces the original organisms in case they are deemed to have better properties than their original counterparts. This whole process would be repeated until a termination criterion is met, at which point the best organism encountered thus far would be returned as the solution to the optimization problem at hand.

Over the years, many variants of the symbiotic organisms algorithm were proposed in the literature, some of which are the discrete symbiotic organisms search algorithm (Ezugwu & Adewumi, 2017a,b), chaos-integrated symbiotic organisms algorithm (Saha & Mukherjee, 2018), oppositional symbiotic organisms algorithm (Chakraborty et al., 2019), and quasi-oppositional-chaotic symbiotic organisms algorithm (Truong et al., 2019), to name a few. That said, the standard symbiotic organisms search algorithm is still a viable option for real-world optimization problems. In fact, the standard symbiotic organisms search algorithm has been successfully used for civil engineering (e.g., Prayogo et al., 2017), computer science (e.g., Ezugwu & Adewumi, 2017a,b), data mining (e.g., Pal et al., 2020; Rajah & Ezugwu, 2020), energy industry (e.g., Prasad & Mukherjee, 2016), financial market (e.g., Cheng et al., 2020), hydrology (e.g., Khalifeh et al., 2020), image processing (e.g., Chakraborty et al., 2020), material engineering (e.g., Do et al., 2019), medical science (e.g., Noureddine et al., 2022), oceanography (e.g., Akbarifard & Radmanesh, 2018), remote sensing (e.g., Jaffel & Farah, 2018), structural engineering (e.g., Kumar et al., 2019), and water resources planning and management (e.g., Bozorg-Haddad et al., 2017a,b). In the following sections, we will explore the computational structure of a standard symbiotic organisms search algorithm.

21.2 Algorithmic Structure of the Symbiotic Organisms Search Algorithm

As a population-based meta-heuristic algorithm enumerates through the search space through a set of search agents. But what distinguishes this algorithm from most meta-heuristic algorithms that we have seen thus far is that there is no explicit transformation from the exploration to the exploitation phase, as there are few parameters to control this transition. However, this algorithm works because the greedy strategy's core principle is embedded within this algorithm's main computational structure. The idea is that the algorithm, for the most part, prevents any non-improving move. So, in a sense, it is a controlled randomized local search that utilizes the idea of the greedy search so that it could constantly improve the properties of the population set.

The symbiotic organisms search algorithm's flowchart is depicted in Figure 21.1. A closer look at the architecture of the symbiotic organisms search algorithm would reveal that it actually consists of three main stages that are the initiation, symbiosis, and termination stages. The symbiosis stage itself consists of three primary operators that are *mutualism*, *commensalism*, and *parasitism*. Using this structure, the algorithm would conduct a thorough search and locate what could be

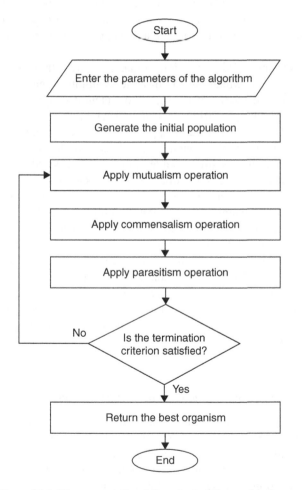

Figure 21.1 The computational flowchart of the symbiotic organisms search algorithm.

the optimum solution to the problem at hand. The following subsection will discuss each of these stages and their mathematical structures.

21.2.1 Initiation Stage

The symbiotic organisms search algorithm is a population-based meta-heuristic optimization algorithm, and as such, it works with multiple search agents, here called organisms, that would enumerate through the search space. As we have seen, in an optimization problem with N decision variables, an N-dimension coordination system could be used to represent the search space. In this case, any point within the search space, say X, can be represented mathematically as a $1{\times}N$ array as follows:

$$X = \left(x_1, x_2, x_3, \ldots, x_j, \ldots, x_N\right) \tag{21.1}$$

where X represents an organism in the search space of an optimization problem with N decision variables, and x_j represents the value associated with the jth decision variable. Note that while these organisms would be labeled differently down the line, they still have the same mathematical structure.

The symbiotic organisms search algorithm starts with randomly placing a series of organisms within the feasible boundaries of the search space. This bundle of arrays, which in the symbiotic organisms search algorithm's terminology is referred to as the population, can be mathematically expressed as $M \times N$ matrix, where M denotes the number of particles or what is technically referred to as the population size. In such a structure, each row represents a single search agent. A population, denoted by *pop*, can be represented as follows:

$$pop = \begin{bmatrix} X_1 \\ X_2 \\ \vdots \\ X_i \\ \vdots \\ X_M \end{bmatrix} = \begin{bmatrix} x_{1,1} & x_{1,2} & \cdots & x_{1,j} & \cdots & x_{1,N} \\ x_{2,1} & x_{1,2} & \cdots & x_{2,j} & \cdots & x_{2,N} \\ & & & \vdots & & \\ x_{i,1} & x_{i,2} & \cdots & x_{i,j} & \cdots & x_{i,N} \\ & & & \vdots & & \\ x_{M,1} & x_{M,2} & \cdots & x_{M,j} & \cdots & x_{M,N} \end{bmatrix} \tag{21.2}$$

where X_i represents the ith organism in the population, and $x_{i,j}$ denotes the jth decision variable of the ith organism.

The initially generated population represents the initial position of organisms. As we progress, the values stored in the *pop* matrix would be altered according to the computational structure of the symbiotic organisms search algorithm's symbiosis stage. By the end of this iterative computation process, when the termination criterion is met, one or possibly multiple organisms could converge to the optimum solution.

21.2.2 Symbiosis Stage

By mimicking the symbiosis relationships that can be found in nature, the algorithm would tend to create a new set of tentative solutions in this stage. The symbiosis stage itself consists of three main pillars that are mutualism, commensalism, and parasitism operators. These mechanisms are used as alteration tools to create a new set of tentative solutions. Using the idea of the greedy strategy, the algorithm would compare these solutions with their original counterparts and replace them in case they show better performance.

So the idea of this stage is that the algorithm would apply all three operators as it goes through all the organisms individually. The operators would randomly pair the organisms with other organisms in the population set and create new tentative solutions in each attempt. These solutions would be compared against the originally selected organism and replaced if the algorithm deems them to be more suitable. The following sections will explore the computation structure of these operators.

21.2.2.1 Mutualism Operator

Mutualism is a relationship where both organisms benefit from the long-term established interaction between two species. To imitate this symbiotic relationship among organisms, the algorithm first needs to identify the best solution within the current population. It goes without saying that the most desirable solution (i.e., identifying the best organism in the population set) has a different interpretation for maximization and minimization problems. For instance, higher objective functions are considered more desirable in a maximization problem, while we are looking for lower values for the objective function in a minimization problem. Here, let $Xbest$ denote the best solution in the set. The said organism could be mathematically represented as follows:

$$Xbest = \left(x_{best,1}, x_{best,2}, x_{best,3}, \ldots, x_{best,j}, \ldots, x_{best,N} \right) \qquad (21.3)$$

where $x_{best,j}$ denotes the position of the best-identified organism in the population in the jth dimension.

The algorithm would continue by randomly coupling two organisms from the population set to emulate an idealized representation of the mutualism relationship. Let us assume that the rth organism, denoted by X_r, has been randomly selected in this stage. The idea here is to establish an interaction between these two organisms that mutually benefits both parties. The mutualism operator uses the following mathematical procedure to create two tentative solutions based on the coupled organisms:

$$\mu_j = \frac{x_{i,j} + x_{r,j}}{2} \qquad \forall j \tag{21.4}$$

$$x'_{i,j} = x_{i,j} + Rand \times \left(x_{best,j} - \mu_j \times \beta_1\right) \qquad \forall j, \beta_1 \in \{1,2\} \tag{21.5}$$

$$x'_{r,j} = x_{r,j} + Rand \times \left(x_{best,j} - \mu_j \times \beta_2\right) \qquad \forall j, \beta_2 \in \{1,2\} \tag{21.6}$$

$$X_i^{new} = \left(x'_{i,1}, x'_{i,2}, x'_{i,3}, \dots, x'_{i,j}, \dots, x'_{i,N}\right) \tag{21.7}$$

$$X_r^{new} = \left(x'_{r,1}, x'_{r,2}, x'_{r,3}, \dots, x'_{r,j}, \dots, x'_{r,N}\right) \tag{21.8}$$

in which, μ_j represents the mutual factor in the jth dimension; $x'_{i,j}$ denotes the new position in the jth dimension for the tentative solution associated with the ith organism; $x'_{r,j}$ is the new position in the jth dimension for the tentative solution associated with the rth organism; *Rand* is a randomly generated value within the range 0–1; X_i^{new} denotes the tentative solution associated with the ith organism; X_r^{new} is the tentative solution associated with the rth organism; both β_1 and β_2 control constants randomly assume the values 1 or 2.

Of course, searching on the uncharted territories of the search agent via a stochastic-based procedure means that there can be no guarantee that the established interaction between these organisms is, in fact, beneficial for both involved parties. In other words, there can be no absolute guarantee that this mechanism will always improve the properties of both organisms, which contradicts the idea of a mutual relationship. The symbiosis organisms' search algorithm would use the greedy strategy to evaluate the search results to create such an effect. This means that the algorithm could not permit non-improving moves to take effect. As such, the tentative solutions would be compared to their original counterparts. The algorithm would replace the original organism with the generated tentative solutions if the mutualism operator were able to improve their properties based on the objective function values of the said solutions.

21.2.2.2 Commensalism Operator

The central theme of the commensalism relationship is that while one organism is definitely getting some benefits, the other involved party is not being helped nor harmed due to this interaction between the two organisms. To implement this idea, the algorithm would need to stage a situation in that one organism could have the potential to be improved via this operation while the other remains intact throughout the whole procedure.

Like what we had in the mutualism operator, the procedure here starts by randomly coupling another organism from the population set to emulate an idealized representation of the mutualism relationship. Let us assume that the *r*th organism, denoted by X_r, has been randomly selected in this stage. The idea here is to establish a type of interaction between these two organisms that could benefit one, while the other is neutral to the interaction. The commensalism operator uses the following mathematical procedure to create a tentative solution based on the coupled organisms.

$$x'_{i,j} = x_{i,j} + Rand \times \left(x_{best,j} - x_{r,j} \right) \qquad \forall j \tag{21.9}$$

$$X_i^{new} = \left(x'_{i,1}, x'_{i,2}, x'_{i,3}, \ldots, x'_{i,j}, \ldots, x'_{i,N} \right) \tag{21.10}$$

Similar to what we have seen earlier, we can safely conclude that there can be no absolute guarantee that this mechanism will always improve the properties of the selected organisms, which contradicts the idea of the commensalism relationship. To create such an effect, the symbiosis organisms search algorithm would use the greedy strategy to evaluate the search result. This means that the algorithm could not permit non-improving moves to take effect. As such, the tentative solution would be compared against its original counterpart. The algorithm would replace the original organism with the generated tentative solution if, and only if, the commensalism operator were able to improve its property based on the objective function value of the said solution.

21.2.2.3 Parasitism Operator

The central theme of the parasitism relationship is that while one organism is definitely getting some benefits, the other involved party is being significantly harmed out of this interaction between the two organisms. To implement this idea, the algorithm would need to stage a situation in one organism that could potentially be improved via this operation while the whole procedure impairs the other.

The algorithm would alter the selected organism via a random alteration to create such an effect. This procedure closely resembles what we had seen in mutation in the genetic algorithm. The idea is that a decision variable would be selected randomly, and the corresponding value for the said variable would be replaced randomly with a value feasible for that dimension. The altered array would be labeled as the *parasite vector*. The parasite vector would then latch on randomly selected organisms from the population set. Let us assume that the *r*th organism, denoted by X_r, has been randomly selected in this stage. If the parasite is deemed more suitable than the host organism, the algorithm will continue by replacing the said host with the created parasite. Otherwise, the parasite would not be able to coexist with the host and, as such, would be removed from the population. Let *l* be the dimension selected randomly for applying these random adjustments. The procedure behind generating the parasite vector can be described as follows:

$$X_i^{new} = \left(x_{i,1}, x_{i,2}, x_{i,3}, \ldots, x_{i,l}', \ldots, x_{i,N} \right) \tag{21.11}$$

$$x_{i,l}' = L_l + Rand \times \left(U_l - L_l \right) \tag{21.12}$$

in which, U_l and L_l represent the upper and lower feasible boundaries of the lth decision variables, respectively.

21.2.3 Termination Stage

Based on the above procedure, the algorithm would go over the organism in the population set one by one and apply all three operations to these organisms. Anytime the algorithm detects an improving move, the position of the search agents would be updated within the search space. However, if the tentative solutions are deemed non-improving, the algorithm would simply toss them aside and move on with the procedure. Like other meta-heuristic algorithms, the sequence of operational structures of this algorithm needs to be executed iteratively until a certain termination criterion is met, at which point the execution of the algorithm would be terminated, and the best organism recorded in the memory would be reported as the solution to the optimization problem. Note that without such a termination stage, the algorithm would potentially be executed in an infinite loop. The termination stage would, in effect, determine whether the algorithm has reached what could be the optimum solution.

As the symbiotic organisms search algorithm is not equipped with an explicitly defined, unique termination mechanism, one could implement the commonly available options, most notably limiting the number of iterations, run time, or perhaps monitoring the improvement made to the best solution in consecutive iterations. Among these options, limiting the number of iterations is arguably the most cited mechanism to create a termination stage for the symbiotic organisms search algorithm. The idea being the process would be executed only for a specified number of times, a parameter known as the maximum iteration. In any case, it should be noted that the selection of the termination mechanism is also considered one of the algorithm's parameters. Bear in mind that in most cases, these termination mechanisms may require setting up additional parameters.

21.3 Parameter Selection and Fine-Tuning the Symbiotic Organisms Search Algorithm

From the *no-free-lunch theorem*, one can conclude that fine-tuning an algorithm is essential to get the best performance out of a meta-heuristic algorithm. This would basically ensure that an algorithm is equipped to handle the unique characteristics of a given optimization problem. Of course, it is possible to use our intuition, experience, and default values suggested for an algorithm's parameters as a good starting point, one should bear in mind that fine-tuning these parameters is more than anything a trial-and-error process. Thus, while it is possible to get a good-enough result by having an educated guess for setting the parameters of these algorithms, to get the best possible performance, it is necessary to go through this fine-tuning process.

In the case of the symbiotic organisms search algorithm, it should be noted that it has been explicitly designed to have the bare minimum number of parameters. In this case, the only parameters used in the structure of the algorithm are the population size (M) and, of course, opting for the termination criterion and all the parameters that are associated with these methods. For instance, if limiting the number of iterations has been selected as a termination criterion, the maximum iteration (T) is another parameter that needs to be defined by the user. As can be seen here, not only the symbiotic organisms search algorithm has very few parameters to begin with, but for the most part, the role of these parameters on the final outcome is easy to deduce intuitively. This makes the fine-tuning process less challenging for those less experienced. That said, to get the absolute best results out of the symbiotic organisms search algorithm, it is best to dabble with these algorithms first to gain some experience and inside knowledge about such parameters. By doing so, you could better understand how to fine-tune these parameters as your initial guesses and parameter selection strategies become more educated. The pseudocode for the symbiotic organisms search algorithm is shown in Figure 21.2.

Begin

    Set the algorithm's parameter and input the data
    Generate the initial population
    Let M denote the population size
    Evaluate the generated organisms
    While the termination criterion is not met
        Let $Xbest$ denote the best solution in the population
        For i in range 1 to M
            Select the rth organism denoted by X_r randomly
            Generate solutions based on mutualism called $??_{??}^{????}$ and $??_r^{????}$
            If generated solutions are better than their counterpart
                Replace the old solutions with the new solutions
            End if
            Select the rth organism denoted by X_r randomly
            Generate a solution based on commensalism called $??_{??}^{????}$
            If the generated solution is better than its counterpart
                Replace the old solution with the new solution
            End if
            Modify the ith organism using parasitism here denoted by $??_{??}^{????}$
            Select the rth organism denoted by X_r randomly
            If $??_{??}^{????}$ is better than the X_r
                Replace the rth organism with $??_{??}^{????}$
            End if
        Next i
    End while
    Report the best solution
End

Figure 21.2 Pseudocode for the symbiotic organisms search algorithm.

21.4 Python Codes

The code to implement the symbiotic organisms search algorithm can be found below:

```python
import numpy as np

def init_generator(pop_size, num_variables, min_val, max_val):
    return np.random.uniform(min_val, max_val, (pop_size, num_
    variables))

def best_solution(pop, obj_func, minimizing):
    results = np.apply_along_axis(obj_func, 1, pop)
    if minimizing:
        index = np.argmin(results)
    else:
        index = np.argmax(results)
    return pop[index]

def mutualism(a, b, xbest, num_variables):
    mutual_factor = np.mean((a, b), axis=0)
    beta1, beta2 = np.random.randint(1,3, 2)
    rand1, rand2 = np.random.uniform(0,1,(2,num_variables))
    a_new = a + rand1*(xbest - (mutual_factor*beta1))
    b_new = b + rand2*(xbest - (mutual_factor*beta2))
    return a_new, b_new

def commensalism(a, b, xbest, num_variables):
    rand = np.random.uniform(0, 1, num_variables)
    a_new = a + rand*(xbest-b)
    return a_new

def parasitism(a, min_val, max_val, num_variables):
    parasite = a.copy()
    index = np.random.randint(num_variables)
    rand = np.random.uniform(min_val, max_val)
    parasite[index] = rand
    return parasite

def evaluator(a, b, obj_func, minimizing):
    of_a = obj_func(a)
    of_b = obj_func(b)
    if minimizing:
        if of_a<of_b:
            return a
        else:
            return b
    else:
        if of_a>of_b:
            return a
        else:
            return b
```

```
def symbiotic_organisms_search(min_val, max_val, num_variables,
                               pop_size, obj_func,
                               iteration, minimizing=True,
                               full_result=False):
    NFE_value = 0
    results = np.zeros(iteration)
    NFE = np.zeros(iteration)
    pop = init_generator(pop_size, num_variables, min_val, max_val)
    NFE_value += pop_size
    for i in range(iteration):
        xbest = best_solution(pop, obj_func, minimizing)
        for j in range(pop_size):
            r = np.random.randint(pop_size)
            organ1, organ2 = mutualism(pop[j], pop[r],
                                       xbest, num_variables)
            pop[j] = evaluator(pop[j], organ1, obj_func,
            minimizing)
            pop[r] = evaluator(pop[r], organ1, obj_func,
            minimizing)
            r = np.random.randint(pop_size)
            oragan1 = commensalism(pop[j], pop[r], xbest, num_
            variables)
            pop[j] = evaluator(pop[j], organ1, obj_func,
            minimizing)
            parasite = parasitism(pop[j], min_val, max_val,
                                  num_variables)
            r = np.random.randint(pop_size)
            pop[r] = evaluator(pop[r], parasite, obj_func,
            minimizing)
            NFE_value+=4
        NFE[i] = NFE_value
        results[i] = obj_func(best_solution(pop, obj_func,
        minimizing))
    if not full_result:
        return best_solution(pop, obj_func, minimizing),
                             obj_func(best_solution(pop, obj_func,
                                      minimizing))
    else:
        return best_solution(pop, obj_func, minimizing),
                             obj_func(best_solution(pop, obj_func,
                             minimizing)), results, NFE
```

21.5 Concluding Remarks

Inspired by the symbiosis relationships observed in nature, the symbiotic organisms search algorithm is a formidable meta-heuristic algorithm for tackling complex real-world problems. A straightforward computational structure is perhaps one of the notable features of this algorithm. But what truly distinguishes this algorithm from most meta-heuristic algorithms we have seen thus far is that not only the structure of the algorithm has very few algorithms to begin with, and for the most part, it is fairly easy to deduce the role of these parameters on the final outcome of the searching process. This could significantly help ease the fine-tuning of the algorithm, even for those with little to no experience with meta-heuristic optimization algorithms. But this seemingly huge advantage comes at a price that lacks enough controlling parameters. This means that there are not many ways to control the properties of the search engine of the algorithm. In other words, the user does not have many options to tailor the algorithm to a specified optimization problem. As a result, the user, for instance, cannot control the transition between the exploration and exploitation phase during the search. This is not to say there are no exploration or exploitation phases during the search procedure. In fact, as the algorithm initiates, the movement is more exploratory, but as the search progresses and the solutions get closer together, the search is more in line with the idea of the exploitation phase. But the critical thing to note here is that the user has no control over the pace of this transition, given that there are no parameters to control the searching properties of the algorithm. As such, the only way to avoid being trapped in local optima or premature convergence is to increase the population size or the number of iterations. Even with these changes, there can be no guarantee that we will avoid these pitfalls altogether. With that said, from a practical standpoint, the symbiotic organisms search algorithm can undoubtedly be considered a viable option when tackling real-world complex optimization problems.

References

Akbarifard, S. & Radmanesh, F. (2018). Predicting sea wave height using symbiotic organisms search (SOS) algorithm. *Ocean Engineering*, 167, 348–356.

Bozorg-Haddad, O., Azarnivand, A., Hosseini-Moghari, S.M., & Loáiciga, H.A. (2017a). Optimal operation of reservoir systems with the symbiotic organisms search (SOS) algorithm. *Journal of Hydroinformatics*, 19(4), 507–521.

Bozorg-Haddad, O., Solgi, M., & Loáiciga, H.A. (2017b). *Meta-heuristic and evolutionary algorithms for engineering optimization*. John Wiley & Sons. ISBN: 9781119386995

Chakraborty, F., Nandi, D., & Roy, P.K. (2019). Oppositional symbiotic organisms search optimization for multilevel thresholding of color image. *Applied Soft Computing*, 82, 105577.

Chakraborty, F., Roy, P.K., & Nandi, D. (2020). Symbiotic organisms search optimization for multilevel image thresholding. *International Journal of Swarm Intelligence Research (IJSIR)*, 11(2), 31–61.

Cheng, M.Y., & Prayogo, D. (2014). Symbiotic organisms search: A new metaheuristic optimization algorithm. *Computers & Structures*, 139, 98–112.

Cheng, M.Y., Cao, M.T., & Herianto, J.G. (2020). Symbiotic organisms search-optimized deep learning technique for mapping construction cash flow considering complexity of project. *Chaos, Solitons & Fractals*, 138, 109869.

Do, D. T., Lee, D., & Lee, J. (2019). Material optimization of functionally graded plates using deep neural network and modified symbiotic organisms search for eigenvalue problems. *Composites Part B: Engineering*, 159, 300–326.

Dobson, A., Lafferty, K.D., Kuris, A.M., Hechinger, R.F., & Jetz, W. (2008). Homage to Linnaeus: How many parasites? How many hosts? *Proceedings of the National Academy of Sciences*, 105(Supplement 1), 11482–11489.

Ezugwu, A.E. & Adewumi, A.O. (2017a). Discrete symbiotic organisms search algorithm for travelling salesman problem. *Expert Systems with Applications*, 87, 70–78.

Ezugwu, A.E. & Adewumi, A.O. (2017b). Soft sets based symbiotic organisms search algorithm for resource discovery in cloud computing environment. *Future Generation Computer Systems*, 76, 33–50.

Jaffel, Z. & Farah, M. (2018). A symbiotic organisms search algorithm for feature selection in satellite image classification. In *Proceeding of 2018 4th International Conference on Advanced Technologies for Signal and Image Processing*, Sousse, Tunisia.

Khalifeh, S., Esmaili, K., Khodashenas, S.R., & Khalifeh, V. (2020). Estimation of nonlinear parameters of the type 5 Muskingum model using SOS algorithm. *MethodsX*, 7, 101040.

Kumar, S., Tejani, G.G., & Mirjalili, S. (2019). Modified symbiotic organisms search for structural optimization. *Engineering with Computers*, 35(4), 1269–1296.

Margulis, L. (1971). Symbiosis and evolution. *Scientific American*, 225(2), 48–61.

Martin, B.D. & Schwab, E. (2012). Symbiosis: "Living together" in chaos. *Studies in the History of Biology*, 4(4), 7–25.

Noureddine, S., Zineeddine, B., Toumi, A., Betka, A., & Benharkat, A.N. (2022). A new predictive medical approach based on data mining and symbiotic organisms search algorithm. *International Journal of Computers and Applications*, 44(5), 465–479.

Pal, S.S., Samui, S., & Kar, S. (2020). A new technique for time series forecasting by using symbiotic organisms search. *Neural Computing and Applications*, 32(7), 2365–2381.

Paracer, S. & Ahmadjian, V. (2000). *Symbiosis: An introduction to biological associations*. Oxford University Press.

Prasad, D. & Mukherjee, V. (2016). A novel symbiotic organisms search algorithm for optimal power flow of power system with FACTS devices. *Engineering Science and Technology, An International Journal*, 19(1), 79–89.

Prayogo, D., Cheng, M.Y., & Prayogo, H. (2017). A novel implementation of nature-inspired optimization for civil engineering: A comparative study of symbiotic organisms search. *Civil Engineering Dimension*, 19(1), 36–43.

Rajah, V. & Ezugwu, A.E. (2020). Hybrid symbiotic organism search algorithms for automatic data clustering. In *2020 Conference on Information Communications Technology and Society*, Durban, South Africa.

Saha, S. & Mukherjee, V. (2018). A novel chaos-integrated symbiotic organisms search algorithm for global optimization. *Soft Computing*, 22(11), 3797–3816.

Truong, K.H., Nallagownden, P., Baharudin, Z., & Vo, D.N. (2019). A quasi-oppositional-chaotic symbiotic organisms search algorithm for global optimization problems. *Applied Soft Computing*, 77, 567–583.

Index